Business Analytics

Walter R. Paczkowski

Business Analytics

Data Science for Business Problems

Springer

Walter R. Paczkowski
Data Analytics Corp.
Plainsboro, NJ, USA

ISBN 978-3-030-87022-5 ISBN 978-3-030-87023-2 (eBook)
https://doi.org/10.1007/978-3-030-87023-2

This Springer imprint is published by the registered company Springer Nature Switzerland AG
The registered company address is: Gewerbestrasse 11, 6330 Cham, Switzerland

Preface

I analyze business data—and I have been doing this for a long time. I was an analyst and department head, a consultant and trainer, worked on countless problems, written many books and reports, and delivered numerous presentations to all levels of management. I learned a lot. This book reflects insights I gained from this experience about *Business Data Analytics* that I want to share.

There are three questions you should quickly ask about this sharing. The first is obvious: "*Share what*?" The second logically follows: "*Share with whom*?" The third is more subtle: "*How does this book differ from other data analytic books*?" The first is about focus, the second is about target, and the third is about competitive comparison. So, let me address each question.

The Book's Focus

My experience has been with practical business problems. When I finished my academic training with a Ph.D. in economics and a heavy statistics exposure, I immediately started my professional career with an AT&T internal consulting group, *The Analytical Support Center* (*ASC*). I quickly learned that I needed both a theoretical, technical understanding of quantitative work—how to estimate a regression model, for example—and an understanding of how to deal with messy data beyond the nice, clean data sets I used as a graduate student. My time at the *ASC* was a great learning experience that I carried throughout my professional career at AT&T, including Bell Labs, and into my own consulting business. The lessons I learned were that good, solid data analysis for practical business problems requires:

1. A theoretical understanding of statistical, econometric, and (in the current era) machine learning methods
2. Data handling capabilities encompassing data organizing, preprocessing, and wrangling
3. Programming knowledge in at least one software language

These three components form a synergistic whole, a unifying approach if you wish, for doing business data analytics, and, in fact, any type of data analysis. This synergy implies that one part does not dominate any of the other two. They work together, feeding each other with the goal of solving only one overarching problem: how to provide decision makers with rich information extracted from data. Recognizing this problem was the most valuable lesson of all. All the analytical tools and know how must have a purpose and solving this problem is that purpose—there is no other.

I show this problem and the synergy of the three components for solving it as a triangle in Fig. 1. This triangle represents the almost philosophical approach I take for any form of business data analysis and is the one I advocate for all data analyses.

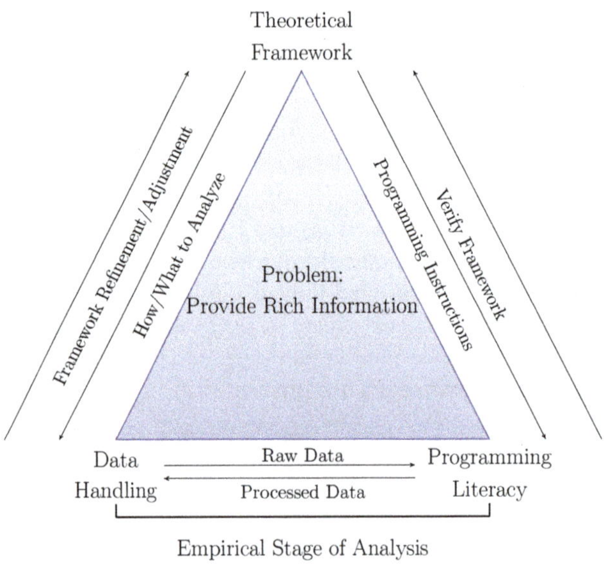

Fig. 1 The synergistic connection of the three components of effective data analysis for the overarching problem is illustrated in this triangular flow diagram. Every component is dependent on the others and none dominates the others. Regardless of the orientation of the triangle, the same relationships will hold

The overarching problem at the center of the triangle is not obvious. It is subtle. But because of its preeminence in the pantheon of problems any decision maker faces, I decided to allocate the entire first chapter to it. Spending so much space talking about information in a data analytics book may seem odd, but it is very important to understand why we do what we do, which is to analyze data to extract that rich information from data.

The theoretical understanding should be obvious. You need to know not just the methodologies but also their limitations so you can effectively apply them to solve a problem. The limitations may hinder you or just give you the wrong answers. Assume you were hired or commissioned by a business decision maker (e.g., a

CEO) to provide actionable, insightful, and useful rich information relevant for their problem. If the limitations of a methodology prevent you from accomplishing your charge, then your life as an analyst will be short-lived, to say the least. This will hold if you either do not know these limitations or simply choose to ignore them. Another methodological approach might be better, one that has fewer problems, or is just more applicable.

There is a dichotomy in methodology training. Most graduate-level statistics and econometric programs, and the newer Data Science programs, do an excellent job instructing students in the theory behind the methodologies. The focus of these academic programs is largely to train the next generation of academic professionals, not the next generation of business analytical professionals. Data Science programs, of which there are now many available online and "in person," often skim the surface of the theoretical underpinnings since their focus is to prepare the next generation of business analysts, those who will tackle the business decision makers' tough problems, and not the academic researchers. Something in between the academic and data science training is needed for successful business data analysts.

Data handling is not as obvious since it is infrequently taught and talked about in academic programs. In those programs, beginner students work with clean data with few problems and that are in nice, neat, and tidy data sets. They are frequently just given the data. More advanced students may be required to collect data, most often at the last phase of training for their thesis or dissertation, but these are small efforts, especially when compared to what they will have to deal with post training. The post-training work involves:

- Identifying the required data from diverse, disparate, and frequently disconnected data sources with possibly multiple definitions of the same quantitative concept
- Dealing with data dictionaries
- Dealing with samples of a very large database—how to draw the sample and determine the sample size
- Merging data from disparate sources
- Organizing data into a coherent framework appropriate for the statistical/econometric/machine learning methodology chosen
- Visualizing complex multivariate data to understand relationships, trends, patterns, and anomalies inside the data sets

This is all beyond what is provided by most training programs.

Finally, there is the programming. First, let me say that there is programming and then there is programming. The difference is scale and focus. Most people, when they hear about programming and programming languages, immediately think about large systems, especially ones needing a considerable amount of time (years?) to fully specify, develop, test, and deploy. They would be correct regarding large-scale, complex systems that handle a multitude of interconnected operations. Online ordering systems easily come to mind. Customer interfaces, inventory management, production coordination, supply chain management, price maintenance and dynamic pricing platforms, shipping and tracking, billing, and collections are just a few components of these systems. The programming for these is complex to say the least.

As a business data analyst, you would not be involved in this type of programming although you might have to know about and access the subsystems of one or more of these larger systems. And major businesses are composed of many larger systems! You might have to write code to access the data, manipulate the retrieved data, and so forth, basically write programming code to do all the data handling I described above. And for this you need to know programming and languages.

There are many programming languages available. Only a few are needed for most business data analysis problems. In my experience, these are:

- *SQL*
- Python
- R

Julia should be included because it is growing in popularity due to its performance and ease of use. For this book, I will use Python because its ecosystem is strongly oriented toward machine learning with strong modeling, statistics, data visualization, and programming functionalities. In fact, its programming paradigm is clear to use, which is a definite advantage over other languages.

The Target Audience

The target audience for this book consists of business data analysts, data scientists, and market research professionals, or those aspiring to be any of these, in the private sector. You would be involved in or responsible for a myriad of quantitative analyses for business problems such as, but not limited to:

- Demand measurement and forecasting
- Predictive modeling
- Pricing analytics including elasticity estimation
- Customer satisfaction assessment
- Market and advertisement research
- New product development and research

To meet these tasks, you will have a need to know basic data analytical methods and some advanced methods, including data handling and management. This book will provide you with this needed background by:

- Explaining the intuition underlying analytic concepts
- Developing the mathematical and statistical analytic concepts
- Demonstrating analytical concepts using Python
- Illustrating analytical concepts with case studies

This book is also suitable for use in colleges and universities offering courses and certifications in business data analytics, data sciences, and market research. It could be used as a major or supplemental textbook.

Since the target audience consists of either current or aspiring business data analysts, it is assumed that you have or are developing a basic understanding of fundamental statistics at the "Stat 101" level: descriptive statistics, hypothesis testing, and regression analysis. Knowledge of econometric and market research principles, while not required, would be beneficial. In addition, a level of comfort with calculus and some matrix algebra is recommended, but not required. Appendices will provide you with some background as needed.

The Book's Competitive Comparison

There are many books on the market that discuss the three themes of this book: analytic methods, data handling, and programming languages. But they do them separately as opposed to a synergistic, analytic whole. They are given separate treatment so that you must cover a wide literature just to find what is needed for a specific business problem. Also, once found, you must translate the material into business terms. This book will present the three themes so you can more easily master what is needed for your work.

The Book's Structure

I divided this book into three parts. In Part I, I cover the basics of business data analytics including data handling, preprocessing, and visualization. In some instances, the basic analytic toolset is all you need to address problems raised by business executives. Part II is devoted to a richer set of analytic tools you should know at a minimum. These include regression modeling, time series analysis, and statistical table analysis. Part III extends the tools from Part II with more advanced methods: advanced regression modeling, classification methods, and grouping methods (*a.k.a., clustering*).

The three parts lead naturally from basic principles and methods to complex methods. I illustrate this logical order in Fig. 2.

Embedded in the three parts are case study examples of business problems using (albeit, fictitious, fake, or simulated) business transactions data designed to be indicative of what business data analysts use every day. Using simulated data

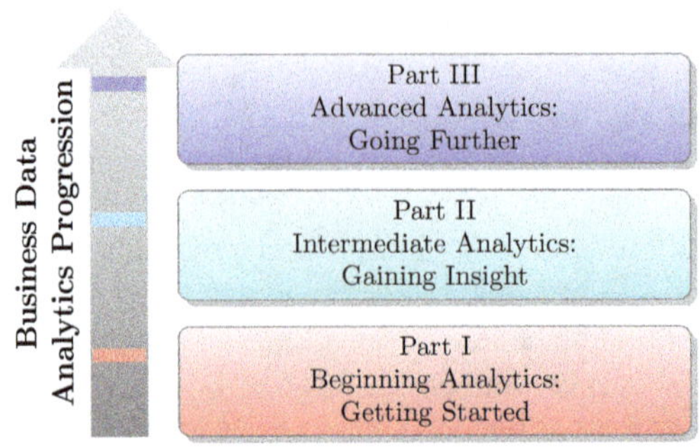

Fig. 2 This is a flow chart of the three parts of this book. The parts move progressively from basics to advanced. At the end of Part I, you should be able to do basic analyses of business data. At the end of Part II, you should be able to do regression and times series analysis. At the end of Part III, you should be able to do advanced machine learning work

for instructional purposes is certainly not without precedence. See, for example, Gelman et al. (2021). Data handling, visualization, and modeling are all illustrated using Python. All examples are in Jupyter notebooks available on Github.

Plainsboro, NJ, USA Walter R. Paczkowski

Acknowledgments

In my last book, I noted the support and encouragement I received from my wonderful wife, Gail, and my two daughters, Kristin and Melissa, and my son-in-law, David. As before, my wife Gail encouraged me to sit down and just write, especially when I did not want to, while my daughters provided the extra set of eyes I needed to make this book perfect. They provided the same support and encouragement for this book, so I owe them a lot, both then and now. I would also like to say something about my two grandsons who, now at 5 and 9, obviously did not contribute to this book but who, I hope, will look at this one in their adult years and say "My grandpa wrote this book, too."

Contents

List of Figures

List of Tables

Part I
Beginning Analytics

This first part of the book introduces basic principles for analyzing business data. The material is at a Statistics 101 level and is applicable if you are interested in basic tools that you can quickly apply to a business problem. After reading this part of the book, you will be able to conduct basic business data analysis.

Chapter 1
Introduction to Business Data Analytics: Setting the Stage

Spoiler-alert: *Business Data Analytics* (*BDA*), the focus of this book, is solely concerned with one task, and one task only: to provide the richest information possible to decision makers.

I have two objectives for this introductory chapter regarding my spoiler alert. I will first discuss the types of problems business decision makers confront and who the decision makers are. I will then discuss the role and importance of information to set the foundations for succeeding chapters. This will include a definition of information. People frequently use the words *data* and *information* interchangeably as if they have the same meaning. I will draw a distinction between them. First, they are not the same despite the fact that they are used interchangeably. Second, as I will argue, information is latent, hidden inside data and must be extracted and revealed which makes it a deeper, more complex topic. As a data analyst, you need to have a handle on the significance of information because extracting it from data is the sole reason for the existence of *BDA*.

My discussion of the difference between data and information will follow with a comparison of two dimensions of information rarely discussed: the quantity and quality of the information decision makers rely on. There is a cost to decision making often overlooked at best or ignored at worst. The cost is due to both dimensions. The objective of *BDA* is not only to provide information (i.e., a quantity issue), but also to provide good information (i.e., a quality issue) to reduce the cost of decision making. Providing good information, however, is itself not without cost. You need the appropriate skill sets and resources to effectively extract information from data. This is a cost of doing data analytics. These two costs—cost of decision making and cost of data analytics—determine what information can be given to decision makers. These have implications for the type and depth of your *BDA*.

W. R. Paczkowski, *Business Analytics*,
https://doi.org/10.1007/978-3-030-87023-2_1

1.1 Types of Business Problems

What types of business problems warrant *BDA*? The types are too numerous to mention, but to give a sense of them consider a few examples:

- Anomaly Detection: production surveillance, predictive maintenance, manufacturing yield optimization;
- Fraud detection;
- Identity theft;
- Account and transaction anomalies;
- Customer analytics:
  - Customer Relationship Management (*CRM*);
  - Churn analysis and prevention;
  - Customer Satisfaction;
  - Marketing cross-sell and up-sell;
  - Pricing: leakage monitoring, promotional effects tracking, competitive price responses;
  - Fulfillment: management and pipeline tracking;
- Competitive monitoring;
- Competitive Environment Analysis (*CEA*); and
- New Product Development.

And the list goes on, and on.

A decision of some type is required for all these problems. New product development best exemplifies a complex decision process. Decisions are made throughout a product development pipeline. This is a series of stages from ideation or conceptualization to product launch and post-launch tracking. Paczkowski (2020) identifies five stages for a pipeline: ideation, design, testing, launch, and post-launch tracking. Decisions are made between each stage whether to proceed to the next one or abort development or even production. Each decision point is marked by a *business case* analysis that examines the expected revenue and market share for the product. Expected sales, anticipated price points (which are refined as the product moves through the pipeline), production and marketing cost estimates, and competitive analyses that include current products, sales, pricing, and promotions plus competitive responses to the proposed new product, are all needed for each business case assessment. If any of these has a negative implication for the concept, then it will be canceled and removed from the pipeline. Information is needed for each business case check point.

The expected revenue and market share are refined for each business case analysis as new and better information –not data– become available for the items I listed above. More data do become available, of course, as the product is developed, but it is the analysis of that data based on methods described in this book, that provide the information needed to approve or not approve the advancement of the concept to the next stage in the pipeline. The first decision, for example, is simply to begin developing a new product. Someone has to say "*Yes*" to the question "*Should we develop a new product*?" The business case analysis provides that decision maker with the information for this initial "*Go/No Go*" decision. Similar decisions are made at other stages.

Another example is product pricing. This is actually a two-fold decision involving a structure (e.g., uniform pricing or price discrimination to mention two possibilities) and a level within the structure. These decisions are made throughout the product life cycle beginning at the development stage (the launch stage of the pipeline I discussed above) and then throughout the post-launch period until the product is ultimately removed from the market. The wrong price structure and/or level could cost your business lost profit, lost market share, or a lost business. See Paczkowski (2018) for a discussion of the role of pricing and the types of analysis for identifying the best price structure and level. Also see Paczkowski (2020) for new product development pricing at each stage of the pipeline.

1.2 The Role of Information in Business Decision Making

Decisions are effective if they solve a problem, such as those I discussed above, and aid rather than hinder your business in succeeding in the market. I will assume your business succeeds if it earns a profit and has a positive return for its owners (shareholders, partners, employees in an employee-owned company) or a sole owner. Information could be about

- current sales;
- future sales;
- the state of the market;
- consumer, social, and technology trends and developments;
- customer needs and wants;
- customer willingness-to-pay;
- key customer segments;
- financial developments;
- supply chain developments; and
- the size of customer churn.

This information is input into decisions and like any input, if it is bad, then the decisions will be bad. Basically, the *GIGO Principle* (*Garbage In–Garbage Out*) holds. This should be obvious and almost trite. Unfortunately, you do not know when you make your decision if your information is good or bad, or even sufficient. You face *uncertainty* due to the amount and quality of the information you have available.

Without any information you would just be guessing, and guessing is costly. In Fig. 1.1, I illustrate what happens to the cost of decisions based on the amount of information you have. Without any information, all your decisions are based on pure guesses, hunches, so you are forced to approximate their effect. The approximation could be very naive, based on gut instinct (i.e., an unfounded belief that you know everything) or what happened yesterday or in another business similar to yours (i.e., an analog business).

The cost of these approximations in terms of financial losses, lost market share, or outright bankruptcy can be very high. As the amount of information increases, however, you will have more insight so your approximations (i.e., guesses) improve and the cost of approximations declines. This is exactly what happens during the business case process I described above. More and better information helps the decision makers at each business case stage. The approximations could now be based on trends, statistically significant estimates of impact, or model-based what-if analyses. These are not "data"; they are information.

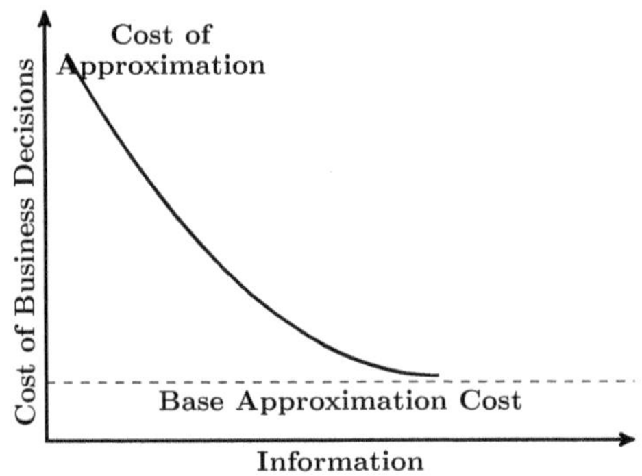

Fig. 1.1 This cost curve illustrates what happens to the cost of decisions as the amount of information increases. The Base Approximation Cost is the lowest possible cost you can achieve due to the uncertainty of all decisions. This is an amount above zero

Adriaans (2019) states that there is an inverse relationship between the amount of information you have and the level of uncertainty you face. Adriaans (2019) cites this as a linear relationship, although I see no reason for linearity because uncertainty is driven to zero at some point under linearity as the amount of information increases. The costs, however, will never decline to zero because you will never have enough information to know everything and to know it perfectly; to know exactly what will happen as a result of your decisions. There will always be some uncertainty associated with any decision. The cost of approximating an outcome will bottom out at a point above zero. You can say it will asymptotically approach a lower limit greater than zero as the amount of information becomes large. The relationship is nonlinear. This is the basis for my cost curve in Fig. 1.1 which shows that as the level of uncertainty declines because of the increased information, the cost of an error will also decline but not disappear.

1.3 Uncertainty vs. Risk

Uncertainty is a fact of life reflecting our lack of knowledge. It is either spatial ("*I don't know what is happening in Congress today.*") or temporal ("*I don't know what will happen to sales next year.*"). In either case, the lack of knowledge is about the *state of the world* (*SOW*): what is happening in Congress and what will happen next year. Business textbooks such as Freund and Williams (1969), Spurr and Bonini (1968), and Hildebrand et al. (2005) typically discuss assigning a probability to different *SOW*s that you could list. The purpose of these probabilities is to enable you to say something about the world before that something materializes. Somehow, and it is never explained how, you assign numeric values representing outcomes, or *payoffs*, to the *SOW*s. The probabilities and associated payoffs are used to calculate an expected or average payoff over all the possible *SOW*s. Consider, for example, the rate of return on an investment (*ROI*) in a capital expansion project. The *ROI* might depend on the average annual growth of real GDP for the next 5 years. Suppose the real GDP growth is simply expressed as declining (i.e., a recession), flat (0%), slow (1%–2%), and robust (>2%) with assigned probabilities of 0.05, 0.20, 0.50, and 0.25, respectively. These form a probability distribution. Let p_i be the probability state i is realized. Then, $\sum_{i=1}^{n} p_i = 1.0$ for these $n = 4$ possible states. I show the *SOW*s, probabilities, and *ROI* values in Table 1.1. The expected *ROI* is $\sum_{i=1}^{4} p_i \times ROI_i = 2.15\%$. This is the amount expected to be earned on average over the next 5 years.

SOW	Real GDP growth	Probability	ROI
Decline	<0%	0.05	−0.01
Flat	0%	0.20	0.01
Slow	1–2%	0.50	0.02
Robust	>2%	0.25	0.04

Table 1.1 For the three *SOW*s shown here, the expected *ROI* is $\sum_{i=1}^{3} ROI_i \times p_i = 0.0215$ or 2.15%

Savage (1972, p. 9) notes that the "world" in the statement "*state of the world*" is defined for the problem at hand and that you should not take it literally. It is a fluid concept. He states that it is "the object about which the person is concerned." At the same time, the "state" of the world is a full description of its conditions. Savage (1972) notes that it is "a description of the world, leaving no relevant aspects undescribed." But he also notes that there is a true state, a "state that does in fact obtain, i.e., the true description of the world." Unfortunately, it is unknown, and so the best we can do until it is realized or revealed to us is assign probabilities to the occurrence of each state for decision making. These are the probabilities in Table 1.1. More importantly, it is the fact that the true state is unknown, and never will be known until revealed that is the problem. No amount of information will ever completely and perfectly reveal this true state before it occurs.

You can, however, mimic that true state prior to its occurrence by creating a model that has some, but not all, of the features of the world you believe will happen. You will not have all the features because the world is just too complicated. So, you have a candidate model of how the world works. You can use a data set to train that model to best mimic the world. If your decision problem is whether or not to extend credit to a class of customers, your training data would have the actual credit worthiness of some customers. You can then test how well your trained model predicts credit worthiness with a separate, independent data set. Once satisfied that your model is well trained, you can deploy it to your population of customers. I discuss ways to develop training and testing data sets in Chap. 9 and how to train, test, and use the models for predictions in succeeding chapters.

Although Table 1.1 is a good textbook example for an introduction to expected values, it has several problems. These are the identification of:

1. *SOW*s;
2. associated *ROI*s; and
3. probabilities.

Where did they come from? I may accept an argument that the *SOW* definitions are reasonable given the past history of business cycles in, say, the U.S. I may also accept an argument for the *ROI* values which may be the averages of investment rates of return for past business cycle periods and for past capital investments. What about the probability distribution? This is an issue. Where did it come from?

The same issue holds for other situations. For example, suppose your credit department has assigned customers a rating denoting their likelihood to default on a payment. The ratings may be "Very Likely", "Somewhat Likely", and "Not at all Likely." The true *SOW* for each customer defaulting or not is unknown until credit is extended. These could be assigned probabilities so you could determine the expected value of payments. But the issue is the same: "*Where did the probability distribution come from*?"

Probabilities are either frequency-based or subjectively-based. Frequency-based probabilities are derived from repeated execution of an experiment (e.g., flipping a fair coin) while subjective probabilities are based on a mental process open to controversy. Experimental results are not controversial (assuming a well-defined and properly executed experimental protocol). Future real GDP growth periods and default ratings cannot be experimentally derived. The probabilities are subjective.

For the default problem, however, there is an alternative. You could classify customers based on their credit history, their current financial standing (perhaps *FICO*® Credit Scores, and sales force input). You could then build a classification model to assign a rating to each customer. In effect, the model would assign a probability of default, hence producing the required probability distribution. The probabilities in a classification problem are referred to as *propensities* and a class assignment is based on these propensities. I discuss classification modeling and probability assignments in Chap. 11.

Knight (1921) distinguished between risk and uncertainty based on knowledge of the subjective probability distribution of the *SOW*s. According to Knight (1921), a situation (e.g., *ROI* payoff on a project) is risky if the distribution is known. The situation is uncertain if it is unknown. The reality, however, is that the distribution is never known. See Zeckhauser (2006) for a discussion and assessment of Knight (1921)'s views. Arrow (Handbook, preface 1) notes that some initial probabilities can be assigned but they are "flimsy" or "flat." Flat means there is an equal chance for each *SOW* whether that *SOW* is the growth in real GDP or the likelihood of a credit default. This distribution is an initial *prior*. As data in the form of news and numeric quantities arrive and are processed to extract information, then the priors are updated and new distributions formed. Better decisions with less uncertainty and thus lower costs can be made. It is not data *per se* that are used for the decisions; it is the information extracted from that data. The uncertainty will never be eliminated; it is just reduced as knowledge of the probability distribution increases.

For a technical, economic discussion of uncertainty, see Hirshleifer and Riley (1996). Spurr and Bonini (1968, Chapters 9 and 10) have an extensive basic discussion of business decision making under uncertainty. Their discussion involves expected profit calculations, opportunity losses, and costs due to uncertainty, all at an elementary level. For a readable discussion of uncertainty (despite the book's subtitle), see Stewart (2019).

1.4 The Data-Information Nexus

To an extent, discussing definitions and terminology is useful for the advancement of scientific and practical solutions for any problem. If you cannot agree on basic terms, then you are doomed at worst and hindered at best from making any progress toward a solution, a decision. You can, however, become so involved in defining terms and so overly concerned about terminology that nothing else matters. Popper (1972, p. 310), cited in Checkland and Howell (1998, p. 88), stated, perhaps a little too strongly, that

> One should never quarrel about words, and never get involved in questions of terminology ... What we are really interested in, our real problem, ... are problems of theories and their truth.

Popper, a philosopher of science, was concerned about scientific problems. The same sentiment, however, holds for practical problems like the ones you face daily in your business. Despite Popper's preeminence, you still need some perspective on the foundational units that drive the *raison d'etre* of *BDA*: data and information.[1] If information is so important for reducing uncertainty, then a logical question to ask is: "*What is information*?" A subordinate, but equally important, question is:

[1] Some writers include knowledge in their discussion so there is a trilogy. I will omit this added component since its inclusion may be too philosophical. See Checkland and Howell (1998) for a discussion.

"*How do you obtain information?*" I will address these two questions in reverse order because information comes from data.

1.4.1 Data and Information Confusion

The words information and data are used as synonyms in everyday conversations. It is not uncommon, for example, to hear a business manager claim in one instance that she has a lot of data and then say in the next instance that she has a lot of information, thus linking the two words to have the same meaning. In fact, the computer systems that manage data are referred to as *Information Systems* (*IS*) and the associated technology used in those systems is referred to as *Information Technology* (*IT*).[2] The C-Level executive in charge of this data and *IT* infrastructure is the *Chief Information Officer* (*CIO*). Notice the repeated use of the word "information."

Even though people use these two words interchangeably it does not mean they have the same meaning. It is my contention, along with others, that data and information are distinct terms that, yet, have a connection. I will simply state that data are *facts*, objects that are true on their face, that have to be organized and manipulated to yield insight into something previously unknown. When managed and manipulated, they become information. The organization cannot be without the manipulation and the manipulation cannot be without the organization. The *IT* group of your business organizes your company's data but it does not manipulate it to be information. The information is latent, hidden inside the data and must be extracted so it can be used in a decision. I illustrate this connection Fig. 1.2. I will comment on each component in the next few sections.

1.4.2 The Data Component

A common starting point for a discussion about data is that they are facts. There is a huge philosophical literature on what is a fact. As noted by Mulligan and Correia (2020), a fact is the opposite of theories and values and "are to be distinguished from things, in particular from complex objects, complexes and wholes, and from relations." Without getting into this philosophy of facts, I will hold that a fact is a checkable or provable entity and, therefore, true. For example, it is true that Washington D.C. is the capital of the United States: it is easily checkable and can

[2] *IT* goes beyond just managing data. It is also concerned with storing, retrieving, and transmitting data. See https://en.wikipedia.org/wiki/Information_technology for a good discussion. Last accessed December 23, 2019.

Fig. 1.2 Data is the base for information which is used for decision making. The *Extractor* consists of the methodologies I will develop in this book to take you from data to information. So, behind this one block in the figure is a wealth of methods and complexities

be shown to be true. It is also a fact that $1 + 1 = 2$. This is checkable by simply counting one finger on your left hand and one finger on your right hand.[3]

You could have a lot of facts on a topic but they are of little value if they are not

1. organized,
2. subsetted,
3. manipulated, and
4. interpreted

in a meaningfully way to provide insight for a recommendation for an action, the action being the problem solution. Otherwise, the facts are just a collection of valueless things. Their value stems from what you can do with them.

Organizing data, or facts, is a first step in any analytical process and the drive for information. This could involve arranging them in chronological order (e.g., by date and time of a transaction), spatial order (e.g., countries in the Northern and Southern Hemispheres), alphanumeric order, size order, and so on in an infinite number of ways. Transactions data, for example, are facts about units sold of a series of products including what products, who bought them, when they were sold, the amount sold, and prices. They are typically maintained in a file without a discernible order: just product, date, and units. There is no insight or intelligence from this data. In fact, it is somewhat randomly organized based on when orders were placed.[4] If sorted by product and date, however, then they are organized and useful, but not much. The best organization is the one most applicable for a practical problem.

In Data Science, statistics, econometrics, machine learning, and other quantitative areas, a common organizational form is a rectangular array consisting of rows and columns. The rows are typically *objects* and the columns *variables* or *features*. An object can be a person (e.g., a customer) or an event (e.g., a transaction). The words *object*, *case*, *individual*, *event*, *observation*, and *instance* are often used interchangeably and I will certainly do this. For the methods considered in this book, each row is an individual case, one case per row and each case is in its own row. The

[3] Believe it or not, a formal mathematical proof is available but which is quite long and intricate. See https://en.wikipedia.org/wiki/Principia_Mathematica, last accessed February 3, 2021, for a discussion as well as the entry at https://math.stackexchange.com/questions/278974/prove-that-11-2, last accessed February 3, 2021. Incidentally, the proof is over 300 pages long!

[4] This statement about randomness is based the data arriving into the transaction system following a Poisson process where the arrival time of an order is a random process.

columns are the variables. Each variable is in one column (with exceptions to be discussed) with one variable per column. In machine learning, variables are called *features*.

In Pandas, a Python package for data management among other functions, the rectangular array is called a *DataFrame*.[5] I will often refer to a data set as a DataFrame since I will illustrate many analytical concepts with data arranged in a DataFrame. One aspect of a DataFrame that makes it powerful is the indexing of the rows. Indexes organize your data. They may or may not be unique, although duplicates add an analytical complication since the same index would identify several rows. At the simplest level, the rows are indexed or numbered in sequential order beginning with 0 (i.e., the integers 0, 1, 2, 3, etc.) because Python itself uses *zero-based indexing*. An object in the first row has index 0, the one in the second row has index 1, and so on. It is useful to think of these index numbers as offsets from the first row: the first row is offset 0 rows from itself; the second row is offset 1 row from the first row, etc.

DataFrame indexing can be changed to begin with 1 or be set to any variable. For example, if your DataFrame contains transactions data measured in time with a date-time variable (called a *datetime*) indicating when a transaction took place, then that datetime variable could be used as the index which is more meaningful than a sequence of (meaningless) integers. If the DataFrame contains customer specific data with a unique customer ID (*CID*) identifying each customer, then the *CID* could be used as the index. These two examples are single-level indexes. A multi-level index, called a *MultiIndex*, is also possible. For example, a DataFrame might contain a combination of temporal and spatial data such as sales data by marketing region and quarter of the year. These two measures could be used to index the DataFrame. A number of multilevel indexes are possible.

You can view multilevel indexing as defining a multi-dimensional object. A *Data Cube* is the highest dimensional object we can visualize or portray (we cannot see or visualize more than three dimensions) enabling a more detailed understanding of data. This also provides more flexibility in handling and managing DataFrames. For example, suppose your business produces six products in four manufacturing plants, one plant in each of four marketing regions. The plants are centered in the regions to minimize travel time to wholesale distributors. When a product is produced, a lot number is stamped on the product to identify when it was produced and at which plant. When customers return defective products, the lot numbers are scanned and the manufacturing date and plant ID are recorded in a returns database. I show an example Data Cube for returns in Fig. 1.3. A cell at the intersection of Month, Plant, and Product has the amount returned for that combination. The associated DataFrame, of which the Data Cube is just a conceptual representation, has three indexes that form a MultiIndex: Month, Product, and Plant. There is only

[5] Note the capitalization. This is standard and helps to differentiate the Pandas concept from the *R* concept of the same name. The Pandas implementation is more powerful and flexible.

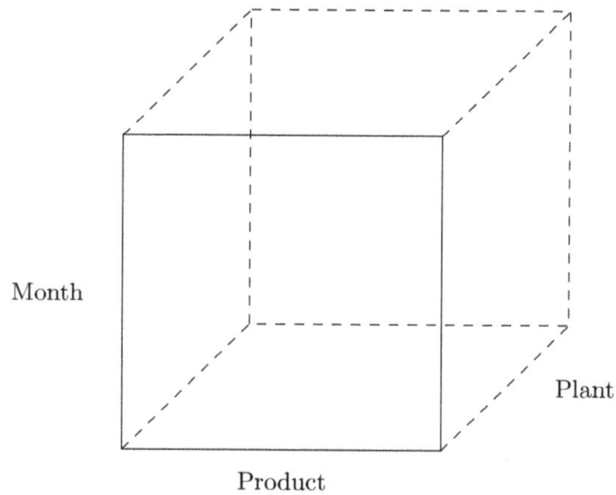

Fig. 1.3 This is an example of a Data Cube illustrating the three dimensions of data for a manufacturer. As I noted in the text, more than three dimensions are possible, but only three can be visualized

			Returns
Product	**Plant**	**Month**	
Product 1	**Plant A**	**January**	1
		February	13
		March	1
		April	4
		May	3

Fig. 1.4 This is a DataFrame version of the Data Cube for the product return example. There are 288 rows. This example has a multilevel index representing the Data Cube. Each combination of the levels of three indexes is unique because each combination is a row identifier, and there can only be one identifier for each row

one variable, the number of returns. I show the first five records, called the *head* in Pandas, of this DataFrame in Fig. 1.4.

A general view of a Data Cube is needed so it can be used for any problem. The dimensions of a Cube are *time*, *space*, and *measure* as I illustrate in Fig. 1.5. All data, whether business data or physical data or scientific data, evolve or are generated and collected in time. For example, sales are generated and recorded daily; payments such as loans, salaries, and interest are made monthly; taxes are paid quarterly; and so on. The temporal domain is everywhere and this domain has a constant forward motion that is universal. In fact, in physics this is called the *Arrow of Time*. See Coveney and Highfield (1990) and Davies (1995) for a detailed discussions of the *Arrow of Time*.

Data also have a spatial aspect. Sales, for example, while generated daily, are also generated in individual stores which may be located in cities which are in states which are in marketing regions which could be in countries. Even online data (e.g., product reviews, advertising, orders) have temporal and spatial features. Anything

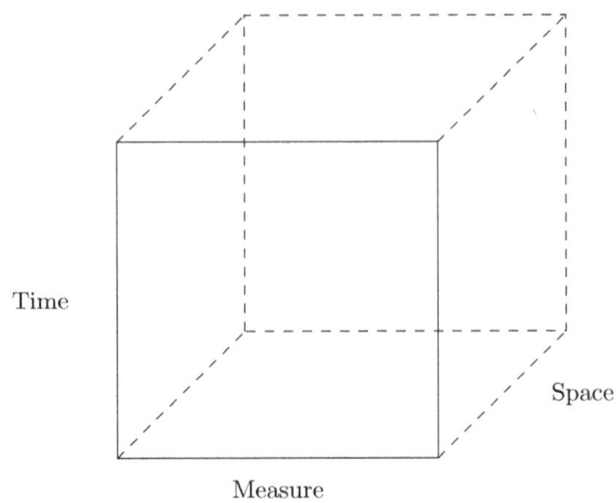

Fig. 1.5 This is a stylized Data Cube illustrating the three dimensions of data

done online is still done in time (e.g., date and time of day) and in space since customers and suppliers live or work in geographic locations.

Finally, the measure could be any business quantity such as unit sales, revenue, net income, cash flow, net earnings, product returns, and inventory levels to mention a few. The time dimension should be obvious while the spatial might be subtle. The measures could vary by regions, facilities, business units, subsidiaries, or franchise units.

I will use the terms "Data Cube", "DataFrame", and "data set", interchangeably to refer to the same concept: an arrangement of complex, multidimensional data. A DataFrame is a two-dimensional flattened version of a (potential) hypercube with multi-indexes representing the dimensions.

This arrangement into a cube (or DataFrame since a cube is only three-dimensional) is just the first step to getting the information from data. It does not mean you have information. Data *per se* are still facts but perhaps a little less meaningless since they are now arranged in some useful fashion. This arrangement allows you to manipulate your data to aid the extraction of information. For instance, you can aggregate over the time dimension (i.e., collapse the Cube) so that you can work with spatial data. Or you can aggregate (i.e., collapse the Cube) over the spatial dimension to work with a time series. I illustrate these possibilities in Fig. 1.6. You could *slice* data out of the cube for a specific time and space. I will discuss advanced handling of DataFrames, including data aggregation and subsetting, in Chap. 9.

Data manipulation goes beyond aggregating and slicing a DataFrame. It also includes *joining* or *merging* two or more DataFrames. This is done more times than you can imagine for any one data analytic problem simply because your data will not be delivered to you in one complete DataFrame. You will have to create the DataFrame you need from several DataFrames. Joining them is a complex issue. I will discuss this in Chap. 3.

Product	Plant	Returns
Product 1	Plant A	36
	Plant B	24
	Plant C	32
	Plant D	27
Product 2	Plant A	49

(a)

Product	Month	Returns
Product 1	January	13
	February	22
	March	6
	April	7
	May	15

(b)

Month	Returns
January	213
February	194
March	109
April	156
May	164

(c)

Fig. 1.6 This illustrates three possible aggregations of the DataFrame in Fig. 1.4. Panel (**a**) is an aggregation over months; (**b**) is an aggregation over plants; and (**c**) is an aggregation over plants and products. There are six ways to aggregate over the three indexes

Manipulation extends even further beyond organizing, joining, and subsetting. Transforming data to another scale such as a standardized, uniform, or share scale for better interpretation and to obtain more meaningful modeling results is a good example. A new scale, especially a standardized one, is often necessary for many statistical and machine learning methods. This, obviously, has to be done before these methods can be used, hence, they are all *preprocessing* steps. I will discuss data preprocessing in Chap. 5.

1.4.3 The Extractor Component

Finally, you have to apply some methods or procedures to your DataFrame to extract information. Refer back to Fig. 1.2 for the role and position of an Extractor function in the information chain. This whole book is concerned with these methods. The interpretation of the results to give meaning to the information will be illustrated as I develop and discuss the methods, but the final interpretation is up to you based on your problem, your domain knowledge, and your expertise.

Due to the size and complexity of modern business data sets, the amount and type of information hidden inside them is large, to say the least. There is no one piece of information–no one size fits all–for all business problems. The same data set can be used for multiple problems and can yield multiple types of information. The possibilities are endless. The information content, however, is a function of the size and complexity of the DataFrame you eventually work with. The size is the number of data elements. Since a DataFrame is a rectangular array, the size is $\#rows \times \#columns$ elements and is given by its *shape* attribute. Shape is expressed as a *tuple* written as $(rows, columns)$. For example, it could be $(5, 2)$ for a DataFrame with 5 rows and 2 columns and 10 elements. The complexity is the types of data in the DataFrame and is difficult to quantify except to count types. They could be floating point numbers (or, simply, *floats*), integers (or *ints*), Booleans, text strings (referred to as *objects*), datetime values, and more. The larger and more complex the DataFrame, the more information you can extract. Let $I = Information$, $S = size$ and $C = complexity$. Then $Information = f(S, C)$ with $\partial I/\partial S > 0$ and $\partial I/\partial C > 0$. For a very large, complex DataFrame, there is a very large amount of information.

The cost of extracting information directly increases with the DataFrame's size and complexity of the data. If I have 10 sales values, then my data set is small and simple. Minimal information, such as the mean, standard deviation, and range, can be extracted. The cost of extraction is low; just a hand-held calculator is needed. If I have 10 GB of data, then more can be done but at a greater cost. For data sizes approaching exabytes, the costs are monumental.

There could be an infinite amount of information, even contradictory information, in a large and complex DataFrame. So, when something is extracted, you have to check for its accuracy. For example, suppose you extract information that classifies customers by risk of default on extended credit. This particular classification may not be a good or correct one; that is, the *predictive classifier* (*PC*) may not be the best one for determining whether someone is a credit risk or not. *Predictive Error Analysis* (*PEA*) is needed to determine how well the *PC* worked. I discuss this in Chap. 11. In that discussion, I will use a distinction between a *training data set* and a *testing data set* for building the classifier and testing it. This means the entire DataFrame will not, and should not, be used for a particular problem. It should be divided into the two parts although that division is not always clear, or even feasible. I will discuss the split into training and testing data sets in Chap. 9.

The complexity of the DataFrame is, as I mentioned above, dependent on the types of data. Generally speaking, there are two forms: text and numeric. Other forms such as images and audio are possible but I will restrict myself to these two forms. I have to discuss these data types so that you know the possibilities and their importance and implications. How the two are handled within a DataFrame and with what statistical, econometric, and machine learning tools for the extraction of information is my focus in this book and so I will deal with them in depth in succeeding chapters. I will first discuss text data and then numeric data in the next two subsections.

1.4.3.1 Text Data

To understand text data, it is helpful to first understand a form of text you encounter in everyday life: news. News is an interesting word because it has three characteristics or properties that are rarely discussed much less thought about. First, it is something that arrives that was not present before. This should be clear since the root part of "news" is "new" and is just the plural form. News could be about a political event, an economic or financial event, a discovery, a technological breakthrough, an international incident, a local event, and so on.

Second, there is no pattern to how news arrives. Consequently, you cannot predict or plan for it. News just happens. You open a newspaper, turn on the TV or radio, look at your favorite news summary service on your smartphone or laptop and you see a headline that was not there a moment before. That headline randomly arrived; you were not waiting for it or expecting it. The only thing you can say is that you know something will arrive but you cannot say what it will be, or even when it will arrive (i.e., the time between the arrival of one piece of news and the next). It is a random process.[6]

Third, your beliefs about the structure of the world, how it works, how its constituent parts interact, and how it evolved are based on and influenced by the news you receive. You have a *belief system*, not a single belief. It is this belief system that is responsible for how you behave and the countless decisions you make every day, some mundane and others significant. At any moment, your belief system forms a *prior* about the world. A prior is a Bayesian concept that refers to a base belief expressed as a subjective probability. I will formally introduce priors in Chap. 11 when I discuss Naive Bayes classification.

You think about the news, mull it over, digest it and its implications at personal, social, economic, political, and global levels. You process the news. You also form opinions about what will happen next and the chances for those happenings. This is the basis for the subjective probabilities I just mentioned.

Random news does not arrive into a vacuum. You have prior insight based on previous random news you already processed. Basically, your knowledge base increases by the amount of each piece of news that randomly arrives and that you process for the insight it contains. The processed news is used to adjust the prior in a Bayesian context to form a *posterior* which is actually a new prior. The posterior is another concept I will introduce in Chap. 11.

In essence, you have a stock or set of information which, in economics, is called an *information set*. Each time you process random news and extract information, that new information is added to your information set. Since news continuously, but nonetheless randomly, arrives and is continuously processed for its information content, your information set continuously increases. Your beliefs about the world continuously expand and change; they evolve.

For my purposes, I will define news as any text-oriented material that randomly arrives to some receiver and which must be stored and processed in order for that receiver to learn about the latent messages contained in that text. This is exactly what you do every morning when you read the newspaper or check a news service on your smartphone.

At one time, defining news as text in a business context would have seemed logical to most business people because there were subscription-based news clipping services, now called *media monitoring services*. Their functions and services are varied as should be expected, but generally, they provide summaries of news from diverse sources allowing subscribers to get an almost panoramic view of issues and developments in a subject area germane to them. This had value to them because

[6] News shocks are studied in economics. See Barsky and Sims (2011), Arezki et al. (2015), and Beaudry and Portier (2006) for examples about news-driven business cycle research.

it allowed them to survey a wider array of events and happenings than they could if they themselves had to peruse a large volume of sources such as newspapers and magazines, which would be very costly in terms of the value of their time. The services, in other words, provided an economic benefit in reduced news gathering time.[7]

The text news I am concerned with extend beyond what the clipping services offered. I am interested in text-based news in the form of company and product reviews, call center logs, emails, warranty claims logs, and so on in addition to competitive advertisements and the traditional "news" about current events and technology, financial, political, regulatory, and economic trends and developments. This form of text-based news is captured and stored *text data*. It is, in fact, no different from any news you might commonly think about. A product review is an example. There is no way to predict before the arrival of a review what a random customer would write and submit. There is also no way to predict beforehand the message, tone, and sentiment of that review. In addition, there is no way to predict the volume of reviews to arrive at the business or any online review site.

Text in a newspaper is logically and clearly written. Journalists receive a lot of training in how to structure and write an article. Their text is structured. Now, however, there is no way to say beforehand what form text will take. The text in product reviews, for example, is *unstructured*. In most instances, it is free-form with abbreviations, foreign words, no punctuation, all upper case, incomplete sentences and even no sentences, just words or symbols.

Text are data just like any other type of data. They have to be extracted from some central collection site or initial repository of the text-data, transformed into a more storable form, and then loaded into a main repository such as a data lake, data warehouse, or data mart. Basically, text data must pass through an *extract-translate-load* (*ETL*) process. The final processing of text data for the informational insight contained within the text messages differ to some extent from the final processing of numeric data which I discuss next, but it is more complicated because of many issues associated with text. See Paczkowski (2020) for an extensive discussion of text processing for new product development. Finally, see Paczkowski (2020) for some discussion of the *ETL* process and data warehousing, data marts, and data lakes.

1.4.3.2 Numeric Data

People most often interpret data as numerics, numbers to which arithmetical operations can be applied and which can be graphed for visualization. This is correct to an extent, but it is short-sighted because there is more to numeric data than meets

[7] Services such as Flipboard and Smartnews are now available online that provides the same service. See the Wikipedia article "Media monitoring service" at https://en.wikipedia.org/wiki/Media_monitoring_service. Last accessed December 15, 2019.

the eye. Numerics can be classified in many different ways. On the one hand, they can be classified, for example, as integers or continuous (also referred to as floating point) numbers. An integer is a whole number without a decimal representation. They appear quite often in computations and can produce problems if they are not correctly identified. For example, in versions of Python prior to 3.x, dividing two integers resulted in an integer even though you might expect a decimal part to the quotient. For example, $5/2 = 2$ which you intuitively might expect is 2.5. Basically, the "floor" of the quotient was returned where the floor is the largest integer not exceeding the correct quotient.[8] Python versions after 3.x do not have this problem since the result is now coerced or cast as a continuous number. A continuous number has a decimal place which can "float" depending how the number is written. For example, you could write 314.159 or 3.14159×10^2 or $3.14159e2$. Hence, these numbers are called *floats*. The numbers, 1.0, 2.5, and 3.14159 are floats. The integer-divide problem does not hold for these (obviously). So, $5/2 = 2.5$ as expected whether the 5 or 2 are integers or floats.

Numerics are classified another way. Stevens (1946) proposed four categories for number which are still used and which appear in most statistics textbooks. In increasing order of complexity, they are

- nominal;
- ordinal;
- interval; and
- ratio.

Nominal and ordinal numbers are integers while interval and ratio are continuous. This scale division has its share of critics. See, for example, Velleman and Wilkinson (1993). I discuss these scales in Chap. 4 for Data Visualization.

Knowing the scale is important for many statistical, econometric, and machine learning applications. For example, a text variable may classify objects in a DataFrame. "Region" is a good example. As text, however, such a variable cannot be used in any computations simply because you cannot calculate anything with words except, perhaps, to count them. The words, however, could be *encoded* and then the encodings used in calculations. If "Region" consists of "Midwest", "Northeast", "South", and "West" (the four U.S. Census regions), then "Region" could be *dummy* or *one-hot* encoded with nominal values 0/1 with one set of values for each region. Or it could be *LabelEncoded* as 0, 1, 2, and 3 (in the alphanumeric order of the regions) where these values are also nominal even though they may, at first glance, appear to be ordinal. Management levels in a company, as another example, could also be dummy encoded or LabelEncoded but for the latter encoding the values would be ordinal (assuming management levels are in order of authority or rank). I will discuss variable encoding in Chap. 5 when I discuss data preprocessing.

[8] See https://en.wikibooks.org/wiki/Python_Programming/Operators#cite_note-1. Last accessed January 28, 2010.

Dates and times are also stored as numerics. Typically, date and time go together and the combination is collectively referred to as a *datetime*. For example, an order is placed on a certain date and at a certain time. The combination date and time is stored as one unit –the *datetime*– which is a numeric value. This characterization of date and time as a numeric is actually a very complex topic and beyond the scope of this book. There is ample documentation on Python's datetime variable online. There is also a multitude of ways to manipulate datatime data to extract year, month, day, day-of-week, hour, and so on. I will review date manipulations in Chap. 7 as part of the discussion about collapsing the spatial dimension of the Data Cube and converting from one calendar frequency to another (e.g., converting from monthly to quarterly data). For a detailed, technical discussion of calendrical calculations, see Dershowitz and Reingold (2008).

Numeric data differ from text data in that numeric data are *structured*, or at least can be structured. By structured, I mean the data are placed into well-defined columns in a DataFrame. If a variable is nominal, then all the values in that column are nominal; if continuous, then all the values in the column are continuous. The same holds for datetime data. There are also usually several different data types in a Pandas DataFrame. The data types can be listed using the *info()* method as I illustrate in Fig. 1.7.

```
##
## Create a Pandas DataFrame using a dictionary
##
dt = {'supplier':[ 'A', 'A', 'A', 'B', 'B', 'C'],
      'averagePrice':[ 1.25, 2.50, 3.10, 2.75, 4.25, 5.00 ],
      'onTime':[ 1, 1, 0, 1, 0, 0 ],
      'dateDelivered':[ '1/11/2020', '1/12/2020', '1/13/2020', \
                        '1/14/2020', '1/15/2020', '1/16/2020' ]}
df = pd.DataFrame( dt )
##
## Convert the delivery date to a datetime variable
##
df.dateDelivered = pd.to_datetime( df.dateDelivered )
##
## Display the information about the DataFrame
##
df.info()
```

```
<class 'pandas.core.frame.DataFrame'>
RangeIndex: 6 entries, 0 to 5
Data columns (total 4 columns):
 #   Column         Non-Null Count  Dtype
---  ------         --------------  -----
 0   supplier       6 non-null      object
 1   averagePrice   6 non-null      float64
 2   onTime         6 non-null      int64
 3   dateDelivered  6 non-null      datetime64[ns]
dtypes: datetime64[ns](1), float64(1), int64(1), object(1)
memory usage: 320.0+ bytes
```

Fig. 1.7 This illustrates information about the structure of a DataFrame. The variable "supplier" is an object or text, "averagePrice" is a float, "ontime" is an integer, and "dateDelivered" is a datetime

1.4.3.3 Data: A Combined View

I will not restrict myself to discussing just numbers or just text in this book. To have a succinct label, I will simply refer to both types as *data*. I will discuss how the specific types of data are handled throughout this book.

This single view of data has an advantage because it fits the Big Data paradigm. There have been numerous attempts to define Big Data but the most common definition focuses on three characteristics: *Velocity*, *Volume*, and *Variety*. Velocity refers to the speed at which data are no collected and added to a database such as a data store, data lake, data warehouse, or data mart. Volume refers to the sheer amount of data. Variety refers to the data types: text, numerics (in all the forms I discussed above), images, audio, and many more. It is the Variety aspect I am referring to when I state that my conception of "data" fits the Big Data paradigm. See Paczkowski (2018) and Paczkowski (2020) for discussions of Big Data.

1.4.4 The Information Component

Information is a word everyone uses every day. Everyone believes its definition is so intuitive and commonsensical that it does not require an elaboration. I was guilty of this until now because I said repeatedly that information is needed, that it is hidden inside data, and that it must be extracted. But I never defined information! If a definition is given by someone, it is quickly derided and dismissed as purely "academic." But define it I must. However, this is not easy. For some discussions about the complexities of defining information, see Floridi (2010), Hoffmann (1980), and Mingers and Standing (2018). For an in-depth historical and philosophical treatment of what is information, see Adriaans (2019) and Capurro and Hjorland (2003). Adriaans (2019), for example, states that

> The term "information" in colloquial speech is currently predominantly used as an abstract mass-noun used to denote any amount of data, code or text that is stored, sent, received or manipulated in any medium. The detailed history of both the term "information" and the various concepts that come with it is complex and for the larger part still has to be written .... The exact meaning of the term "information" varies in different philosophical traditions and its colloquial use varies geographically and over different pragmatic contexts. Although an analysis of the notion of information has been a theme in Western philosophy from its early inception, the explicit analysis of information as a philosophical concept is recent, and dates back to the second half of the twentieth century. At this moment it is clear that information is a pivotal concept in the sciences and humanities and in our everyday life. Everything we know about the world is based on information we received or gathered and every science in principle deals with information. There is a network of related concepts of information, with roots in various disciplines like physics, mathematics, logic, biology, economy and epistemology.

Rather than try to define information, it might be better to note its characteristics or attributes. Adriaans (2019) notes that information has a major characteristic. It is additive. This means that "the combination of two independent datasets with the same amount of information contains twice as much information as the separate individual datasets." Economists refer to this as increasing returns to scale: doubling the inputs yields more than double the output. This is very important because it indicates that one data set will not suffice for solving a business problem and that several data sets may have to be used. As I noted above, however, these data sets have to be joined or merged since statistical, econometric, and machine learning methods require just one data set. The merged data set reflects the information contained in the individual data sets, but it is actually more than the separate sets. For example, a DataFrame that contains transactions data by *CID* can only tell you about those transactions but nothing about the customers involved. A separate DataFrame on the customers (e.g., their locations, tenure as a customer, customer satisfaction, and credit worthiness) is of limited use because it tells you nothing about their buying patterns over time. Combine the two DataFrames, however, and you have a lot of potential information that you can extract. The size and complexity of what you have increases. For example, you can extract buying patterns by marketing regions. The additivity effect is, in fact, the main reason for merging two or more DataFrames: to take advantage of it.

The additivity effect, in part, reflects a quantity characteristic to information. The cost curve in Fig. 1.1 implies this quantity characteristic. But there is also a quality characteristic that cannot be ignored. Information can also be viewed as varying from *Poor* to *Rich*. I illustrate these two extreme types along an *Information Quality Continuum* in Fig. 1.8. See Paczkowski (2020) for a discussion of the Continuum for new product development. The two adjectives, "Poor" and "Rich", imply a quality aspect to information indicated by the form the information takes. I also show the forms in Fig. 1.8.

Poor Information usually takes the form of simple means, proportions, and their ilk that tell you something about your problem, but not much. They lack depth and the ability to imply any significance or substance action. Knowing the mean sales for your product, for instance, is not very actionable. Graphs and tabulations (i.e., tabs) are a slight improvement and do better at summarizing means and proportions, but they are still simplistic. The graphs I am referring to are the pie and bar charts ubiquitous in business presentations. They actually form the basis for *infographics*. Tabs are in the same category as graphs in this regard. Scientific graphs, which are better referred to as *data visualizations*, are a different breed of graphic displays that are more powerful and insightful. I will discuss scientific data visualization in Chap. 4.

Rich Information takes the form of complex interactions and cause and effect relationships. It goes beyond the obvious or mildly informative Poor Information to instill a "Wow!" factor in you. For example, a form of analysis called correspondence analysis can reveal patterns and relationships in a cross-tabulation table that the table itself just does not show. The table actually hides as much information as the raw data. The results from this type of analysis are more insightful. I provide an example in Chap. 8. *Key Driver Analysis* (i.e., the explanation of why business results are they way they are; for example: what determines customer satisfaction) and regression analysis to calculate price elasticities (essential for optimum product and service pricing) are other examples. See Paczkowski (2018) and Paczkowski (2020) for discussion about pricing analytics and elasticities.

The forms comprising Poor Information result from *Shallow Data Analytics*. This type of analytics makes the simplest, almost primitive, use of data in part because advanced methodologies are unknown or their use and applicability to a problem are not well understood. *Deep Data Analytics* provides a different, more sophisticated, insightful, and actionable level of information that is Rich Information.

The forms of information at the extremes of the Continuum, and anything in between, are input into meeting three information objectives:

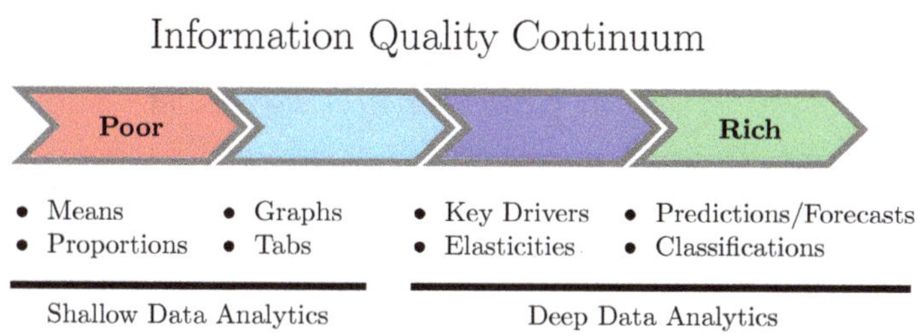

Fig. 1.8 Not only does information have a quantity dimension that addresses the question "*How much information do you have*?', but it also has a quality dimension that addresses the question "*How good is the information*?" This latter dimension is illustrated in this figure as varying degrees from Poor to Rich

1. describing what did or is currently happening in your business;
2. predicting what will happen under different conditions; and
3. classifying objects (e.g., customers, loan applicants, suppliers, agents).

The first objective is met with tools at the shallow side of data analytics. This is usually referred to as *Business Intelligence*. The second and third objectives are met by tools at the deep side of data analytics. These are in the domain of *BDA*: *the Deep Data Analytics for extracting a lot of Rich Information from a complex cubical organization of data, whether numeric or text*. Business Intelligence, except for a casual mention every now and then in this book, is not my concern. Shallow Analytics—involving means and proportions—is also not my focus, although they do play a role, albeit minor, in *BDA* as you will see. In terms of Fig. 1.2, *BDA* is the Information Extractor process. The quality of information extracted depends on whether your objective is to *predict* or *classify*. To predict means to say what will happen to a quantity or object of interest (e.g., unit sales) under different conditions (e.g., price is lowered 1%, 2%, or 3%). The methodologies are ordinary or logistic regression. To classify means to assign objects to groups. The methods are logistic regression, naive Bayes, Support Vector Machines, and clustering. Table 1.2 summarizes the extraction methods and where I discuss them in this book.

Example problem	Extractor method	Chapters
Price elasticity	*OLS* regression	6, 10
	Logistic regression	11
Geographic effect	OLS regression	6, 10
	Logistic regression	11
	Decision trees	11
Segment customers/suppliers	Decision trees	11
	Logistic regression	11
	Support vector machines	11
	Naive Bayes	11
	Clustering	12
Credit default risk	Decision trees	11
	Logistic regression	11
	Support vector machines	11
	Naive Bayes	11
Sales prediction/what-if	*OLS* regression	6, 10
	Logistic regression	11
Key drivers	*OLS* regression	6, 10
	Logistic regression	11
	Decision trees	11
Price segmentation	Logistic regression	11
	Decision trees	11
	Clustering	11
Forecasting	Time series analysis	7
Data visualization	Graphs and tables	4, 8
Dimensionality reduction	*PCA*	10

Table 1.2 Information extraction methods and chapters where I discuss them

1.5 Analytics Requirements

Let me reconsider the cost curve in Fig. 1.1. There is another cost of analysis that, when combined with the cost of approximation, determines the level of information you will have for your decisions. It is the *Cost of Analytics* which I illustrate in Fig. 1.9. This reflects the skill set of your data analysts, now called *data scientists*, as well as the amount and quality of your data. The more data you have, the higher the cost of working with it and extracting the most information from it simply because the data structure is more complex as I discussed above. I discuss data structure in more detail in Chap. 2. At the same, time, the more data you have the higher the needed skill level of your data scientists because they need to have more expertise and knowledge to access, manipulate, and analyze that data. Analytics is not cheap. The more Rich Information you want, the higher the cost of getting it. The Cost of Analytics curve is upward sloping.

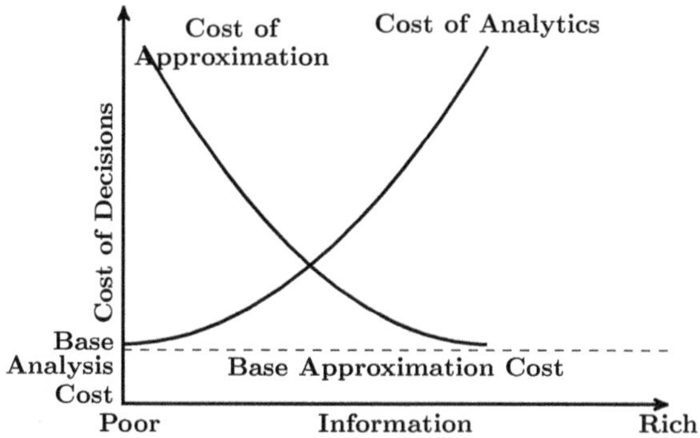

Fig. 1.9 Cost curves for Rich Information extraction from data

Although the cost of analytics rises with the level of desired Rich Information, the costs can be controlled. You, even as a data scientist, are as responsible for managing these costs of analytics as any other manager in your business. Good, solid, cost effective data analysis requires that you have a

1. theoretical understanding of statistical, econometric, and (now in the current era) machine learning techniques;
2. data handling capabilities encompassing data cleaning and wrangling; and
3. data programming knowledge in at least one software language

so that you can effectively provide the Rich Information your decision makers need. These three requirements are not independent but are components of a synergistic whole I illustrate in Fig. 1.10.

The three components are displayed on the vertices of an equilateral triangle. An equilateral triangle is used because no matter how the triangle is rotated, the message is the same. The implication is that one component is not more important than the other two. All three contribute equal weight to solving a business problem. In addition, there is a two-way interconnection of each vertex to the next as shown in Fig. 1.10. I discuss these interconnections in the next subsections.

1.5.1 Theoretical Framework

The need for a theoretical framework for solving a business problem may seem obvious: the framework is a guide for analysis, any analysis, whether it be for a business problem such as my focus, or an academic or basic research focus. You need to know how to organize thoughts, identify key factors, and eliminate impossible or trivial relationships. Basically, a theoretical framework keeps you out of trouble. The framework may be, and probably will be, incomplete and maybe

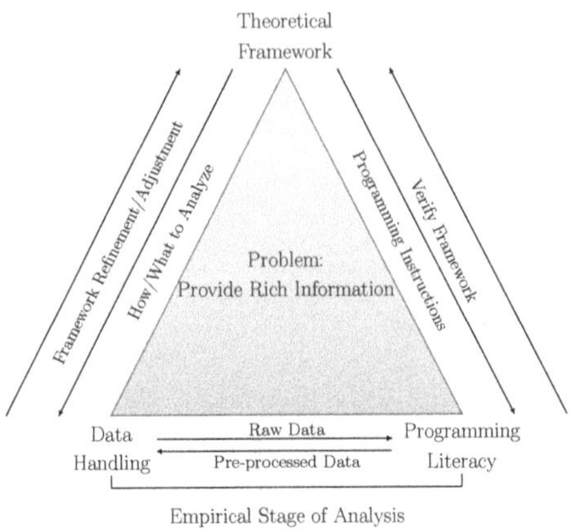

Fig. 1.10 The synergistic connection of the three components of effective data analysis for business problems is illustrated in this triangular flow diagram. Every component is dependent on the others and none dominates the others. Regardless of the orientation of the triangle, the same relationships will hold

questionable in some regards, but a framework nonetheless helps. A famous quote by Leonardo da Vinci[9] sums up the concept:

> He who loves practice without theory is like the sailor who boards ship without a rudder and compass and never knows where he may cast.

There are two aspects to a theoretical framework: *domain theory* and *methodological theory*. Domain theory is concerned with subject matter concepts and principles, such as those applicable to business situations. Economic theory, financial theory, and management science theory, just to name three, are relevant examples. The methodological theory is concerned with the mathematical, statistical, econometric, and machine learning concepts and principles relevant for working with data to solve a problem such as a business problem. Ordinary least squares (*OLS*) theory is an example. Domain theory and methodological theory often work together as do, for example, economic theory and econometrics. My focus in this book in not on domain theory, but methodological theory for solving business problems.

You need to know not just the methodologies but also their limitations so you can effectively apply the tools to solve a problem. The limitations may hinder you or just give you the wrong answers. Assume you were hired or commissioned by a business executives (e.g., a *CEO*) to provide actionable, insightful, and useful Rich Information relevant for a supply chain problem. If the limitations of a methodology prevent you from accomplishing your charge, then your life as an analyst would be short-lived to say the least. This will hold if you do not know these limitations or

[9] Source: https://www.leonardodavinci.net/quotes.jsp. Last accessed November 14, 2019.

choose to ignore them even if you do know them. Another methodological approach might be better, one that has fewer problems, or is just more applicable.

There is a dichotomy in methodology training. Most graduate level statistics and econometric programs, and the newer *Data Science* programs, do an excellent job instructing students in the theory behind the methodologies. The focus of these academic programs is largely to train the next generation of academic professionals, not the next generation of business analytical professionals. Data Science programs, of which there are now many available online and "in person," often skim the surface of the theoretical underpinnings since their focus is to prepare the next generation of business analysts, those who will tackle the business decision makers' tough problems, and not the academic researchers. Something in between the academic and data science training is needed for the successful business data analysts.

1.5.2 Data Handling

Data handling is not as obvious since it is infrequently taught and talked about in academic programs. In those programs, beginning students work with clean data with few problems and that are in nice, neat, and tidy data sets. They are frequently just given the data. More advanced students may be required to collect data, most often at the last phase of training for their thesis or dissertation, but these are small efforts, especially when compared to what they will have to deal with post-training. The post-training work involves:

- identifying the required data from diverse, disparate, and frequently disconnected data sources with possibly multiple definitions of the same quantitative concept;[10]
- dealing with data dictionaries;
- dealing with samples of a very large database—how to draw the sample and determine the sample size;[11]
- merging data from disparate sources;
- organizing data into a coherent framework appropriate for the statistical/econometric/machine learning methodology chosen; and
- visualizing complex multivariate data to understand relationships, trends, patterns, and anomalies inside the data sets.

This is all beyond what is provided by most training programs.

[10] Revenue is a good example. Students just learn about "revenue" which is price times quantity. There is, however, gross revenue, revenue net of returns, revenue before discounts but net of returns, revenue net of discounts and returns, billed revenue, booked revenue, and the list goes on.

[11] At AT&T, detailed call records were maintained for each customer: call origin, call destination, call length in minutes, billing information, etc. which just intuitively would amount to millions of records of information per day. The sheer volume made it impractical to use all the data so samples, say 5% or even 1%, were used.

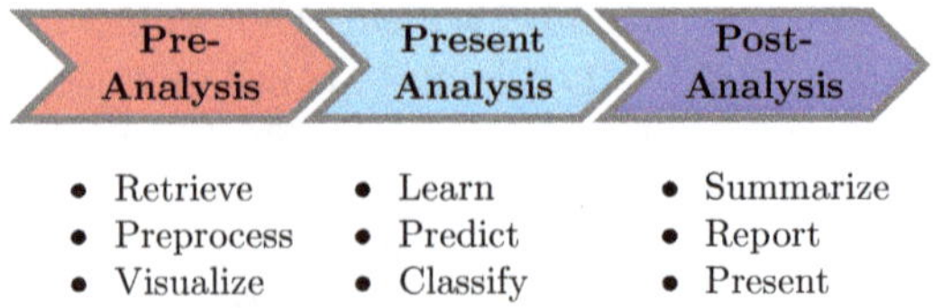

Fig. 1.11 Programming roles throughout the Deep Data Analytic process

1.5.3 Programming Literacy

Finally, there is programming. First, let me say that there is programming and then there is programming. The difference is scale and focus. Most people, when they hear about programming and programming languages, immediately think about large systems, especially ones needing a considerable amount of time (years?) to fully specify, develop, test, and deploy. They are correct regarding large scale, complex systems that handle a multitude of interconnected operations. Online ordering systems easily come to mind. Customer interfaces, inventory management, production coordination, supply chain management, price maintenance and dynamic pricing platforms, shipping and tracking, billing, and collections are just a few components of these systems. The programming for these is complex to say the least and generally require a team effort to implement let alone maintain.

Business analysts may not be involved in this type of programming although they might have to know about and access the data from the subsystems of one or more of these larger systems. And major businesses are composed of many large systems! The business data analyst might have to write a program (*a.k.a.*, code) to access the data,[12] manipulate the retrieved data, and so forth; basically, write programming code to do all the data handling described above. Even after the data are collected and manipulated for a particular problem, the programming effort may still not be over. There are other programming tasks required as part of the analysis process. Programming, in other words, has a *pre-analysis*, *present analysis*, and *post-analysis* role that I illustrate in Fig. 1.11. For this, you need to know programming and languages. As an example, suppose a regression model is estimated for a product demand problem. Programming may be required to further explore the distributional properties of key measures predicted from the estimated model. Monte Carlo and Bootstrap simulations could be used, but these require some level of programming to be effective.

[12] The code could be structured query language (*SQL*) code, for example.

• Data visualization	• Statistical modeling
• Probabilistic programming	• Data manipulation
• Machine learning	• Numerical calculations
• Deep learning frameworks	• Natural language processing

Table 1.3 These are some major package categories available in Python

There are many programming languages available.[13] Only a few are needed for most business data analysis problems. In my experience these are:

- Python
- *SQL*
- R

Julia should be included because it is growing in popularity due to its performance and ease of use.

Notice that I do not include spreadsheets. They are well entrenched, but this does not mean they are the correct tool to use. They have problems that greatly hinder their use in any serious business data analytic work, and actually may make that work impossible. Yet, people try to make spreadsheets do the work. Here are seven reasons why you should not use spreadsheets for business data analytics:[14]

1. They lack database management functionality.
2. They do not handle large data sets.
3. They lack data identification methods.
4. They often enable data to be entered in multiple worksheets.
5. They cannot handle complex data structures.
6. They lack basic data manipulations such as joining, splitting, and stacking.
7. They have limited data visualization methods.

The fact that there are seven reasons indicates a problem. Software such as Python and R have definite advantages for *BDA*. They are designed to handle and manage large data sets and do this efficiently! They have many add-on packages that extend their power. Python, for example, has 130,000+ packages,[15] and growing. Some major package categories are listed in Table 1.3. R has a similar package structure and active support community.

What about the Structured Query Language (*SQL*), which is the foundational language for accessing, and, by default, organizing vast quantities of data? *SQL*,

[13] Wikipedia lists 693. The list includes "all notable programming languages in existence, both those in current use and historical ones.... Dialects of BASIC, esoteric programming languages, and markup languages are not included."" Source: https://en.wikipedia.org/wiki/List_of_programming_languages. Last accessed November 2, 2019.

[14] Source: Paczkowski (2020).

[15] https://www.quora.com/How-do-I-know-how-many-packages-are-in-Python. Last accessed January 12, 2021.

introduced and standardized in the 1980s, enabled the development of data stores, data lakes, data warehouses, and data marts so common and prevalent in our current business environment. These are the major data sources you will work with, so knowing *SQL* will help you access this data more efficiently and effectively: efficiently because you can access it faster with less resources; more effectively because you will be able to access exactly the data needed and in the format needed. I will not discuss *SQL* since it is outside the scope of this book although I do make some comments about it in Chap. 2.

A subtle shortcoming of spreadsheets is their inability to aid documenting workflow for reproducibility. These are two separate but yet connected tasks important in any research project; they should not be an afterthought. Trying to recall what you did is itself a daunting task usually subordinated to other daunting tasks of research. Researchers typically document after their work is done but at this point the documentation is incomplete at best and error prone at worse.

Documentation is the logging of steps in the research process including data sources, transformations, and steps to arrive at an answer to management or a client's business question. *Reproducibility* is the ability to rerun an analysis. Quite often, analysts produce a report only to have management or the client call, even months later, requesting either a clarification of what was done, perhaps for a legal reason, or to request further analysis. This means a business data analyst must recall exactly what was done, not to mention what data were used. This is where reproducibility comes in.

The tool I recommend as a front-end statistical/econometric/machine learning and programming framework is Jupyter. With a notebook paradigm, it allows documentation and reproducibility because of a cell-orientation. There are two cells:

1. a code cell where code is entered; and
2. a "markdown" cell where text is entered.

Jupyter is also an ecosystem that it will handle multiple languages, Python being just one of them. It will also handle R, Fortran, Julia, and many more. In fact, about 120!

1.5.4 Component Interconnections

The three components of the triangle in Fig. 1.10 are not independent of each other, but are intimately interconnected. The connecting arrows in the figure are self-explanatory.

Chapter 2
Data Sources, Organization, and Structures

I stated in Chap. 1 that information is hidden, latent inside data so obviously you need data before you can get any information. But just saying, however, that you need data first is too simplistic and trivial. Where data originate, how you get them, and what you do with them, that is, how you manipulate them, before you begin your work is important to address. You need to ask four questions:

1. Where do I get my data?
2. How do I organize my data?
3. How is my data structured?
4. How do I extract information from my data?

I will address the first three questions in this chapter and the fourth question throughout this book.

The first question is not easy to answer so I will discuss it at length. There are many sources for data, both internal and external to your business, which make the first question difficult to answer. Internal data should be relatively easy to obtain, although internal political issues may interfere. In particular, different internal organizations may maintain data sets you might need, but jealously safeguard them to maintain power. This is sometimes referred to as creating data *silos*. See Wilder-Jame (2016) for a discussion about the need to break down silo barriers. External data would not have this problem although they may be difficult to locate, nonetheless.

Addressing the first question is only part of your problems. You must also store and organize that data. This means putting them into a format suitable for the type of analysis you plan to do. That organization may, and most likely will, change as your analysis proceeds. This will become clear as I develop organization concepts. Part of organizing data is documenting them, so I will also touch on this topic in this chapter.

Once you have your data, you must then begin to understand its structure. Structure is an overlooked topic in all forms of data analysis so I will spend some

W. R. Paczkowski, *Business Analytics*,
https://doi.org/10.1007/978-3-030-87023-2_2

time on it. Knowing structure is part of "looking" at your data, a time-honored recommendation. This is usually stated, however, in conjunction with discussions about graphing data to visualize distributions, relationships, trends, patterns, and anomalies. Sometimes, you are told to print your data and physically look at it as if this will reveal something you did not previously know. This may help you identify large missing value patterns or coding errors, although there are other more efficient ways to gain this insight such as using software to identify and portray missing value patterns. If your data set is large, especially with thousands of records and hundreds of variables, physically "looking" at it is impractical to say the least. Visualization, whether by graphs or physical inspection, while important, will not reveal anything about the types of structure I am concerned with, structure in which, for example, cases are subsets of larger entities so that they are nested under these larger entities; there is a hierarchy. This understanding can only come from knowing what your data represent and how they were collected. I discuss data visualization in Chap. 5.

2.1 Data Dimensions: A Taxonomy for Defining Data

There are five aspects or dimensions to data that bear on how they can be analyzed and with what tool. There is a taxonomy consisting of:

1. Source
2. Domain
3. Level
4. Continuity
5. Measurement Scale

Each of these introduces its own problems. Together, these dimensions define a *Data Taxonomy* which I illustrate in Fig. 2.1. I will discuss the taxonomy's components in the following subsections.

2.1.1 Taxonomy Component #1: Source

Data do not just magically appear. They come from somewhere. Specifically, they come from one of two sources: primary data collection and secondary data collection. Knowing your data's source is important because poorly controlled data collection can mask errors that lead to the outliers, thus jeopardizing any analysis for decision making. All data are subject to errors, sometimes due to human and mechanical measurement failures, other times to unexplainable and surprising random shocks. These are data *anomalies* or *outliers*. The outliers can be small or large, innocuous or pernicious. It is the large pernicious ones that cause the grief.

The two sources most business people immediately mention are transactions and surveys. But this is shortsighted. Depending on your business and the nature of your

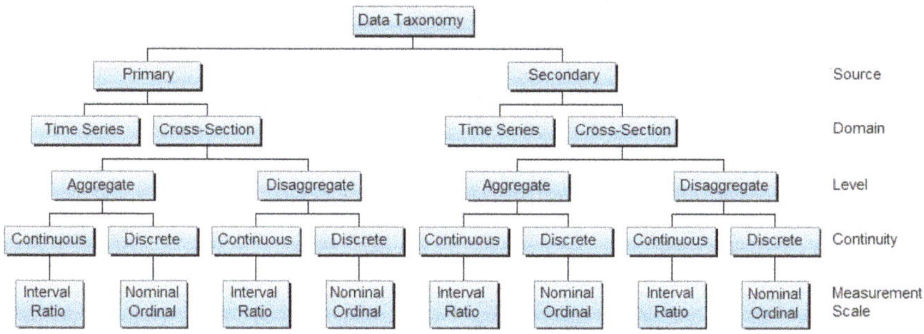

Fig. 2.1 A data taxonomy. Source: Paczkowski (2016). Permission to use granted by SAS Press

problem, experiments should also be on the list and, in the modern technology era, sensor data on processes and actions must be included.

I further classify data sources by their *endogeneity* and *exogeneity*. *Endogenous data* are generated by, and are intimately connected with, your business. They are generated by your business itself; that is, by the processes, functions, and decisions made for your operations and market presence. This includes sales, price points, messages and claims, purchases (i.e., inputs), taxes paid or owed, interest payments paid on debt, and production flows. They result from decisions you made in the past, since it should be obvious that you cannot collect data on actions that have neither been decided nor enacted.

Exogenous data are generated outside your business by decisions and forces beyond your control. Examples are numerous: economic growth and business cycle patterns; monetary and fiscal policy; federal and state tax policies; political events; international developments, events, and issues; new technology introductions; regulatory actions; and competitive moves and new market entrants just to mention a few. And you certainly cannot ignore pandemics such as the 2020 COVID-19 pandemic that had such as devastating business, economic, political, and personal impact.

Consider business cycle patterns. There are two phases to a business cycle: growth and recession.[1] During a growth phase, real GDP, the measure of the size of the economy, increases, so the economy expands; during a recession, it shrinks. These phases, however, are too simplistic because they could also be distinguished by their severity and length. A growth phase, for instance, could be mild or robust as well as prolonged. A recession could be mild or deep and severe, such as for the

[1] Some people define four phases which includes the peak and trough which are two turning points. See Stupak (2019) for example. Two phases will suffice for my discussion. Also see Zarnowitz (1992) for a historical treatment of business cycles.

2007–2009 recession, as well as short or prolonged.[2] All industries are affected in one way or another by the severity and length of these two phases.

Exogenous factors are further divided into two groups: *universal exogenous factors* and *local exogenous factors*. Universal exogenous factors affect all businesses in all industries, so the scope of impact is large and encompassing. The degree of the effect of a universal factor, however, varies. The business cycle is again an example. No business can escape its impact regardless of its severity or length. Berman and Pfleeger (1997) found very high and positive correlations between sector employment and real GDP in household furniture; miscellaneous plastics products; personnel supply services; plumbing and nonelectric heating equipment; stone, clay, and miscellaneous mineral products; and electric lighting and wiring equipment to mention a few. The positive correlations indicate that as real GDP rises, so does sector employment, and employment falls as real GDP falls.[3] They also found mediocre negative correlations in agricultural production; pipelines, except natural gas; crude petroleum, natural gas, and gas liquids; Federal government enterprises, *n.e.c.*; and private hospital sectors.[4] The effects are clearly different, but they are present nonetheless.

Local exogenous factors affect a specific firm or industry so their scope is narrow. These factors include competitive actions such as pricing, messaging, and new products. It also includes technological advances that have the potential to eliminate a whole industry or sector of the economy. The buggy whip industry is a commonly used example. See, however, Stross (2010) for an interesting view of this example.

Most business people focus on local exogenous data by asking questions such as "*What is my competition up to*?" or "*What are regulators planning*?" Many businesses have organizations responsible for tracking and collecting data, for example, on competitive prices and advertisements.[5] The importance of universal exogenous data is recognized, but is often of less concern from a data collection perspective, albeit depending on the industry. Berman and Pfleeger (1997) found that household furniture industry employment is highly correlated with movements in real GDP with a correlation coefficient of $r = 0.9591$. They also found that this sector's correlation between real GDP and final demand is $r = 0.7713$, also high and positive. So, a firm in the household furniture industry should track real GDP as well as other macroeconomic data (e.g., unemployment, industrial production index) that have a relationship with real GDP. On the other hand, they found that educational services and personal services, *n.e.c.* have very low correlations with

[2] See the National Bureau of Economic Research (*NBER*) September 20, 2010 announcement of the end of 2007–2009 recession: "Business Cycle Dating Committee, National Bureau of Economic Research" at https://nber.org/cycles/sept2010.html. Last accessed January 15, 2020. Also see the *NBER* calendar of peaks and troughs in the business cycle at https://nber.org/cycles/cyclesmain.html. The 2007–2009 recession lasted 18 months and was the longest in the post-World War II period.

[3] Recall that correlation just shows the degree of association, not cause-and-effect.

[4] "n.e.c." is a standard abbreviation for "Not Elsewhere Classified."

[5] I use to do this myself for AT&T's Computer System division in the early 1990s.

real GDP ($r = 0.0343$ and $r = 0.0186$, respectively) and equally low correlations with final demand ($r = 0.3000$ and $r = 0.2572$, respectively). Firms in these industries most likely will not track macroeconomic data as closely as the others.

Tracking exogenous data means that internal databases must be constructed and maintained by updating the data elements as new data become available and making these databases internally available in the business. These data, however, could also be accessed and downloaded from outside sources that undertake these tasks. Accessing and downloading could be free if the provider is a government agency. Many U.S. macroeconomic time series, for example, can be freely downloaded from the *Federal Reserve Economic Database* (*FRED*).[6] The World Bank is an equally free source for international economic times series and a host of other data.

What about endogenous data? There are five sources: transactions, experiments, surveys, sensors, and internal. Some elementary statistics books list only four endogenous data sources: observational, experimental, surveys, and census. See Moore and Notz (2017), for example.

Transactions are probably the most commonly thought of source for *BDA*. This is because they are the heart of any business. Without a transaction, your business would cease to exist since, after all, a transaction is a sale. Transactions data could come from company-owned and maintained web sites, traditional brick and mortar store fronts, and purchases by wholesalers who order through a dedicated sales force. Regardless of where these data come from, a central feature is that they flow into the business from an outside source (i.e., a customer) and this flow is beyond your control, but not influence. You cannot control when customers will make a purchase, or how much they will purchase, or even if a long-standing customer is still or will remain a customer. Some customers "die" (i.e., cease buying from you) with you unaware of this. See Schmittlein et al. (1987) on this issue. You can influence the flow of transactions through pricing strategies (which includes discount structures), messages, and claims. Basically, by manipulating the *Marketing Mix*.

Regardless of the source, a transactions database contains (at a minimum):

- order number (*Onum*);
- customer ID (*CID*);
- transaction date;
- product ID (*PID*);
- units sales;
- price (list and pocket);[7]
- discounts;
- order status (fulfilled or not); and
- sales rep ID (*SID*) if applicable

[6] See https://fred.stlouisfed.org/. Last accessed January 23, 2020.

[7] See Paczkowski (2018) on pocket prices.

This database, however, would be linked to other data bases that have data on the customers (e.g., location, longevity as a customer, credit history), products (e.g., description), and fulfillment (e.g., shipped, back-ordered, or in process). Linking means there is a common identifier or "key" in two or more databases that allow you to *merge* or *join* them. This is how you gain the additivity effect I mentioned in Chap. 1. I discuss data organization in Sect. 2.2 and merging in Sect. 3.3.

Unlike transactions data, experimental data arise from controlled procedures following well defined and articulated protocols. The controls are under the direction of, or are determined by, a market researcher or an industrial statistician. Experiments could be conducted for market tests of new products in the new product pipeline or an industrial experiment to determine the right settings for a manufacturing process or product development. The latter is outside the scope of this book. See Box et al. (1978) for the classic treatment of experimental design with examples of industrial experiments. Also see Diamond (1989).

Market experiments result in data that business data analysts could use; they probably have little need for industrial experiments. Market experiments are typically *discrete choice experiments* designed to elicit responses from customers or key stakeholders (e.g., employees) about a variety of business issues. Pricing, product design, competitive positioning, and messaging are four examples. A discrete choice experimental study involves defining or creating a series of *choice sets*, each consisting of an optimal arrangement of the levels of *attributes*. The optimal arrangement is based on *design of experiment* (*DOE*) procedures and principles which are the protocols I mentioned above. Each arrangement in a choice set represents an object of interest such as a potential new product prototype. Study subjects (customers, employees, or stakeholders) are shown a series of choice sets from which they select one object from each set. If the objects are prototypes for a new product, then each choice set consists of several prototypes and potential customers are asked to select one from each set. Models, estimated from the choice responses, are used to predict *choice probabilities* (a.k.a., *take rates*), or shares such as *market share*, *share of wallet*, or *share of preference*. See Paczkowski (2018) for an extensive discussion of choice study design and estimation for product pricing. Also see Paczkowski (2020) for a discussion of the use of these models in the testing phase of new product development and for optimal messaging for new product launching.

Choice experiments are in the context of a survey. This is the only way to elicit choice responses from customers or key stakeholders. Surveys, however, are more broadly used. Examples are many: customer satisfaction, purchase intent, purchase behavior identification, and attitudes/opinions/interests (*AIO*) identification to mention a few. In jewelry customer surveys, for example, it is common to collect data on where jewelry customers currently shop (e.g., local store, malls, online), what they typically purchase (e.g., watches, broaches, pendants, etc.), how much they typically spend on a jewelry purchase, who they typically buy for (e.g., spouse, fiancée, friend, relative, or self), and the reason for the purchase (e.g., engagement, wedding, graduation, birth, or a personal reward for reaching a milestone). In addition, extensive batteries of demographics are collected including gender, age, employment, education, and so on.

All this data can be, and should be, compiled into one survey database for use in analyses beyond the original purpose of the surveys. This database is now a rich source of data and resulting information. For example, again consider jewelry surveys. Surveys spanning several years, each one containing data on the type of jewelry purchased, can be used to identify shifts in purchasing behavior in the industry that may not be evident in the company's transactions databases. These databases contain local endogenous data while a survey database contains universal exogenous data. A decline in the local endogenous sales data cannot be interpreted as a general shift in buying behavior (e.g., a shift to online purchases) because customers may simply be reacting to Marketing Mix changes. The universal exogenous data, however, indicate that an industry-wide shift is occurring perhaps due to lifestyle changes or age distribution shifts. Incidentally, the analysis of surveys conducted over time is a *tracking study*. See Paczkowski (2021b) for survey data analysis methods.

Sensors are becoming a prominent source of data. They are everywhere. They are in our homes, appliances, in cars, on street corners, in medical facilities and devices, in manufacturing plants, and in many more places and used in more applications than I could list here. Sensor data are measures of processes or actions. Sensors in automobiles measure speed and driving behavior (e.g., frequency of braking, breaking distances, veering outside a lane). Sensors on a manufacturing assembly line measure product throughput and product assembly at various stages of the production process (e.g., filling containers to the right height or amount). In medical devices, they measure patients' vital signs. See Paczkowski (2020) for discussions of sensors.

Sensor data can be used to identify otherwise undetectable problems. And they certainly can be used to predict future issues if data collected in real-time begin to deviate from a long-term trend. This is the case, for example, in a manufacturing setting where sensor readings that deviate from trend indicate a production problem that must be immediately addressed. Sensor data, being so extensive, contain a treasure trove of information, probably more so now than, say, a decade ago. We are only now beginning to develop ways to manage and analyze this data. See Paczkowski (2020) for some discussion of sensor data for new product development.

Internal data sources are everything else not covered by the previous categories. This includes *HR* employee records; financial accounts; manufacturing supplier data; production downtime data; and so forth.

Primary data are data you collect for a specific purpose. They are collected for a particular study and not for any other use. The collection techniques, criteria specifying what to collect, and data definitions are all designed for a particular research objective. The key is that you have control over all aspects of your data's collection.

Secondary data are collected for someone's purpose or research objective but then used directly or adapted for another study. If your study is the "other study" then you have little to no control over data generation short of copying it from one location to another. Your only control is over which series you use and how you use it; not the collection. Real GDP from *The Economic Report of the President* is an example. Data stored in a data warehouse or data mart that you access and use in a *BDA* study is another example of secondary data.

Since someone else specified all aspects of the data you are secondarily using, then you cannot control its quality which limits what you can say about the data. For example, you cannot say anything about the existence and magnitude of measurement errors without conducting your own detailed study of the data. See Morgenstern (1965) for a classic discussion of measurement errors in secondary economic data. Control is the key to the data. Some key questions you might ask are:

1. Where did the data come from?
2. What are its strengths and weaknesses?
3. How were variables defined?
4. What instruments (e.g. questionnaire) were used to collect the data and how good were they?
5. Who collected the data and how good was that person or organization (i.e., was he/she conscientious, properly trained)?

2.1.2 Taxonomy Component #2: Domain

There are two data domains: *spatial* (i.e., *cross-sectional*) and *temporal* (i.e., *time series*). Cross-sectional data are data on different units measured at one point in time. Measurements on sales by countries, states, industries, and households in a year are examples. The label "spatial" should not be narrowly interpreted as space or geography; it is anything without a time aspect. Time series, or *temporal domain* data, are data on one unit tracked through time. Monthly same-store sales for a 5-year period is an example.

Time series data have complications that can become very involved and highly technical. I discuss some of these below and further discuss time series in Chap. 7.

2.1.3 Taxonomy Component #3: Levels

There are two levels to data:

1. disaggregate, and
2. aggregate.

Disaggregate data are the most fundamental level although the boundary between disaggregate and aggregate is blurry. Data on consumer purchases collected by *point-of-sale* (*POS*) scanners is an example of disaggregate data while sales by stores and marketing regions are examples of *aggregate data.* Summation or averaging of disaggregate units, sometimes with weights applied to account for varying importances of the underlying measurement units, produce the aggregate data. Collapsing the Data Cube involves moving from disaggregate to aggregate data.

2.1.4 Taxonomy Component #4: Continuity

The continuity of data refers to their smoothness. There are two types:

1. continuous; and
2. discrete.

Continuous data have an infinite number of possible values with a decimal representation, although typically only a finite number of decimal places are used or are relevant. Such data are floating point numbers in Python. An example is a discount rate (e.g., a quantity discount). Discrete data have only a small, finite number of possible values represented by integers. Integers are referred to as *ints* in Python. They are often used for classification and so are categorical. Gender in consumer surveys is typically encoded with discrete numeric values such as "1 = Male" and "2 = Female" although any number could be used. The values are just for classification purposes and are not used in calculations, although there are exceptions.

The notion of a decimal place is important. Computers and associated visualization software interpret numbers differently from how we humans interpret them. For example, to the average person the number 2 has meaning in the context of "2 items" or "2 books"; so 2 is just 2. To a computer, 2 could be 2 as an integer or 2.0 as a float. Integers and floats are different. A float conveys precision while an integer does not, although the integer is understood to be exact. To say the distance between two objects is 2.1 feet is more precise than to say it is 2 feet, but yet 2 feet sounds very definite. In Python, integers are not restricted to a fixed number of *bits* where a bit is the basic unit used in computer technology. A bit is represented as a 0 or 1 which is a binary or *base 2* representation of numbers. There is no limit to the number of bits that can be used to represent an integer, except, of course, the size of the computer's memory. The implication is that an integer can be represented exactly in binary form. A floating-point number, however, uses a maximum of 53 bits for a double precision computer. Floats, therefore, cannot be represented exactly. This often leads to surprising results for calculations. For example, if you hand-calculate $1.1 + 2.2$, you get 3.3. However, using Python (and many other software languages) you get 3.3000000000000003. The reason is the internal representation of the floats. See the Python documentation "Floating Point Arithmetic: Issues and Limitations" for a detailed discussion.[8]

[8] This is for Python 3.8.2. It is available at https://docs.python.org/3/tutorial/floatingpoint.html. Last accessed March 12, 2020.

Most analysts use and are familiar with floating point numbers. Integers, are also common but require recognition. In survey research, for example, it is not uncommon to ask a person their gender. This is usually encoded as 1 or 2, perhaps "1 = Male" and "2 = Female". These are integers. They are also nominal. They are, by their nature, discrete so they can be used for categorization; floating point numbers cannot be used for categorization. Floats could be binned or grouped for categorization, but then the bins are encoded with integers.

In many software packages, integers and floats have to be pre-identified or *typed* if they are to be handled correctly. These languages *statically type* all objects. An example is C++. This is not the case with Python which *dynamically types* all objects. It determines the type from the context of the object at run-time. This is a great saving to the user because the time otherwise spent designating an object's type is reallocated to more useful work. Plus, there is less chance for error since the Python interpreter does all the work, not you. See VanderPlas (2017, pp. 34–35) for a discussion about object typing.

2.1.5 Taxonomy Component #5: Measurement Scale

There are four measurement scales attributed to Stevens (1946). They are controversial but generally most practicing data analysts adhere to them, more or less. See Velleman and Wilkinson (1993) for some discussion. The scales are:

1. nominal;
2. ordinal;
3. interval; and
4. ratio.

The nominal scale is the most basic since it consists of a numeric encoding of labels using discrete values. The exact encoding is immaterial and arbitrary. This is very common in market research surveys with a "*Buy/Don't Buy*" question as a good example. Statistically, only counts, proportions, and the mode can be calculated. Ordinal data, as the name suggests, are data for which order is important and must be preserved. This is also common in market research. A Likert Scale for purchase intent (e.g., *Not at all Likely, Somewhat Unlikely, Neutral, Somewhat Likely, Very Likely*) is an example. Three levels of management (e.g., Entry-level, Mid-level, and Executive) for *HR* data is another example. Counts, proportions, the mode, the median, and percentiles can be calculated. Means and standard deviations do not make sense. What is the meaning of the average level of three management levels? Sometimes, means and standard deviations are calculated for ordinal data collected on a Likert Scale, but this is controversial. See Mangiafico (2016) and Brill (2008) for discussions. I am on the side that believes Likert Scale data are ordinal and so means should not be calculated.

Both nominal and ordinal scaled data are discrete; interval and ratio are continuous. Interval data do not have a fixed zero as an origin which means the distance between two values is meaningful but the origin is meaningless since it can be changed. This implies that a ratio is meaningless. A thermometer scale is the classic example. Suppose you take two temperature readings on the Fahrenheit scale: 40 °F and 80 °F so the difference has meaning (80 °F is 40 °F hotter). But are you justified in saying that 80° is twice as hot as 40°? That is, does $^{80\,°\mathrm{F}}/_{40\,°\mathrm{F}} = 2$? Consider the same temperatures converted to the Celsius scale by the formula: $Celsius = (Fahrenheit - 32) \times {}^{5}/_{9}$. So, 40 °F = 4 °C and 80 °F = 27 °C. Clearly the higher temperature is not twice the lower on this scale (i.e., ($^{27\,°\mathrm{C}}/_{4\,°\mathrm{C}} \neq 2$)). You get a different answer by changing the scale, yet the sense of "hotness" is the same. Statistically, you can calculate counts, proportions, mode, median, mean, and standard deviation.

Ratio scaled data are the most commonly known type. Most economic and business data are ratio scaled. They are continuous with a fixed zero as an origin so the distance between values is meaningful but the origin is also meaningful. If you have zero sales in a calendar quarter, no change of scale (i.e., the origin) will suddenly give you non-zero sales. The same holds for different currencies if you have international sales. Sales of \$50 in one quarter and \$100 in the next quarter means the second quarter is twice the first regardless if dollars or Francs or Yen are used to measure sales: $^{k \times \$100}/_{k \times \$50} = 2$ where k is the exchange rate. So the origin acts as the reference point for making all calculations and comparisons. You can calculate counts, proportions, mode, median, mean, and standard deviation.

I show the four measurement scales in Fig. 2.2 including their complexity relationship and the allowable statistics for each. The complexity of data increases as you move up the scale. At the same time, the allowable statistics cumulatively build from simple counts and proportions to arithmetic means and standard deviations.

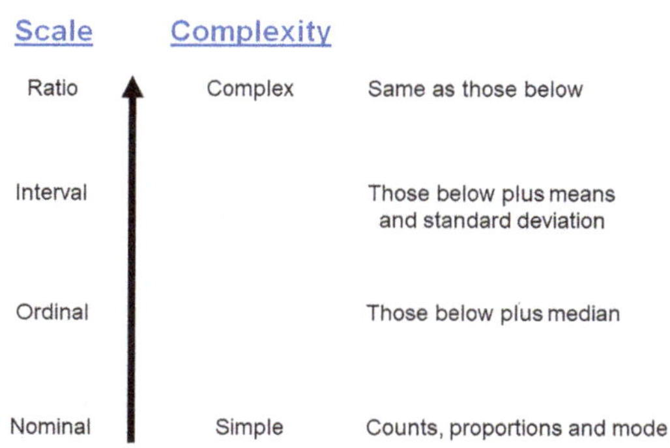

Fig. 2.2 Measurement scales attributed to Stevens (1946). Source for this chart: Paczkowski (2016). Permission to use granted by SAS Press

2.2 Data Organization

Data Organization is a complex topic, but one that has to be addressed even if to a small extent, because without an understanding of how your data are organized, you will be unable to access and understand them. In fact, you will have to depend on someone else to access data you need which is inefficient and ineffective. I should note that the organization I am referring to is outside your control and is independent of how you organize your final study data in a DataFrame. The former is determined, perhaps, by your *IT* department while the latter is determined by you for your specific purpose. The *IT* data structure will remain unchanged except for the obvious addition of new data, while your personal structure will change from problem to problem as well as during your analysis.

Data organization is two-fold. First, it refers to how data are organized in large databases. Your *IT* department is responsible for maintaining and organizing data for efficient storage and delivery to end-users such as you. There is a process, *extract-translate-load* (*ETL*), in which data are extracted from numerous data collection systems, translated to a more usable and understandable form, and finally loaded into databases accessible by end-users. This is an *external data structure*: it is external to you as the analyst.

Second, data, finally used by you as the end-user, have an internal structure that represents how data concepts are related to each other. The external structure is important to understand because you will eventually have to interact with these external databases to obtain the data you require for a particular problem. The internal structure refers to what you will actually use for your problem once you have accessed and downloaded data from the externally structured databases. The internal structure of your data tells you what is possible from an analytical perspective. You determine this by arranging your data. I will review internal data structures below.

2.2.1 External Database Structures

This topic is very complex. See Lemahieu et al. (2018) for an excellent, in-depth textbook treatment. Typically, data are stored in *relational databases* comprised of data tables which are rectangular arrays comparable to DataFrames. The difference is that they are more complex and detailed. Every table in a relational database is linked to one or more other tables by *keys*: *primary keys* and *foreign keys*. A primary key is associated with a table itself and uniquely identifies each record in that table. They are comparable to DataFrame indexes. In an orders data table, a unique order ID number (*Onum*) identifies each order record. The record itself has the order date and time (i.e., a timestamp), the amount ordered, the product ordered, and perhaps some other data about the order. The *Onum* and its values form a *key:value* pair.[9]

Another table could have data regarding the status of an order. It would have the order number plus indicators about the status: in-fulfillment, fulfilled, on back-ordered. The *Onum* thus appears in both tables which allows you to link them. The order number is a primary key in both tables.

A foreign key is not a unique identifier of a record in the table, but instead allows you to link to other tables for supplemental data. For an orders data table, the product ordered is indicated by a product ID number (*PID*). The actual name and description of the product are typically not shown or used because they are too long, thus taking up too much storage space, making it inefficient. This inefficiency is compounded if the product appears in multiple orders. It is more efficient to have a separate product table that has the *PID* and its description. The *PID* and its associated name/description appear only once in the table. The *PID* and description form a key:value pair.

In my example, the *PID* appears in both the orders table and the product table. In the orders table, it is a foreign key while in the product table it is a primary key. Reports on an order can be created by linking the two tables using the *PID*. This linking produces the additive effect I discussed in Chap. 1. The *Structured Query Language* (*SQL*) is the programming language typically used to manage, summarize, and link data tables. This is a human language-like programming language that has a simple syntax which can, of course, become complicated. Everyone involved in *BDA* must know something about *SQL*.

An *SQL* statement is referred to as a *query*. The simplest query consists of three statements, each starting with a verb:

1. *Select* statement that specifies what is to be selected;
2. *From* statement that specifies the data table (or tables) to use; and
3. *Where* statement that specifies a condition for the selection.

The *Select* and *From* statements are required; the *Where* statement is optional. If *Where* is not included, then everything listed in the *Select* statement is selected. The *Select* statement can include summary functions such as average or sum. These three key words are verbs. Another verb used when summary values are calculated is *Group by* which controls the groups for which the summary statistics are calculated. See Celko (2000) and Hernandez and Viescas (2000) for good background instruction.

As an example, suppose you have a small supplier table showing the names of suppliers of a raw material for your production process. In addition to the name, the table also has the last delivery amount (in tons) and an indicator showing whether or not the delivery was on time (e.g., 1 = Yes, 0 = No). As a side note, the 0/1 integers are more efficient to store (i.e., they take less hard drive or memory space)

[9] The "value" in the key:value pair could be a list of items as in my example or a single item. If a single item, then the list is a singleton list.

in a computer compared to the strings "No" and "Yes" or "Late" and "On Time", respectively. In Python, a DataFrame is created using the code in Fig. 2.3 and a *SQL* query statement in Python to select all the on-time suppliers is shown in Fig. 2.4.

```
##
## Create a Pandas DataFrame using a dictionary
##
dt = {'supplier':[ 'A', 'A', 'A', 'B', 'B', 'C'],
      'amount':[ 1, 2, 3, 2, 4, 5],
      'onTime':[ 1, 1, 0, 1, 0, 0 ] }
df = pd.DataFrame( dt )
df.style.set_caption( 'Supplier DataFrame' ).\
    set_table_styles( tbl_styles ).hide_index()
```

Supplier DataFrame

supplier	amount	onTime
A	1	1
A	2	1
A	3	0
B	2	1
B	4	0
C	5	0

Fig. 2.3 This is the Pandas code to create the supplier on-time DataFrame. The resulting DataFrame is shown

```
##
## Specify the query as a string.
## Select all suppliers who were on time.
##
qry = """
    Select supplier
    from df
    where onTime = 1
"""
##
## Run the query
##
sqldf( qry ).style.set_caption( 'SQL Query Result' ).\
    set_table_styles( tbl_styles )
```

SQL Query Result

	supplier
0	A
1	A
2	B

Fig. 2.4 This is the SQL code to select the on-time suppliers. The resulting DataFrame is shown. Notice that the query string, called "qry" in this example, contains the three verbs I mentioned in the text

A query can be expanded to include joining multiple tables, all linked on an appropriate key. Specific variables can be selected from each table as well as aggregated in different ways. See Hernandez and Viescas (2000) and Celko (2000) for more details and examples on writing queries.

Pandas has a query method that allows you to query a DataFrame in an almost *SQL* fashion. I will show you how to use this method for comparable queries in succeeding chapters. See the Jupyter notebook for this chapter for examples

2.2.2 Internal Database Structures

Once you have identified the data you need and created a DataFrame, perhaps using *SQL* queries, you still need to take an additional step to understand your data. You have to understand the *internal structure* of your DataFrame and perhaps manipulate it to your advantage. Knowing the structure of a DataFrame enables you to apply the right toolset to extract latent information. The more complex the structure, the more information is inside, the more difficult it is to extract that information from all the data "stuff," and the more sophisticated the tools needed for that extraction.

For an analogy, consider two books: Dick & Jane and War & Peace. Each book is a collection of words that are data points no different than what is in a dataset. The words *per se* have no meaning, just as data points (i.e., numbers) have no meaning. But both books have a message (i.e., information) distilled from the words; the same for a data set.

Obviously, Dick & Jane has a simple structure: just a few words on a page, a few pages, and one or two simple messages (the information). War & Peace, on the other hand, has a complex structure: hundreds of words on a page, hundreds of pages, and deep thought-provoking messages throughout. You would never read War & Peace the way you would read Dick & Jane: the required toolsets are different. And if you could only read at the level of Dick & Jane, you would never survive War & Peace. You would never approach War & Peace the way you would approach Dick & Jane. Yet this is what many do regarding their data: they approach a complex data set the way they approach a Stat 101 data set, a simplistic data set with a handful of observations and variables used to illustrate concepts and not meant for serious analysis.

When you read War & Peace, or a math book, or a physics book, or a history book, or an economics book, anything that is complex, the first thing you (should) do is look at its structure. This is given by the Table of Contents with chapter headings, section headings, and subsection headings all in a logical sequence. The Index at the back of the book gives hints about what is important. The book's cover jacket has ample insight into the structure and complexity of the book and even about the author's motive for writing it. Even the Preface has clues about the book's content, theme, and major conclusions. You would not do this for Dick & Jane. See Adler and Doren (1972) for insight on how to read a book.

Just as you would (or should) look at the above items for a complex book, you should follow these steps for understanding a data set's structure. A data dictionary is one place to start; a questionnaire is obvious; missing value patterns a must; groupings, as in a multilevel or hierarchical data set, are more challenging. The analysis is easier once the structure is known. This is not to say that it will become trivial if you do this, but you will be better off than if you do not.

Data structure is not the number of rows and columns in a DataFrame or which columns come first and which come last. This is a relatively unimportant physical layout although I will discuss physical layouts below. The real structure is the organization of columns relative to each other so that they tell a story. For example, consider a survey data set. Typically, case ID variables are at the beginning and demographic variables are at the end; this is a physical structure and is more a convenience than a necessity. The real structure consists of columns (i.e., variables) that are conditions for other columns. So, if a survey respondent's answer is "No" in one column, then other columns might be dependent on that "No" answer and contain a certain set of responses, but contain a different set if the answer is "Yes." The responses could, of course, be simply missing values. For a soup preference study, if the first question is "*Are you a vegetarian*?" and the response is "Yes", then later columns for types of meats preferred in the soups have missing values. This is a structural dependency.

The soup example is obviously a simple structure. In a DataFrame, a simple structure is just a few rows and columns, no missing values, and no structural dependencies. Very neat and clean, and always very small. This is the Stat 101 data set. All the data needed for a problem are also in that one data set. Real world data sets are not neat, clean, small, and self-contained. Aside from describing them as "messy" (i.e., having missing values, structural and otherwise), they also have complex structures. Consider a dataset of purchase transactions that has purchase locations, date and time of purchase (these last two making it a *panel* or *longitudinal* dataset), product type, product class or category, customer information (e.g., gender, tenure as a customer, last purchase), prices, discounts, sales incentives, sales rep identification, multilevels of relationships (e.g., stores within cities and cities within sales regions), and so forth. This data may be spread across several data bases so they have to be joined together in a logical and consistent fashion using primary and foreign keys. And do not forget that this is Big Data with gigabytes! This is a complex structure different from the Stat 101 structure as well as the survey structure.

A simple data structure is a rectangular array or *matrix* format as I noted in Chap. 1. A simple structure also has just numeric data, usually continuous variables measured on a ratio scale. Sometimes, a discrete nominal variable is included but the focus is on the continuous ratio variable. A nominal variable is categorical with a finite number of values, levels, or *categories*. The minimum is, of course, two; with one category it is just a constant. For this simple structure, there is a limited number of operations and analyses you can do and, therefore, a limited amount of information can be extracted. This is Poor Information and the analysis is Shallow Analytics.

As an example, suppose your business wants to build new research facilities. You have to study the potential employment pool by state to determine the best location. The first five records of a state DataFrame indexed on *State* that has data on unemployment and median household income are shown in Fig. 2.5.[10]

```
##
## Import tech data
##
file = '020620_1 stateData01.csv'
##
df = pd.read_csv( path + file, index_col = 'State' )
df.head().style.set_caption( 'Simple DataFrame' ).\
    set_table_styles( tbl_styles )
```

Simple DataFrame

State	medianHHIncome	unemployRate
Alabama	49936	2.700000
Alaska	68734	6.100000
Arizona	62283	4.600000
Arkansas	49781	3.600000
California	70489	3.900000

Fig. 2.5 This is a simple DataFrame for state data

This structure is a 50×2 rectangular array with each state in one row and only one state per row. There are only two numeric/continuous variables. The typical analysis is simple: means for the unemployment rate and household income calculated across the states. Graphs, such as bar charts are created to show distributions. I will discuss more enhanced data visualizations in Chap. 4.

Now consider a more complex data structure. It is the above data set but with one additional variable: a score for the presence of high-end technical talent. A high score indicates a high concentration of technology talent in the state. The score is based on a composite set of indexes for the concentration of computer and information scientist experts, the concentration of engineers, and the concentration of life and physical scientists.[11] The tech-concentration score was dummified to two levels for this example by assigning states with a score above 50 to a "Tech" group and all others to a "NonTech" group. I show the distribution in Fig. 2.6.

There is more to this data structure because of this technology index. This makes it a little more complex, although not by much, because states can be divided into two groups or clusters: technology talented and non-technology talented states. The unemployment and income data can then be compared by the technology talent. This invites more possible analyses and potentially useful Rich Information so

[10] Only 5 out of 50 states shown. Sources: Unemployment rate:https://www.bls.gov/web/laus/laumstrk.htm; Income: Table H-8. Median Household Income by State https://www.census.gov/data/tables/time-series/demo/income-poverty/historical-income-households.html.

[11] See http://statetechandscience.org/statetech.taf?page=overall-ranking&composite=tswf for the data and http://statetechandscience.org/statetech.taf?page=outline for the index methodology. Last accessed February 7, 2020.

```
##
## Tech group distribution
##
x = df.tech_grp.value_counts( normalize = True )
pd.DataFrame( x ).style.format( '{:.1%}' ).\
    set_caption( 'Tech Group Distribution' ).\
    set_table_styles( tbl_styles )
```

Tech Group Distribution

	tech_grp
NonTech	68.0%
Tech	32.0%

Fig. 2.6 States are categorized as technology talented or not. This shows that only 32% of the states are technology talented

the information content extracted is larger. One possibility is a comparison of the unemployment rate by the technology indicator using a two-sample t-test of the means. I show the test and results in Fig. 2.7. The p-values show there is a difference between the household income for states with technology talent vs those without technology talent for any alternative hypothesis. A similar test could be done for the unemployment rate.

```
##
## Create Tech and NonTech DataFrames
##
tech = df.query( " tech_grp == 'Tech' " )
nontech = df.query( " tech_grp == 'NonTech' " )
##
## Calculate income difference
##
diff = tech.medianHHIncome.mean() - nontech.medianHHIncome.mean()
##
## t-Test of income difference: tech - nontech
##
x = stats.ttest_ind( tech[ 'medianHHIncome' ], nontech[ 'medianHHIncome' ],
                     equal_var = False )
##
print( 'T-Test Results' )
print( f'\nDifference: ${diff:.2f}' )
print( f't-value: {x[ 0 ]:.4f}' )
print( f'Prob > |t|: {x[ 1 ]:.4f}' )
print( f'Prob > t: {1 - x[ 1 ]/2:.4f}' )
print( f'Prob < t: {x[ 1 ]/2:.4f}' )
```

```
T-Test Results

Difference: $12680.61
t-value: 5.1357
Prob > |t|: 0.0000
Prob > t: 1.0000
Prob < t: 0.0000
```

Fig. 2.7 A two-sample t-test for a difference in the median household income for tech vs non-tech states shows that there is a statistical difference. Notice my use of the query statements

Adding the one extra variable increased the DataFrame's dimensionality by one so the structure is now slightly more complex. There is now more opportunity to extract Rich Information. The structure is important since it determines what you can do. The more complex the structure, the more the analytical possibilities and the more information that can be extracted. Andrew Gelman referred to this as the blessing of dimensionality[12] as opposed to the curse of dimensionality.

Some structures across the rows of a DataFrame are explicit while others are implicit. The explicit structures are obvious based on the variables in the data table. Technology is the example I have been using. A variable on technology is in the data table so a division of the data by it is clear. How this variable is actually used is a separate matter, but it can and should be used to extract more and richer information from the whole data table. Regions of the country, however, are implicit in the DataFrame. There is no variable named "Region" in the table, yet it is there in the sense that the states comprising regions are there; states are nested in regions. States are mapped to regions by the U.S. Census Bureau, and this mapping is easy to obtain.[13] Using this mapping, a region variable could be added enabling further cuts of the data leading to more detailed and refined analysis. For example, you could study the unemployment rate by technology by region.

The explicit structural variables are clear-cut: they are whatever is in the DataFrame. The implicit structural variables also depend on what is already in the DataFrame, but their underlying components have to be found and manipulated (i.e., wrangled) into the new variables. This is what I did with mapping states to regions. Variables that are candidates for a mapping include, but are certainly not limited to, any of the following:

1. Telephone numbers to extract:
   - international dialing codes;
   - domestic US area codes; and
   - toll-free numbers.
2. ZIP codes and other postal codes.
3. Time/Date stamps to extract:
   (a) Day-of week
   (b) Work day vs weekend
   (c) Day-of-month
   (d) Month
   (e) Quarter
   (f) Year
   (g) Time-of-day (e.g., Morning/afternoon/evening/night)
   (h) Holidays (and holiday weekends)
   (i) Season of the year
4. Web addresses

[12] See (http://andrewgelman.com/2004/10/27/the_blessing_of/.

[13] See www2.census.gov. Last accessed February 3, 2020.

5. Date-of-Birth (*DOB*)
   (a) Age
   (b) Year of birth
   (c) Decade of birth
6. *SKU*s which are often combinations of codes
   (a) Product category
   (b) Product line within a category
   (c) Specific product

You could also bin or categorize continuous variables to create new discrete variables to add further structure. For example, people's age may be calculated from their date-of-birth (*DOB*) and then binned into pre-teen, teen, adults, and seniors.

In each of these cases, a single explicit variable can be used to identify an implicit variable. The problem is compounded when several explicit variables can be used. Which ones and how? This is where several multivariate statistical methods come in. Two are *cluster analysis* and *principal components analysis* (*PCA*). The former is used to group or cluster rows of the DataFrame based on a number of explicit variables. The result is a new implicit variable that can be added to the DataFrame as a discrete variable with levels or values identifying the clusters. This new discrete variable is much like the region and technology variables I previously discussed. I will discuss *PCA* in Chap. 5 and cluster analysis in Chap. 12.

If there are many variables in the DataFrame, then it may be possible to collapse several of them into one (or several) new summary variable that captures or reflects the essence of them. The original variables could then be deleted and only the summary variable kept. This will reduce the dimensionality of the DataFrame and, perhaps, increase its information content because the summary variable may be more revealing. Cluster analysis could be used for this purpose but the clustering is by variables rather than by rows of the DataFrame. *PCA* would accomplish the same task but have the added benefit that the new summary variables, called *principle components*, is extracted and have the property that they are linearly independent of each other. This independence is important for linear modeling as I will discuss later.

A feature of explicit and implicit structural variables is that they define structure across the records within a single data table. So, the explicit structural variable "technology" tells you how the rows of the DataFrame are divided. This division could be incorporated in a regression model by including a technology *dummy variable* to capture a technology effect. I discuss dummy variables and their issues in Chap. 5.

There could be structure across columns, again within a single DataFrame. On the simplest level, some groupings of columns are natural or explicit (to reuse that word). For example, a survey DataFrame could have a group of columns for demographics and another group for product performance for a series of attributes. A *Check-all-that-apply* (*CATA*) question is another example since the responses to this type of question are usually recorded in separate columns with nominal values (usually 0/1). As for the implicit variables for the rows of the data table, new variables can be derived from existing ones.

Combining variables is especially important when you work with a large DataFrame with hundreds if not thousands of variables which makes the DataFrame "high dimensional." A high-dimensional data table is one that not only has a large number of variables, but this number far outstrips the number of cases or observations. If N is the number of cases and P is the number of variables, then the DataFrame is high dimensional if $P >> N$. Standard statistical methods (e.g., regression analysis) become jeopardized because of this. Regression models, for example, cannot be estimated if $P >> N$. Some means must be employed, therefore, to reduce the dimensionality. See Paczkowski (2018) for a discussion of high dimensional data for pricing analytics.

As I mentioned above, just as there is cluster analysis for the rows of a data table, an approach most analysts are readily familiar with, so there is cluster analysis for the variables. This form of cluster analysis and *PCA* both have the same goal: dimension reduction. *PCA* is probably better known, but variable clustering is becoming more popular because the developing algorithms are handling different types of data plus it avoids a problem with *PCA*. The data types typically found in a data table are quantitative and qualitative (also known as *categorical*). The former is just numbers. The latter consists of classifiers such as gender, technology, and region. These may be represented by numbers in the DataFrame or words such as "Male" and "Female." If words are used, they are usually encoded with values. Ordinal values are also possible if there is an intuitive ordering to them such as "Entry-Level Manager", "Mid-Level Manager", and "Executive" in a *HR* study. Technically, *PCA* is designed for continuous quantitative data because it tries to find components based on variances which are quantitative concepts, so technically *PCA* should be avoided. Variable clustering algorithms do not have this issue.

There is more to data structure than the almost obvious aspects I have been discussing. Consider, for example, clustering demographic variables to create segments of consumers. The segments were implicit (i.e., hidden or latent) in the demographics. A clustering procedure reveals what is there all the time. Collapsing the demographics to a new single variable that is the segments adds more structure to the data table. Different graphs can now be created for an added visual analysis of the data. And, in fact, this is frequently done—a few simple graphs (the ubiquitous pie and bar charts) are created to summarize the segments and a few key variables such as purchase intent by segments or satisfaction by segments.

In addition to the visuals, sometimes a simple *OLS* regression model is estimated with the newly created segments as an independent variable. Actually, the segment variable is dummified and the dummies are used since the segment variable *per se* is not quantitative but categorical and so it cannot be used in a regression model. Unfortunately, an *OLS* model is not appropriate because of a violation of a key independence assumption required for *OLS* to be effectively used. This assumption is independence of the observations. In the case of segments, almost by definition the observations are not independent because the very nature of segments says that all consumers in a single segment are homogeneous. So, they should all behave "the same way" and are therefore not independent.

The problem is that there is a hierarchical or multilevel structure to the data that must be accounted for when estimating a model with this data. There are micro, or first level, units nested inside macro, or second level, units. Consumers are the micro units embedded inside the segments that are the macro units. The observations used in estimation are on the micro units. The macro units give a context to these micro units and it is that context that you must accounted for. There could be several layers of macro units so my example is somewhat simplistic. I illustrate a possible hierarchy consisting of three macro levels for consumers in Fig. 2.8, Panel (a). Consumer traits such as their income and age are used in modeling but so are higher level context variables. In the case of segments there is just a single macro level. Business units also could have a hierarchical structure. I show a possibility in Fig. 2.8, Panel (b).

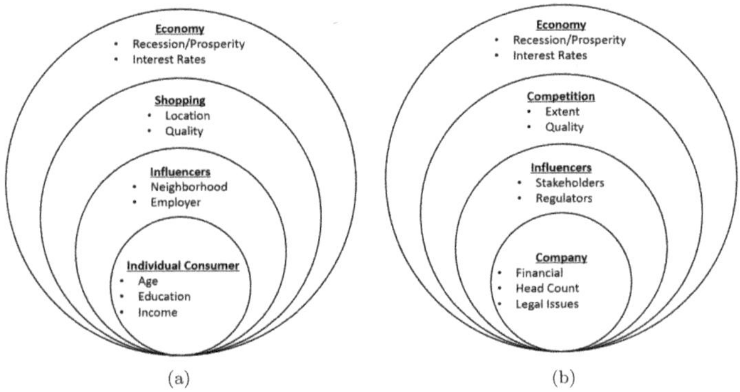

Fig. 2.8 This is a hierarchical structure of consumers and businesses. (**a**) Consumer structure. (**b**) Business structure

The macro levels usually have key driver variables that are the influencers for the lower level. For consumers, universal exogenous data such as interest rates, the national unemployment rate, and real GDP in one time period influence or drive how consumers behave. Interest rates in that time period are the same for all consumers. Over several periods, the rates change but they change the same way for all consumers. Somehow, this factor, and others like it, have to be incorporated in a model. Traditionally, this has been done either by aggregating data at the micro level to a higher level or by disaggregating the macro level data to a lower level.

Both data handling methods have aggregation and disaggregation issues.

Aggregation problems:

1. Information is more hidden/obscure. Recall that information is buried inside the data and must be extracted.
2. Statistically, there is a loss of power for statistical tests and procedures.

Disaggregation problems:

1. Data are "blown-up" and artificially evenly spread.
2. Statistical tests assume there are independent draws from a distribution, but they are not since they have a common base thus violating the key assumption.
3. The sample size is affected since measures are at a higher level than what the sampling was designed for.

This last point on sample size has a subtle problem. The sample size is too large because the measurement units (e.g., consumers) are measured at the wrong level so the standard errors are too small. Test statistics are then too large so you would too often reject the Null Hypothesis of no effect.

Examples of nested structures abound. The classic example usually referred to when nested or multilevel data are discussed is students in classes which are in schools which are in school districts. The hierarchical structure is inherent to the data simply because students are in classes which are in schools which are in districts. A hierarchical structure could result, by the way, if cluster sampling is used to select schools in a large metropolitan district and then used again to select classes. Whichever method is used, the nesting is self-evident which is why it is usually used to illustrate the concept. Examples for *BDA* include:

- segments;
- stores;
- marketing regions;
- states;
- neighborhoods;
- organization membership; and
- brand loyalty.

Consumers are nested in segments; they are nested in stores; they are nested in neighborhoods, and so forth. This nesting must be accounted for in modeling behavior. But how?

The basic *OLS* model, which I review in Chap. 6, is inappropriate. Sometimes analysts will add dummy variables to reflect group membership (or some other form of encoding of groups such as effects coding), but this does not adequately reflect the nesting, primarily because there may be key drivers that determine the higher-level components of the structure. For example, consider a grocery store chain with multiple locations in urban and suburban areas. Consumers are nested within the stores in their neighborhood. Those stores, and the neighborhoods they serve, have their own characteristics or attributes just as the consumers have their own. The stores in urban areas may, for instance, have smaller footprints because real estate is tighter in urban areas but have larger ones in suburban areas where real estate is more plentiful. Stores in Manhattan are smaller than comparable stores in the same chain located in Central New Jersey. The store size determines the types of services that could be offered to customers, the size of stock maintained, the variety of products, and even the price points.

An extension to the basic *OLS* model is needed. This is done by modeling the parameters of the *OLS* model where those models reflect the higher level in the structure and these models could be functions of the higher-level characteristics. This is a more complicated, and richer, model. Subtly, the random component for the error is a composite of terms, not just one term as in an *OLS* model as you will see in Chap. 6. A dummy variable approach to modeling the hierarchical structure would not include this composite error term which means the dummy variable approach is incorrect; there is a model misspecification—it is just wrong. The correct specification has to reflect random variations at each level and, of course, any correlations between the two. Also, the composite error term contains an interaction between an error and a lower level predictor variable which violates *OLS* assumptions. A dummy variable *OLS* specification would not do this.

Many variations of this model are possible:

- Null Model: no explanatory variables;
- Intercept varying, slope constant;
- Intercept constant, slope varying; and
- Intercept varying, slope varying.

There is one more aspect to data structure. This is the physical arrangement of the data. I stated in Chap. 1 that data are stored in rectangular arrays. It is the form of the array that varies depending on the analysis. If the rectangular array has a large number of variables so $P >> N$, then the DataFrame is said to be in *wide-form*. If, however, it has more observations than variables, then it is in *long-form*. To be more explicit about the long-form arrangement, consider a DataFrame that has data on survey respondents' answers to a five-point Likert Scale question about jewelry purchases: "*Which of these six brands of watches did you last purchase*?" Each respondent is assigned to one of four marketing segments. The question about brands could be a check-all-that-apply (*CATA*) question recorded as a "No" or "Yes" for each watch. Assume that the "No"/"Yes" answers are dummy coded as 0/1, respectively. There are then seven columns: six for the brands and one for the segment. This is a wide-form structure. You could create a simple summary table showing the sum of the 0/1 values for each brand and segment, with, say, the segments as rows and brands as columns. This is still a wide-form structure. If you now stack each row (i.e., each segment) under each other, so that there are three columns (segments, brands, counts of responses) then the data have a long-form structure.

Which structure you use depends on the analysis you will do. Some statistical analyses require the data to be in long-form. Correspondence analysis is an example. Others require a wide form. Correlation analysis is an example. I will illustrate the two forms and the reshaping of DataFrames from one form to the other throughout this book.

2.3 Data Dictionary

Given the volume of data in most businesses, *IT* departments have established and maintain *data dictionaries* as a best practice. A data dictionary contains *metadata* about the data. According to Wikipedia,[14]

> Metadata means "data about data". Although the "meta" prefix ... means "after" or "beyond", it is used to mean "about" in epistemology. Metadata is defined as the data providing information about one or more aspects of the data; it is used to summarize basic information about data which can make tracking and working with specific data easier.

Metadata can be anything that helps you understand and document your data. This could include:

- means of creation;
- purpose of the data;
- time and date of creation;
- creator or author of data;
- placement on a network (electronic form) where the data was created;
- what standards were used

and so on.[15] I will restrict the metadata in data dictionaries used in this book to include only

- variable name;
- possible values;
- source; and
- mnemonic.

The variable name is just a brief description of the variable. It could be, for example, "Revenue net of taxes", "List Price", "Dealer Discount", "Customer ID (*CID*)", "Supplier ID (*SID*)", just to mention a few possibilities. The possible values include ranges and perhaps a brief description of the type of data such as "Nominal", "Ordinal", "Dates in MM/DD/YYYY format", and so on. The source is where the data came from. In this book, some illustrative data will come from the marketing department, other data from the sales department, and still other data from the financial organization (*CFO*). It helps to know the source to better understand your data. Finally, the mnemonic (i.e., a memory aid) is the variable abbreviation or acronym you will use in formulas or in visualization function calls. The variable name *per se* may be, and often is, too complex or long to be used so the mnemonic

[14] http://en.wikipedia.org/wiki/Metadata. Last accessed February 3, 2020.

[15] http://en.wikipedia.org/wiki/Metadata. Last accessed February 3, 2020.

substitutes for it. The data dictionary provides a useful look-up feature to find what the mnemonic stands for.

The data dictionary can be created using a Python dictionary and the Python module *tabulate*. Their use is illustrated in this chapter's Jupyter notebook. The main Python code is shown in Fig. 2.9. This data dictionary could also be created in a Jupyter notebook Markdown cell. See this chapter's Jupyter notebook for examples. To install *tabulate*, use *pip install tabulate* or *conda install -c conda-forge tabulate*.

```
##
## Define a Python dictionary with
## variables as the keys and the other three
## items as the values.
##
dt = { "Order Number":[ 'Nominal Integer', 'Order Sys', 'Onum' ],
       "Customer ID":[ 'Nominal Integer', 'Customer Sys', 'CID' ],
       "Transaction Date":[ 'MM/DD/YYYY', 'Order Sys', 'Tdate' ],
       "Product Line ID":[ 'Five rooms of house', 'Product Sys', 'Pline' ],
       "Product Class ID":[ 'Item in line', 'Product Sys', 'Pclass' ],
       "Unit Sales":[ 'Number of units per order', 'Order Sys', 'Usales' ]
        }
##
## Flatten the dictionary
##
lst = [ [ k ] + v for k,v in dt.items() ]
##
## Specify headers and display table
##
headers = ['Variable', 'Values', 'Source', 'Mnemonic']
display( HTML( tb.tabulate( lst, tablefmt = 'html', headers = headers ) ) )
```

Variable	Values	Source	Mnemonic
Order Number	Nominal Integer	Order Sys	Onum
Customer ID	Nominal Integer	Customer Sys	CID
Transaction Date	MM/DD/YYYY	Order Sys	Tdate
Product Line ID	Five rooms of house	Product Sys	Pline
Product Class ID	Item in line	Product Sys	Pclass
Unit Sales	Number of units per order	Order Sys	Usales

Fig. 2.9 This is a Python script to generate a data dictionary

Chapter 3
Basic Data Handling

Small, tidy data sets are used to illustrate concepts and techniques in typical textbook treatments of statistics and econometrics. They always have just a few rows or *records* and a few columns called *variables* or *features*. The data are neat, clean, and available, meaning you are never told about the complexities involved in finding the data let alone importing them into a statistical or programming package. In addition, there is only one dataset. The use of several that may have to be merged or joined is not discussed. Consequently, learners must determine from other sources how to handle "messy" and large amounts of data. These are, in fact, typical operations in Business Data Analytics. They require *preprocessing* before any analysis can begin.

Preprocessing includes dealing with massive amounts of data, far more than what is used in a statistics or econometrics course. Importing data then becomes complex because there could be more than your computer can handle, especially if programs and operations (e.g., statistical, econometric, and machine learning) are performed mostly in your computer's finite memory.

Your data could be in a *flat data file*. A flat file has just two dimensions: rows and columns. In essence, it is the Data Cube I discussed in Chap. 1 but flattened to a rectangle. If the number of columns is large, then the file is *high dimensional*. This is typical of many *BDA* data sets.

In addition, the data you need may be spread across several datasets. This is another level of complexity you have to deal with, not from the perspective of getting your data but from one of merging the data sets into one containing just those data elements you need for your analysis. And once you have the needed data, its organizational form may not be appropriate for many procedures you want to use. Your data may be in *wide-form*—many variables and few rows—while statistical procedures require that your data be in *long-form*—many rows, few variables. So your data have to be reorganized or *reshaped*.

Finally, your data may have missing values, different scales, or inappropriate scales. Missing data is a common problem, no matter the ultimate data source. The

W. R. Paczkowski, *Business Analytics*,
https://doi.org/10.1007/978-3-030-87023-2_3

impact of missing values depends on the extent of the missingness as well as its basis. Different scales could bias results because variables measured on one scale could overpower the influence of other variables measured on another scale simply due to their scales. This suggests that a common base is required for all variables. At the same time, a transformation of a variable, which goes beyond just changing its base, may be needed to change the variable's distribution, linearize it, or stabilize its variance. All this comes under the label of *preprocessing*.

In this chapter, I will focus on four tasks for managing or handling your data and its complexities. This includes importing data into Pandas, especially large amounts of data; identifying some preliminary statistical aspects of your data; merging several data sets into one; and reshaping your data set. In the next chapter, I will address the remaining preprocessing issues such as changing scales and dimension reduction.

3.1 Case Studies

There are two case studies used in this and succeeding chapters. Both are based on a different aspect of a business. One deals with orders by customers so it is transactions data and the other with measures of order fulfillment. I describe these Case Studies in the next two subsections.

3.1.1 Case Study 1: Customer Transactions Data

A fictitious household furniture manufacturer sells to locally owned, boutique retailers scattered throughout the U.S. which is divided into four marketing regions consistent with U.S. Census regions: Midwest, Northeast, South, and West. The company has 43 products in six product lines for the major rooms in a house: Den, Dining Room, Kids' Room, Kitchen, Living Room, and Master Bedroom. Each product line is divided into product classes such as Chairs, Tables, and Baker's Racks for the Kitchen product line. The sales force is regionally dispersed and managed by a regional executive vice president. The sales reps offer four types of discounts, at their discretion, to the retailers to incent them to buy their furniture for resale. The discounts are: Competitive Discount, Order Size Discount, Dealer Discount, and a Pick-up Discount (offered if the retailer picks up an order).

In addition to taking orders and offering discounts, the sales force rates each customer on the ease of doing business with them. A three-point scale is used: Poor/Good/Excellent. They also administer a Loyalty Program although not all retailers are members. Finally, the sales force collects customer satisfaction data from the retailers to judge how satisfied they are with the product lines. This information is sent to the product managers for each product line.

There are three databases:

Group	Class	Example
Grocery	Mass Merchandise	Target, Walmart
	Supermarkets	Krogers, Shoprite
	Club Stores	BJs, Costco, Sams Club
Convenience	Convenience Stores	7-Eleven
	Dollar Stores	Dollar Store
	Drug Stores	CVS, Walgreen, Rite Aid
Restaurants	Restaurants	Dinners, etc.

Table 3.1 This is a listing of the bakery's customers by groups and classes within a group

1. Orders: contains an order number (*ONUM*), a customer ID (*CID*), a timestamp for each order, the amount ordered, list price, the discounts offered, a return flag if the product is returned, the returned amount, and material cost for the product;
2. Customers: the *CID*, where the customer is located; and
3. Marketing: the *CID*, Loyalty Program membership, the Buyer Rating, and the buyer satisfaction score.

The product manager for the Living Rooms product line wants to understand the key drivers of sales of living room window treatments, specifically blinds. She specified the following objectives:

1. Identify sales patterns by:
   - Marketing Region
   - Customer Loyalty
   - Buyer Rating
2. Estimate the price elasticity for living room blinds.
3. Check for statistical differences among the four discounts.
4. Provide a tool for predicting sales for a customer.

The company's data scientist (i.e., you) has been charged with answering her questions.

3.1.2 Case Study 2: Measures of Order Fulfillment

A large national bread-products company supplies loaves of bread, hot dog rolls, and hamburger rolls to different types of store fronts. It has seven classes of customers divided into three groups as I show in Table 3.1. I will use the classes in Chap. 8 and the groups until then. Its customers are located in urban and rural areas in four marketing regions that correspond to the four U.S. Census regions (i.e., Midwest, Northeast, South, and West). The baked goods must be delivered fresh each morning, usually before 6 AM local time, to each customer.

The company contracted with local bakeries to actually produce and deliver the loaves and rolls. There are 400 bakeries or *baking facilities*, each identified by a facility ID (*FID*). These 400 bakeries are located in the same urban and rural areas as the customers, allowing the national company to deliver fresh products each day.

The size of each day's order varies based on how much a customer sold the day before and, therefore, how much remains from a previous order which are put on a "day old" shelf. Any bread products that are more than 2 days old are automatically removed from store or restaurant shelves and discarded. The customer places an order at the end of each day using an electronic order placement system developed by the national bakery company. The order goes into an ordering system and then forwarded electronically to an appropriate local baker for fulfillment. This system allows the national company to monitor orders and performance.

When the order is delivered by the contract baker, someone from the customer's staff receives it, counts the loaves and rolls delivered, checks the quality of the order (damaged bags for bread products, crushed bread and rolls, and water damaged breads that cannot be sold), and verifies the charges for the order. This verification is all done through an Internet connection via a tablet app specifically designed for tracking order fulfillment. Using this app, the receiving customer indicates if the order is complete, damage free, delivered on time (i.e., before 6 AM local time), and the invoice documentation is correct. The responses are submitted electronically via the app to a main data collection facility which is a data store. An *ETL* protocol processes and loads the data from the data store into a data warehouse or data mart.

The amount for data collected is large. For each contract baker, there are several daily delivery orders. For each order, there is a measure on completeness, on time, damage free, and correct documentation, all recorded in the database as 1 = "Yes" or 0 = "No". In addition, for each order there is the baking facility ID (*FID*) and delivery date. Another database has information about each *FID*: marketing region, local market name, and if the baker is in an urban or rural area.

You, as the company's data scientist, were asked to develop and analyze a *Perfect Order Index* (*POI*) as a measure of order fulfillment. The *POI* is simply calculated as the product of the percentage of *Yes* responses to each of the four order fulfillment measures. That is,

$$POI = complete\% \times damage\% \times ontime\% \times document\% \tag{3.1.1}$$

$$= \prod_{i=1}^{4} measure_i \tag{3.1.2}$$

where $measure_i$ is the ith measure in percent terms. I show an example in Table 3.2. I will expand this Case Study in later chapters.

Measure	Yes (%)
Order complete	97.1
Damage free	98.2
Delivered on time	98.0
Documents correct	99.4
POI	92.9

Table 3.2 This illustrates the calculation of the *POI*

3.2 Importing Your Data

An obvious first step in any analytical process, aside from locating the right data, is to import your data into an analytical framework. This is actually more complicated than imagined. Some issues to address are the current data format, the size of the dataset to import, and the nature of the data once imported. I address these in the next few subsections.

3.2.1 Data Formats

Several data formats in common use are easily read by Pandas, the Python data manipulation package. Pandas provides a set of very flexible import functions. Which one you should use depends on your data format. I provide some typical formats and relevant functions in Table 3.3.

Data format	Extension	Import function	Write function
CSV	.csv	read_csv	to_csv
Excel	.xls, .xlsx	read_excel	to_excel
Clipboard		read_clipboard	to_clipboard
JSON	.json	read_json	to_jason
HDF5	.hdf	read_hdf	to_hdf
SAS	.ss7bdat	read_sas	NA
Stata	.dta	read_stata	to_stata
SQL		read_sql	to_sql

Table 3.3 Pandas has a rich variety of read and write formats. This is a partial list. The complete list contains 18 formats. An extended version of this list is available in McKinney (2018, pp. 167–168). Notice that there is no SAS supported write function. The clipboard and *SQL* extensions vary

The *Comma Separated Value* (*CSV*) and Excel formats are probably the most commonly used formats in *BDA*. *CSV* is a simple format with each value in a record separated by a comma and text strings often, but not always, denoted by quotation marks. This format is supported by almost all software packages because of its simplicity.[1] Excel is also very popular because many analysts mistakenly believe that Excel, or any spreadsheet package, is sufficient for data analytical work. This is far from true. Nonetheless, they store their data in Excel workbooks and worksheets.

Java Script Object Notation (*JSON*) is another popular format that allows you to transfer data, software code, and more from one installation to another. Jupyter notebooks, for example, are *JSON* files.[2]

HDF5 (*Hierarchical Data Format*, Version 5) is a format used with very large datasets, a size typical for *BDA*. It has three characteristics: groupings of data, attributes for the groups, and measures on the items in the groups. These data sets are hierarchically organized into groups that store related items. Attributes are arbitrary *metadata* for the groups that are stored directly with the groups. The arbitrariness means the metadata could be added without following fixed rules and it could vary from one group to the next; it is whatever is relevant for documenting the measures in a group. The measures are just the data themselves. For the *POI* data, a logical grouping is all the data by *FID*. The attributes could be the delivery date and time. The measures are the order measures (i.e., on time, correct documentation, complete, and damage free). For the transactions data, a logical grouping could be based on the sales representatives. The attributes could be the type of discount offered and the measures the actual discount values. This hierarchical structuring has efficiency advantages for very large data sets that actually distinguish it from an *SQL* structure. See Collette (2014) for an extensive discussion of hierarchical data structures and how *HDF5* can be used with this type of data.

SAS is, from my perspective, the oldest and most comprehensive statistical package available and one that is well entrenched in quantitative and data processing organizations in corporate and government agencies world-wide. Its extensive library of functions have many high-powered routines that have been developed, maintained, and expanded upon over several decades. The libraries are called *PROCs*, short for "procedures." They cover data processing and management, data visualization, basic statistical operations (e.g., hypothesis testing, regression, *ANOVA*), reporting, and time series analysis and forecasting to mention a few. If you are in a data analytic shop, in a major corporation such as a Fortune 1000 company, then the chances are high that your company uses SAS. SAS is by license only.

Stata is a powerful econometric-oriented software package that has a wide array of state-of-the-art econometric routines, data visualization capabilities, matrix operations, and programming functionality. The programming functionality allows users to develop their own methods and contribute them through a wide user community, thus expanding Stata's capabilities.

[1] I have not done an exhaustive search of all software packages, so this claim is just based on my experience.

[2] See https://en.wikipedia.org/wiki/JSON for information about JSON. Last accessed January 12, 2021.

SQL is the query language *par excellence* I mentioned in Chap. 2. There are many dialects of *SQL*, each with an added functionality to differentiate that dialect from others to gain a competitive advantage. Fundamentally, all dialects have the same basic, core set of human-like verbs to enable "easy" querying of a data base or data set. I list these verbs in Table 3.4.

Verb	Description
Select	Select variable(s); aggregate
From	Data source(s)
Where	Selection condition(s); filters row of a table
Group By	Groups table rows for aggregation in the Select clause
Having	Filters groups created by Group By
Order By	Sort results

Table 3.4 These are the basic, core verbs used in a *SQL* query statement. Just the *Select* and *From* verbs are required since they specify what will be returned and where the data will come from. Each verb defines a clause with all clauses defining a query. The *Where* clause must follow the *From* clause and the *Having* clause must follow the *Group By* clause. There are many other verbs available

The data sources for the *SQL From* verb are *SQL*-ready data tables and the result of the *Select* verb is a data table satisfying the query. A powerful and useful feature of *SQL* is the use of a returned table in the *From* clause so that, in effect, you could embed one query in another. Figure 2.4 illustrates a simple query of a DataFrame. Although it is not necessary to know the *SQL* query language for *BDA*, I highly recommend that you gain some proficiency in its basics.

The list of formats I show in Table 3.3 is actually a mix of *pure formats* (*PF*) and *associated formats* (*AF*). Pure formats are independent of any analytical framework or engine so they can be used with any analytical software. They have to be translated to a software's own internal data format, but this does not change the fact that they are not directly associated with a particular software. *CSV*, *HDF5*, and *JSON* are in this group. The associated formats are part of a software package's framework. *SAS* and Stata's formats are examples. Every *SAS* PROC reads and manipulates its format. Any *PF* imported into *SAS* is translated to SAS's data format. This also holds for *Stata*. *SQL* is a query language but also a database structure.

3.2.2 *Importing a CSV Text File into Pandas*

I show an example in Fig. 3.1 of importing a *CSV* file into Pandas. The basic import or read command consists of four parts:

1. the package where the function is located: Pandas identified by its alias *pd*;

2. the read function: *read_csv*;
3. the location or path to the data as an argument: *path*; and
4. the file name as an argument: *file*.

The package alias must be "chained" to the *read_csv* import function, otherwise the Python interpreter will not know where to find the function. Both the path and file name can be separately defined for convenience and cleaner coding; I consider this a Best Practice. You must always specify the file path so Pandas can find the file, unless the data file is in the same directory as your Jupyter notebook. Then a path is unnecessary since Pandas always begins a search in the same directory as the notebook. Otherwise, you have to specify it as I show in Fig. 3.1. This function has

```
##
## Import a small test DataFrame
##
file = 'chunkTestData.csv'
##
df_chunk = pd.read_csv( path + file )
df_chunk.style.set_caption( 'Data Read by Chunks' ).\
    set_table_styles( tbl_styles )
```

Data Read by Chunks

	x1	x2	x3
0	1	3	2
1	2	2	3
2	4	4	1
3	3	5	3
4	2	6	2
5	4	2	4
6	5	1	5
7	6	3	3
8	2	4	2
9	6	5	1

Fig. 3.1 Importing a *CSV* file. The path for the data would have been previously defined as a character string, perhaps as *path* = '../Data/'. The file name is also a character string as shown here. The path and file name are string concatenated using the plus sign

Argument	Purpose	Default	Example
sep	Separates values in a row	','	sep = ';'
header	Row to use as header	'infer'	header = 0
names	Column names to use	None (set header = 0)	names = ['A', 'B']
index_col	Column to use as index	None	index_col = 'Year'
skiprows	Rows to skip reading (0 based)	None	skiprows = 5
skip_blank_rows	Skip blank rows	True	skip_blank_rows = True
na_values	Strings to recognize as missing	See text explanation	
parse_dates	Parse a string as a date variable	False	parse_dates = ['Date']
chucksize	Read chucks of data	See text explanation	

Table 3.5 This is just a partial listing of arguments for the Pandas *read_csv* function. See McKinney (2018, pp. 172–173) for a complete list

a large number of arguments that extend its reading capabilities and flexibility. I list a few key arguments in Table 3.5.

3.2.3 Importing Large Files in Chunks

The data files you need for a *BDA* problem are typically large, perhaps larger than what is practical for you to import at once. In particular, if you process a large file after importing it, perhaps to create new variables or selectively keep specific columns, then it is very inefficient to discard the majority of the imported data as unneeded. Too much time and computer resources are used to justify the relatively smaller final DataFrame needed for an analysis. This inefficiency is increased if there is a processing error (e.g., transformations are incorrectly applied, calculations are incorrect, or the wrong variables are saved) so it all has to be redone. Importing chunks of data, processing each separately, and then concatenating them into one final, albeit smaller and more compact, DataFrame is a better way to proceed. For example, one small chuck of data could be imported as a test chunk to check if transformations and content are correct. Then a large number of chunks could be read, processed, and concatenated.

The Pandas *read_csv* command has a parameter *chunksize* that specifies the number of rows to read at one time from a master *CSV* file. This produces an *iterable* which allows you to *iterate* over objects, chunks of data in this case, processing each chuck in succession. I provide examples in Figs. 3.2, 3.3, and 3.4 for the same example DataFrame in Fig. 3.1.

```
chunks = pd.read_csv( path + file, chunksize = 5 )
##
## Process the chunks
##
for chunk in chunks:
    chunk[ 'rowSum' ] = chunk.sum( axis = 1 )
    print( chunk )
```

```
   x1  x2  x3  rowSum
0   1   3   2       6
1   2   2   3       7
2   4   4   1       9
3   3   5   3      11
4   2   6   2      10
   x1  x2  x3  rowSum
5   4   2   4      10
6   5   1   5      11
7   6   3   3      12
8   2   4   2       8
9   6   5   1      12
```

Fig. 3.2 Reading a chunk of data. The chunk size is 5 records. The columns in each row in each chunk are summed

```
chunks = pd.read_csv( path + file, chunksize = 5 )
##
## Process the chunks
##
for chunk in chunks:
    chunk[ 'rowSum' ] = chunk.sum( axis = 1 )
    chunk.drop( [ 'x1', 'x2' ], axis = 1, inplace = True )
    print( chunk )
```

```
   x3  rowSum
0   2       6
1   3       7
2   1       9
3   3      11
4   2      10
   x3  rowSum
5   4      10
6   5      11
7   3      12
8   2       8
9   1      12
```

Fig. 3.3 Processing a chunk of data and summing the columns, but then deleting the first two columns after the summation

```
chunks = pd.read_csv( path + file, chunksize = 5 )
##
## Create an empty DataFrame
##
df_chunk = pd.DataFrame()
##
## Process the chunks and concatenate them
##
for chunk in chunks:
    chunk[ 'rowSum' ] = chunk.sum( axis = 1 )
    chunk.drop( [ 'x1', 'x2' ], axis = 1, inplace = True )
    df_chunk = pd.concat( [ df_chunk, chunk ] )
df_chunk.style.set_caption( 'Processed Chunks' ).\
    set_table_styles( tbl_styles ).hide_index()
```

Processed Chunks

x3	rowSum
2	6
3	7
1	9
3	11
2	10
4	10
5	11
3	12
2	8
1	12

Fig. 3.4 Chunks of data are processed as in Fig. 3.3 but then concatenated into one DataFrame

This *chunksize* argument is available for *read_csv*, but, unfortunately, there is no comparable argument for *read_excel*. A possible way to handle large chucks when reading an excel file is to use another parameter for the function that allows you to read in a specified number of lines.

3.2.4 Checking Your Imported Data

Once you have imported your data, you should perform five checks of them before beginning your analytical work:

Check #1 Display the first few records of your DataFrame. Ask: "*Do I see what I expect to see?*"
Check #2 Check the shape of your DataFrame. Ask: "*Do I see all I expect to see?*"
Check #3 Check the column names in your DataFrame. Ask: "*Do I have the correct and cleansed variable names I need?*"
Check #4 Check for missing data in your DataFrame. Ask: "*Do I have complete data?*"
Check #5 Check the data types of your variables. Ask: "*Do I have the correct data types?*"

Notice that I do not have data visualization on my list. You might suppose that it should be part of Check #1: *Look at your data*. You would be correct. Data visualization, however, is such a large and complex topic that I devote Chap. 4 to it.

3.2.4.1 Check #1: Display the First Few Records

Students are always advised (i.e., taught) to look at their data as a Best Practice. This is vague advice because it is never clear what it means. Look at them how and for what? But this is also impractical advice for large, or even moderately large, DataFrames. Nonetheless, one way to "look" at your data is to determine if they are in the format you expect. For example, if you expect floating point numbers but you see only integers, then something is wrong. Also, if you see character strings (e.g., the words "True" and "False") when you expect integers (e.g., 1 for True and 0 for False), then you know you will have to do extra data processing. Similarly if you see commas as thousands separators in what should be floating point numbers, then you have a problem because the numbers will be treated and stored as strings since a comma is a character.[3]

You only have to view the first few records to make a quick assessment. You can view them using the DataFrame's *head( )* method. A *method* is a function attached to or associated with a DataFrame the moment the DataFrame is created. It is applicable to that DataFrame; it is not a stand-alone function that can apply to any object. A method has a parameter, in this case the number of records to display. The default is five. As a method, it requires opening and closing parentheses. The *head( )* method is *chained* to the DataFrame name using "dot" notation. For example, *df.head( )*. I show an example in Fig. 3.5. Incidentally, you could look at the last five records using *tail( )* where five is also the default.

[3] In European countries, a dot is used as a separator.

```
##
## Set data path
##
path = '../Data/'
```

```
##
## Import the POI data and display the head
##
file = '030320_1 poi.csv'
df = pd.read_csv( path + file, parse_dates = [ 'dates' ] )
df.head().style.set_caption( 'POI Data' ).\
    set_table_styles( tbl_styles )
```

POI Data

	fids	dates	ontime	complete	damage	document
0	100	2016-01-01 00:00:00	1	1	1	1
1	100	2016-01-02 00:00:00	1	1	1	1
2	100	2016-01-03 00:00:00	1	1	1	1
3	100	2016-01-04 00:00:00	1	1	1	1
4	100	2016-01-05 00:00:00	1	1	1	1

Fig. 3.5 Display the *head( )* of a DataFrame. The default is $n = 5$ records. If you want to display six records, use *df.head(6)* or *df.head(n = 6)*. Display the tail with a comparable method. Note the "dot" between the 'df" and "head(). This means that the *head( )* method is chained or linked to the DataFrame "df"

Pandas has a *style* method that can be chained to DataFrames when they are displayed that make the display more readable and documented. There are several styles I use in this book:

- *set_caption* for adding a caption (actually, a title) to the DataFrame;
- *bar* for adding a bar chart to displayed columns;
- *format* for formatting displayed columns; and
- *hide_index* to hide the DataFrame index.

You could also define a table style that you might use often. As an example, you could define one for the caption so it displays in 18-point font. I show one possible definition in Fig. 3.6. You can see an example in Fig. 3.7 and many other examples throughout this book.

```
##
## DataFrame styles
##
tbl_styles = [ {
    'selector': 'caption',
    'props': [
        ('color', 'darkblue'),
        ('font-size', '18px')
    ] } ]
```

Fig. 3.6 This is a style definition for setting the font size for a DataFrame caption

```
df_orders.head().style.set_caption( 'Orders Data' ).\
    set_table_styles( tbl_styles ).\
    hide_index()
```

Fig. 3.7 This is an example of using a style for a DataFrame

3.2.4.2 Check #2: Check the Shape of the DataFrame

The *shape* of a DataFrame is a *tuple* (actually, a 2-tuple) whose elements are the number of rows and the number of columns, in that order. A tuple is an immutable list which means it cannot be modified. The shape tuple is an *attribute* of the DataFrame so it is an automatic characteristic of a DataFrame that you can always access. To display the shape, use *df.shape*. I provide an example in Fig. 3.8.

```
##
## Check the shape
##
file = '030320_1 poi.csv'
df = pd.read_csv( path + file, parse_dates = [ 'dates' ] )
df.shape
```

```
(730000, 6)
```

Fig. 3.8 Display the *shape* of a DataFrame. Notice that the *shape* does not take any arguments and parentheses are not needed. The shape is an attribute, not a method. This DataFrame has 730,000 records and six columns

Although a tuple is immutable, this does not mean you cannot access its elements for separate processing. To access an element, use the square brackets, [ ], with the element's index inside. For example, to access the number of rows, use *df.shape[0]*. Remember, Python is zero-based so indexing starts with zero for the first element.

3.2.4.3 Check #3: Check Column Names

Checking column (i.e., variable) names is a grossly overlooked step in the first stages of data analysis. A name will certainly not impact your analysis in any way, but failure to check names could impact the time you spend looking for errors rather than doing your analysis. Column names, which are also attributes of a DataFrame, could have stray characters and leading and trailing white spaces. White spaces are especially pernicious. Suppose a variable's name is listed as 'sales' with a leading white space. When you display the *head* of the DataFrame, you will see 'sales' displayed without the white space, but the white space is really there. You will naturally try to use 'sales' (notice there is no white space) in a future command, say a regression command. The Python interpreter will immediately display an error message that 'sales' is not found and the reason is simply that you typed 'sales' (notice the lack of a white space) rather than 'sales' (notice the leading white space).

You will needlessly spend time trying to find out why. Checking column names up-front will save time later.

You can display column names by using the *columns* attribute attached to the DataFrame. Simply use *df.columns*. See Fig. 3.9 for an example. You remove leading or trailing white spaces using *df.columns = df.columns.str.strip( )* where *str* is a string *accessor* in Pandas that operates on a string. There are four accessors in Pandas and more can be written by users. I list these four in Table 3.6. In my example, a list of column names returned by *df.columns* is passed or chained to the *str* accessor which is chained to the *strip()* function. You may also want to convert all column names to lower case as Best Practice using *df.columns = df.columns.str.lower( )* for consistency across names and to reduce the chance of typing errors since the Python interpreter is case sensitive. You could do both operations at once using *df.columns = df.columns.str.strip( ).str.lower( )*. Notice the use of *str* twice in this expression.

```
##
## Check column names
##
file = '030320_1 poi.csv'
df = pd.read_csv( path + file, parse_dates = [ 'dates' ] )
df.columns
```

```
Index(['fids', 'dates', 'ontime', 'complete', 'damage', 'document'], dtype='object')
```

Fig. 3.9 Display the column names of a DataFrame using the *columns* attribute

Accessor	Operates on
dt	datetime variables
str	strings
cat	categorical variables
sparse	sparse variables

Table 3.6 These are four accessor methods available in Pandas. The text illustrates the use of the *str* accessor which has a large number of string functions

3.2.4.4 Check #4: Check for Missing Values

Missing values are a headache in statistics, econometric, and machine learning. You cannot estimate a parameter if the data for estimation are missing. Some estimation and analysis methods automatically check for missing values in the DataFrame and then delete records with at least one missing value. If the DataFrame is very large, this automatic deletion is not worrisome. If the DataFrame is small, then it is worrisome because the degrees-of-freedom for hypothesis testing could be reduced enough to jeopardize the validity of a test. It is also troublesome if you are working with time series because the deletion will cause a break in the continuity of the series. Many time series functions require this continuity.

```
##
## NaN value ignored
##
x = pd.Series( [1, 2, 3, np.nan, 4 ] )
print( f'Sum: {x.sum()}\nCount: {x.count()} \
\nProduct: {x.prod()}\nMean: {x.mean()}' )

Sum: 10.0
Count: 4
Product: 24.0
Mean: 2.5
```

Fig. 3.10 These are some examples where an *NaN* value is ignored in the calculation

```
##
## NaN value not ignored
## It appears as a break in the series
##
x = pd.Series( [1, 2, 3, np.nan, 4 ] )
print( f'Cumulative Sum:\n{x.cumsum()}\
\nCumlative Product:\n{x.cumprod()}' )

Cumulative Sum:
0     1.0
1     3.0
2     6.0
3     NaN
4    10.0
dtype: float64
Cumlative Product:
0     1.0
1     2.0
2     6.0
3     NaN
4    24.0
dtype: float64
```

Fig. 3.11 These are some examples where an *NaN* value is not ignored in the calculation

Missing values are also either ignored or converted to zero by some Pandas methods and functions before a calculation is done. The *sum( )*, *count( )*, and *prod( )* methods are examples. I illustrate this in Fig. 3.10. In some situations, missing values cannot be ignored. The cumulative methods such as *cumsum( )* and *cumprod( )* are examples. I illustrate this in Fig. 3.11. It is important to identify the columns in a DataFrame that have missing values. If there are any, then you have to decide what to do with them.

Missing values are indicated in various ways in Pandas (and in other software). Pandas uses the symbol *NaN*, which stands for "Not a Number."[4] This is actually a floating-point number stored in memory at a specific location. I illustrate this in Fig. 3.12. Note that I referred to "x" and "y" as *symbols* in Fig. 3.12 and not variables. This is because they refer to, or point to, the memory location of an object, which is the *NaN* float in this case. They are not equal to the *NaN* float but identify where it is in memory.

[4] It is the same value as the Numpy *NaN* value since Pandas uses Numpy as a base.

```
##
## Check memory Location of NaN
##
x = np.nan
y = np.nan
##
## Print memory Locatiosn using id( ) function
##
print( f'Location for x: {id( x )}\nLocation for y: {id( y )}' )

Location for x: 2819840139152
Location for y: 2819840139152
```

Fig. 3.12 Two symbols are assigned an *NaN* value using Numpy's *nan* function. The *id( )* function returns the memory location of the symbol. Both are stored in the same memory location

Since *NaN* is a float, it only applies to float variables. Integers, Booleans, and strings (which have the data type, or *dtype*, *object*), are not floats so *NaN* does not apply to them. If *NaN* appears in one location in an object variable, then the whole variable is recast as a float. So, if a variable has all integer values with one *NaN*, then all the values are recast as floats.

You can also represent a missing value with *None*, but this is dangerous because *None* is special. It literally means—none; not zero, False, empty, or missing.

While *NaN* applies to missing values for floats, *NaT* applies to missing *datetime* values. A datetime value is a special representation of a date and time, such as April 1, 2020 at 1 PM EST. You could receive a data set with a date stamp representing when an order was placed. It is possible that a computer failure interrupted the data recording so the date stamp is missing. This is indicated by *NaT*. The concerns about *NaN* equally apply to *NaT*.

The Pandas *info( )* method attached to each DataFrame is a way to detect missing values. Calling *df.info( )* will return a display that contains the name of each column in the DataFrame and the count of the number of non-null values. Columns with a count less than the total number of records in the DataFrame indicated by the *shape* attribute have missing values. The problem with the *info( )* method is that it only returns a display, so you cannot access the values directly. The *count( )* method, also attached to the DataFrame, is an alternative. It returns the count of the number of non-null records for each variable which you could save in a separate DataFrame for further analysis. I provide an example in Fig. 3.13.

You could use the *isna( )* and *notna( )* Pandas methods to check for missing values. Both return True or False, as appropriate, for each element in the object. For example, if you have a Pandas data series object, x, then *x.isna( )* returns an equal length series of True/False elements; this returned series has datatype *Boolean*. I show the relationship between *isna()* and *notna()* in Table 3.7. The advantage of using *is.na( )* or *notna( )* is that you can count the number of missing values. You do this by chaining the *sum( )* function to the method you use. For example, *x.isnull( ).sum( )* returns the number of missing values in the variable x. You can then use this in other calculations or a report.

```
##
## Checking missingness
##
## Use the info() method
print( df.info() )
##
## Use the count() method
##
x = df.count()
pd.DataFrame( x, columns = [ 'Count' ] ).style.\
    set_caption( 'Missing Counts' ).set_table_styles( tbl_styles )
```

```
<class 'pandas.core.frame.DataFrame'>
RangeIndex: 730000 entries, 0 to 729999
Data columns (total 6 columns):
 #   Column    Non-Null Count   Dtype
---  ------    --------------   -----
 0   fids      730000 non-null  int64
 1   dates     730000 non-null  datetime64[ns]
 2   ontime    730000 non-null  int64
 3   complete  730000 non-null  int64
 4   damage    730000 non-null  int64
 5   document  730000 non-null  int64
dtypes: datetime64[ns](1), int64(5)
memory usage: 33.4 MB
None
```

Missing Counts

	Count
fids	730000
dates	730000
ontime	730000
complete	730000
damage	730000
document	730000

Fig. 3.13 This illustrates counting missing values by the columns of a DataFrame. The top portion of the output shows the display from the *info()* method while the bottom portion shows the results from the *count()* in a DataFrame

Element state	isna()	notna()
Missing	True	False
Not missing	False	True

Table 3.7 The two Pandas missing value checking methods return Boolean indicators as shown here for the state of an element in a Pandas object

Another approach for checking missingness is to create a graph of a set of variables by record number and have a black stripe wherever there is a missing value for each variable-record combination. I show an example in Fig. 3.14. This particular display is called a *heatmap* which I discuss in Chap. 4.

Using the Pandas *describe( )* method and domain-specific knowledge is a final way to identify missing values. Domain-specific knowledge is prior knowledge of the data and the possible range of values based on experience and data documentation. A specific variable may have a range of values that does not include zero, but yet zero may be in the data set for some records. The zero may indicate missing. If the *describe()* method reports that the minimum value is zero, then you have a clue that there is at least one missing value. This is an example of how missing values are coded in your original data set. A data set could contain special codes to indicate a missing value and quite often different codes to represent different situations or causes of missing values. For example, a business database could contain a *Gender* variable for customers with "1 = Male" and "2 = Female." The numerics would be stored in a customer's record as opposed to the words "Male" and "Female" because it is more efficient to store integers. Suppose, however, that the gender is unknown. The database protocol may require the missing gender numeric be coded as 9, which

```
##
## Create a Heatmap of Missing Values
##
lst = [ 'ontime', 'complete', 'damage', 'document' ]
n = len( lst )
ax = sns.heatmap( x[ lst ].notnull().T, cbar = False )
ax.hlines( range( 1, n ) , *ax.get_xlim() )  ## Places horizontal line between variables
ax.set_title( 'Heatmap of Missing Data', fontsize = font_title )
ax.set( xlabel = 'Record Number', ylabel = 'Variable' )
base = 'Note:\nStripes indicate missing values'
footer(ax);
```

Fig. 3.14 This illustrates a possible display of missing values for the four *POI* measures. The entire DataFrame was subsetted to the first 1000 records for illustrative purposes. Missing values were randomly inserted. This map visually shows that "documentation" had no missing values while "ontime" had the most

is not uncommon. Similarly, customer income intervals could be represented by numerics ranging from 1 to 10 to cover values from less than $10,000 annually to more than $1 Million annually. Some customers, however, may not have provided their income out of privacy concerns so a 99 could be used to represent the missing value. The values 9, 99, 999, and 9999 are common. Other cases are certainly obvious.

This coding may reflect a disconnect between the *IT* department that designs and maintains the database and you as the analyst who will use this data. If you are unaware of the missing value codes they specify, then any results you produce will be nonsensical if these codes (e.g., 9, 99, etc.) are included. Not only must you know the list of *IT* missing values codes, but you must also let Pandas know them. The Pandas *read_csv* function has an argument to specify a list or dictionary of missing value codes. By default, Pandas recognizes a number of values such as "NA", "<NA>", "NULL" and many others, all of which are translated to the universal Pandas NaN code for floating point numbers.

How can you minimize the disconnect between the *IT* department and you? Ask them for a data dictionary. The dictionary's composition and detail vary by the extent of the *IT* department's databases, but generally it contains information on a variable's name, a mnemonic as a short-hand name, dates for creation and modification, ownership, and special coding to mention just a few. The special coding is where you find information about missing value codes.

There is a problem with having just one universal missing value code, NaN, for floats. This is the loss of information that multiple codes otherwise convey. For example, consider the income example I mentioned above. A customer may refuse to provide information about his/her income or just overlooked providing it. In both cases, there is a reason for the missingness but a single *NaN* will not tell you that reason. The *IT*'s data dictionary should be a source for these.

There are two more final issues regarding missing data. *What is the cause of the missingness*? and *What do you do about missing values*? I address these issues in Chap. 5 when I discuss preprocessing data. I treat them as part of data preprocessing because dealing with missingness is one of the overall preprocessing steps.

3.2.4.5 Check #5: Check the Data Types

Pandas manages different types of data which I list in Table 3.8. There are counterparts for most of these in Python and Numpy, but these are the key ones you will encounter in *BDA*.

Type	Designation (dtype)	Example
String or text	object	"Satisfied", "$1.99"
Integer number	int64	1, 2, 3
Floating point number	float64	3.14159
Boolean	bool	0 or 1
Date and time	datetime64	05/06/2020
Categorical string or text	category	"Small", "Medium", "Large"

Table 3.8 This is a partial listing of the data types available in Pandas

Strings are text enclosed in either single or double quotation marks. Numbers could be interpreted and handled as strings if they are enclosed in quotation marks. For example, 3.14159 and "3.14159" are two different data types. The first is a number *per se* while the second is a string. An integer in a number without a decimal; it is a whole number that could be positive or negative. A float is a number with a decimal point that can "float" among the digits depending on the values to represent. Integers and floats are treated differently and operations on them could give surprising, and unexpected results.

Boolean variables are simply nominal variables with values 0 and 1. Almost all software and programming languages interpret 0 as *False* and 1 as *True*. Boolean values are returned from comparison operations. I list some of these operations in Table 3.9.

Dates and times, combined and referred to as *datetime*, is a complex object treated and stored differently than floats, integers, and strings. Their use in calculations to accurately reflect dates, times, periods, time between periods, time zones, Daylight/Standard Saving time, and calendars in general is itself a complex topic. See Dershowitz and Reingold (2008) for a detailed discussion of calendrical calculations and different types of calendars. Pandas has a plethora of functions to work with datetime variables plus an accessor, *dt*, for extracting specific times from

Comparison	Operator	Example
Equality	$==$	$X == Y$
Inequality	$!=$	$X! = Y$
Greater than	$>$	$X > Y$
Less than	$<$	$X < Y$
Greater than/equal to	$>=$	$X >= Y$
Less than/equal to	$<=$	$X <= Y$

Table 3.9 These are the standard comparison operators that return a Boolean value: 1 if the statement is True; 0 otherwise

a datetime variable. For example, you could use the *dt* accessor to extract the year from a datetime variable in a Pandas DataFrame. I discuss datetime variables and the *dt* accessor in Chap. 7.

Categorical variables are a convenient way to store and manage a concept variable with a finite number of levels. A concept variable, which I more fully discuss in Chap. 5 with data encoding, is a non-numeric variable that divides a data set into parts, the minimum is, of course, two. The concept does not actually exist but is artificially designated, sometimes arbitrarily, to help identify groups. By artificial, I mean they do not exist in nature; they are invented by humans for some purpose. By arbitrary, I mean their definition can be changed and what constitutes the concept can be changed. For example, marketing regions is a concept for where different marketing activities are handled. They are meant to improve an organization's inefficiencies by dividing the overall target market into submarkets, each with a different management structure. Regions can be, and often are, redefined as marketing, economic, and political situations change. They are just concepts. As such, they have discrete levels that are mutually exclusive and completely exhaustive. For example, a business may have four marketing regions consistent with the U.S. Census Regions: Midwest, Northeast, South, and West. These four completely divide up the U.S. territory so they are completely exhaustive. A customer is in one and only one of these regions, so they are mutually exclusive. The specific regions could be viewed as levels or *categories*. The same holds for other concepts such as gender, income, education, and so forth.

The levels for the categorical concept are usually designated by character strings. Sometimes, it is more efficient for computer storage and processing, as well as other operations such as sorting, to treat the level designations differently, not as character strings but as numerics. The category data type does just this. For a management study, for example, there could be three levels for the concept variable "Manager": "Entry-level", "Mid-level", and "Executive." Clearly, these are arbitrary labels. These labels are strings. This manager variable could be designated a categorical variable and stored by its three levels rather than by the constant repetition of the three strings. When storing the manager variable this way, the order of the levels, which is purely artificial, could also be stored: Entry-level < Mid-level < Executive.

This cannot be done if the levels are maintained as strings. I will return to the category data type in Chap. 5.

The data type for any variable can be found using the *dtype* method. It is chained to a DataFrame name and returns a series indexed by each variable name and its corresponding data type. You can change from one data type to another using the Pandas *astype( )* method. For example, if *X* in the DataFrame *df* is an integer, you use *df.X.astype(float)* to cast *X* as a float variable.

3.3 Merging or Joining DataFrames

You will often have data in two (or more) DataFrames so you will need to *merge* or *join*[5] them to get one complete DataFrame. As an example, a second DataFrame for the baking facilities has information on each facility: the marketing region where the facility is located (Midwest, Northeast, South, and West), the state in that region, a two character state code, the customer location served by that facility (urban or rural), and the type of store served (convenience, grocery, or restaurant). This DataFrame must be merged with the *POI* DataFrame to have a complete picture of the baking facility. The merge is done on the facility ID (*FID*) which is a primary key in both DataFrames. One of these DataFrames in on the left side of the merge command and the other is on the right side. The one on the left is sometimes called the *main* DataFrame. The *POI* DataFrame is the main one in this example.

There are many types of joins but I will only describe one: an *inner join*, which is the default method in Pandas because it is the most common. I illustrate others in Fig. 3.15. The inner join operates by using each value of the primary key in the DataFrame on the left to find a matching value in the primary key on the right. If a match is found, then the data from the left and right DataFrames for that matching key are put into an output DataFrame. If a match is not found, then the left primary key is dropped, nothing is put into the output DataFrame, and the next left primary key is used. This is repeated for each primary key on the left.

You might recognize the inner join as the intersection of two circles or bubbles in a Venn Diagram as I show in Fig. 3.15. I show an example of merging two DataFrames in Fig. 3.16. This illustrates an inner join on a common primary key called "key." Merging DataFrames is a complex topic. I illustrate merging throughout this book.

[5] The two terms are used interchangeably.

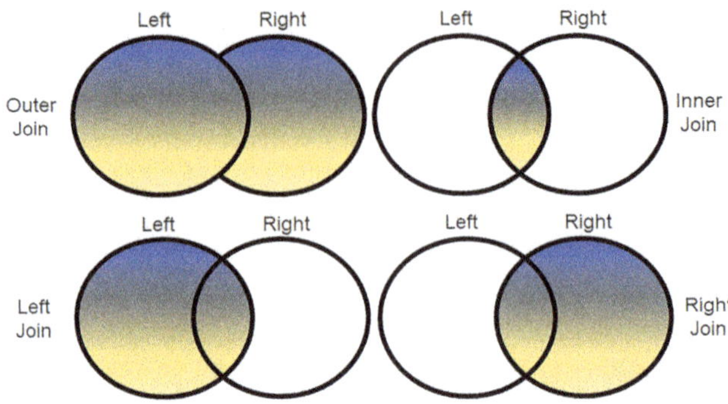

Fig. 3.15 This illustrates several different types of joins using Venn Diagrams. Source: Paczkowski (2016). Used with permission of SAS

```
##
## Define the Left or main Dataframe
##
dt_L = { 'key':[ 'A', 'B', 'C', 'D' ], 'L1':[ 1, 0, 0, 1 ], 'L2':[ 0, 0, 1, 0 ] }
df_L = pd.DataFrame( dt_L )
df_L.style.set_caption( 'Left or Main DataFrame' ).set_table_styles( tbl_styles )
```

Left or Main DataFrame

	key	L1	L2
0	A	1	0
1	B	0	0
2	C	0	1
3	D	1	0

```
##
## Define the right Dataframe
##
dt_R = { 'key':[ 'A', 'C', 'E' ], 'R1':[ 2, 4, 6 ], 'R2':[ 3, 1, 5 ] }
df_R = pd.DataFrame( dt_R )
df_R.style.set_caption( 'Right DataFrame' ).set_table_styles( tbl_styles )
```

Right DataFrame

	key	R1	R2
0	A	2	3
1	C	4	1
2	E	6	5

```
##
## Merge the Left and right DataFrames on 'key'
##
df = pd.merge( df_L, df_R, how = 'inner', on = 'key' )
df.style.set_caption( 'Merged DataFrame' ).set_table_styles( tbl_styles )
```

Merged DataFrame

	key	L1	L2	R1	R2
0	A	1	0	2	3
1	C	0	1	4	1

Fig. 3.16 This illustrates merging two DataFrames on a common primary key: the variable "key." Notice that the output DataFrame has only two records because there are only two matches of keys in the left and right DataFrames: key "A" and key "C". The non-matches are dropped

3.4 Reshaping DataFrames

I previously described a DataFrame as a rectangular array. This description is correct but overly simplistic because it does not specify the shape of that rectangle. It could be wide and shallow or narrow and deep. The width is the number of columns and the depth is the number of rows. A wide and shallow DataFrame is in *wide-form*; a narrow and deep one is in *long-form*.

A DataFrame's *shape* attribute provides information on the number of rows and columns. Sometimes this shape is inappropriate for a specific form of analysis and so it must be changed, the DataFrame must be *reshaped* to make its shape more appropriate. Changing a DataFrame from wide- to long-form involves stacking the rows vertically on top of each other (basically transposing each row) into a new column in a new DataFrame and using the original DataFrame's column names as values in yet another new column alongside the transposed rows. Those column names will repeat for each transposed row. As an example, a DataFrame could have 5 years of monthly sales data with one column for the year and a separate column for each month. There are then be 13 columns (one for year and 12 for months) and five rows for the 5 years. The shape is the tuple (5, 13). This is a wide-form. A simple regression analysis for sales as a function of a year and month effect requires a different data arrangement: one column for sales, one for year, and one for month. This is a long-form. So, the wide-form must be reshaped to long-form by stacking.

There are two Pandas methods for reshaping a DataFrame from wide- to long-form: *stack* and *melt*. They basically do the same thing, but the *melt* method is slightly more versatile in providing a name for the new columns of transposed rows. The *stack* method is better for operating on MultiIndexed DataFrames. The reverse reshaping operation of going from long- to wide-form is *unstacking*.[6] I show an example of melting a DataFrame in Fig. 3.17 that uses the merged DataFrame from Fig. 3.16. I show the reverse operation of unstacking this melted DataFrame Fig. 3.18. I illustrate reshaping throughout this book.

I used the *pivot_table* function in Fig. 3.18 to convert a DataFrame from long- to wide-form. This function is very powerful and versatile because it not only allows you to reshape a DataFrame but also summarize it. It has an argument, *aggfunc*, that accepts a list of functions for aggregating the entries in the pivoted table. The default function is the mean so the entries in the pivoted table will be the means of the corresponding rows and columns. I often use this function in subsequent chapters when data have to be aggregated and reshaped. I discuss pivot tables in Chap. 8.

---

[6] See the Pandas User Guide at https://pandas.pydata.org/pandas-docs/stable/user_guide/reshaping.html. Last accessed on March 25, 2020.

```
##
## Stack or melt a DataFrame
##
lst = ['L1', 'L2', 'R1', 'R2' ]
df_long = pd.melt( df, id_vars = [ 'key' ], value_vars = lst )
df_long.sort_values( by = [ 'key', 'variable' ], inplace = True )
df_long.reset_index( inplace = True )
df_long.style.set_caption( 'Long-Form DataFrame' ).set_table_styles( tbl_styles )
```

Long-Form DataFrame

	index	key	variable	value
0	0	A	L1	1
1	2	A	L2	0
2	4	A	R1	2
3	6	A	R2	3
4	1	C	L1	0
5	3	C	L2	1
6	5	C	R1	4
7	7	C	R2	1

Fig. 3.17 This illustrates melting a DataFrame from wide- to long-form using the final merged DataFrame from Fig. 3.16. The rows of the melted DataFrame are sorted to better show the correspondence to the DataFrame in Fig. 3.16

```
##
## Unstack or pivot a DataFrame
##
df_wide_2 = df_long.pivot_table( index = 'key', \
                columns = 'variable', values = 'value' )
df_wide_2.reset_index( inplace = True )
df_wide_2.style.set_caption( 'Wide-Form DataFrame' ).\
    set_table_styles( tbl_styles )
```

Wide-Form DataFrame

variable	key	L1	L2	R1	R2
0	A	1	0	2	3
1	C	0	1	4	1

Fig. 3.18 This illustrates the unstacking of the DataFrame in Fig. 3.17 from long- to wide-form

3.5 Sorting a DataFrame

Sorting is the next most common and frequently used operation on a DataFrame. This involves putting the values in a DataFrame in a descending or ascending order based on the values in one or more variables. The baking facilities could be sorted in ascending order (i.e., starting with the lowest value and proceeding to the largest) based on the *POI* measure. This would place those facilities with the lowest *POI* at the beginning of the DataFrame so the worst performers could be easily identified. Customers in a transactions DataFrame could be sorted by the date of last purchase to identify those customers with the most recent purchase or they could be sorted by the size of their purchase to identify the largest customers. I illustrate a simple sort in Fig. 3.17 for the DataFrame I previously created by melting or reshaping a wide-form DataFrame.

In some instances, sorting the DataFrame to identify the largest customers in terms of purchases or those who purchased most recently, or to identify the poorest performing facilities may not have to be done. A *query* of the DataFrame may be all that is need in these cases. I discuss queries in the next section. You could also use the *nlargest* or *nsmallest* methods. Each takes an argument for the number of records to return and a list of the columns to search on. For example, *df.nlargest(10, 'X')* returns the 10 largest values from the *X* column in the DataFrame *df*.

3.6 Querying a DataFrame

A DataFrame, whether small or large, contains a lot of latent information as I discussed before. One way you can get some information out of it is by literally asking it a question. That is, querying the DataFrame. You do this with *Boolean operators* and *query statements*. I will cover Boolean operators and query statements in the next subsections.

3.6.1 Boolean Operators and Indicator Functions

In everyday arithmetic, the equal sign has a meaning we all accept: the term or value on the left is the same as that on the right, just differently expressed. If the two sides are the same, then the expression is taken to be true; otherwise, it is false. So, the expression $2 + 2 = 4$ is true while $2 + 2 = 5$ is false. In Python, and many other programming languages, the equal sign has a different interpretation. It assigns a symbol to an object, which could be numeric or string; it does not signify equality as in everyday arithmetic. The assignment names the object and is said to be bound to the object. The object is on the right and the name on the left. So, the expression $x = 2$ does not say that x is the same as 2 but that x is the name for, and is bound to, the value 2. As noted by Sedgewick et al. (2016, p. 17), "*In short, the meaning of = in a program is decidedly not the same as in a mathematical equation.*"[7]

This assignment is even deeper than it seems. An object, whether numeric or string or a function or anything Python considers to be an object, is stored in a location in memory. The statement $x = 2$ names the memory location of 2 so this object can be retrieved. In some languages, x is referred to as a pointer to the object 2 in memory.

The object, 2 in this case, is stored once in memory but there could be several names bound to that memory location. So $x = 2$ and $y = 2$ have the symbols x and y naming (or pointing to) the same object in the same memory location. There is only one 2 in that memory location but two names or pointers to it. This has

[7] Emphasis in the original.

implications for changing names. See Sedgewick et al. (2016) for a programming-level discussion of assignments, names, and pointers in Python. Also see VanderPlas (2017, pp. 35–36) for some insight regarding how data are stored in computer memory.

If the equal sign does not indicate mathematical equality but names an object, then what does signify equality? The answer is a *Boolean Operator.* A Boolean Operator returns a true or false result where "True" is also represented by the integer 1 and "False" by the integer 0. The Boolean Operator symbol for equality is a double equal sign: "==". So, the arithmetic statement $2 + 2 = 4$ is written as the Boolean statement $2 + 2 == 4$ and "True" is returned. The Boolean statement $2 + 2 == 5$ returns "False." The Boolean statement is, therefore, a test which is either true or false. I show a collection of Boolean Operators in Table 3.9.

Several Boolean statements could be connected using the *logical and* connector ("&") and the *logical or* connector ("|"). For example, you could test if x is greater than 4 and less than 8 using "$(x > 4)$ & $(x < 8)$". You could also use $(4 < x < 8)$. You could test if x is greater than 4 or less than 8 using "$(x > 4) \mid (x < 8)$". The & symbol represents logical "and" while the | symbol (called a "pipe") represents logical "or". You could substitute the words "and" and "or" for "&" and "|", respectively. A *truth table* from fundamental logic summarizes possibilities. See Table 3.10. See Blumberg (1976, Chap. 5) for an extensive discussion of truth tables.

A	B	A & B	A | B
T	T	T	T
T	F	F	T
F	T	F	T
F	F	F	F

Table 3.10 This is a truth table for two Boolean comparisons: logical "and" and logical "or." See Sedgewick et al. (2016) for a more extensive table for Python Boolean comparisons

A Boolean statement could be compactly written in mathematical notation using an *indicator function*. An indicator function returns a 0 or 1 for a Boolean statement, usually for the data or a subset of the data. Suppose you have a list of six values for a variable X: [1, 2, 3, 4, 5, 6] and the Boolean statement "$x > 3$". This is written in mathematical notation as

$$\mathbb{I} = \begin{cases} 1 & \text{if } x > 3 \\ 0 & \text{otherwise} \end{cases}$$

or more simply as $\mathbb{I}(x > 3)$. The $\mathbb{I}$ is the indicator function. It returns the list $[0, 0, 0, 1, 1, 1]$ for this example. If you consider a subset of X, say the first three entries, $A = [1, 2, 3]$, then the indicator function is written as $\mathbb{I}_A(x > 3)$ which returns [0, 0, 0]. Indicator functions will be used in this book. See Paczkowski (2021a) for a discussion of the indicator function.

3.6.2 Pandas Query Method

Pandas has a *query* method for a DataFrame that takes a Boolean expression and returns a subset of the DataFrame where the Boolean expression is True. This makes it very easy to query a DataFrame to create a subset of the data for more specific analysis. I show two examples in Fig. 3.19. You will see applications of this query method in later chapters. The *query* method is chained to the DataFrame. The argument is a string with opening and closing single or double quotation marks (they must match). If a variable is used in the Boolean expression, you must enclose it in quotation marks. For example, you could write "$x >$ 'sales'". Notice the single and double quotation marks. You could also define a variable with a value before the query but then use that variable in the query. In this case, you must use an @ before the variable so that the Python interpreter knows the variable is not in the DataFrame. For example, you could define $Z = 3$ and then use "$x > @Z$".

```
##
## Query the POI DataFrame for all fids == 100
##
df_subset = df.query( 'fids == 100' )
print( f'Shape of the subset:\n{df_subset.shape}' )
print( f'\nList of the unique fids:\n{df_subset.fids.unique()}' )
```

```
Shape of the subset:
(1825, 6)

List of the unique fids:
[100]
```

```
##
## Query the POI DataFrame for all fids >= 100 and
## fids < 300.  The result should have just two fids:
## fids = 100 and fids = 101.
##
df_subset = df.query( ( 'fids == 100' ) and ( 'fids < 102' ) )
print( f'Shape of the subset:\n{df_subset.shape}' )
print( f'\nList of the unique fids:\n{df_subset.fids.unique()}' )
```

```
Shape of the subset:
(3650, 6)

List of the unique fids:
[100 101]
```

Fig. 3.19 These are two example queries of the POI DataFrame. The first show a simple query for all records with a *FID* equal to 100. There are 1825 of them. The second show a more complex query for all records with a *FID* between 100 and 102, but excluding 102. There are 3650 records

Chapter 4
Data Visualization: The Basics

Data visualization issues associated with the graphics used in a presentation, not in the analysis stage of developing the material leading to the presentation, are discussed in many books. I focus on data visualization from a practical analytical point-of-view in this chapter, not their presentation. This does not mean, however, that they cannot be used in a presentation; they certainly can be used. The graphs I describe are meant to aid and enhance the extraction of latent Rich Information from data.

4.1 Background for Data Visualization

Business Data Analytics is a combination of advanced and sophisticated statistical, econometric, and machine learning tools and methods for extracting Rich Information from data. You probably think of these as regression models, clustering algorithms, correspondence maps, decision trees, and many more. You spend a great deal of time learning their intricacies and how to apply them. One tool, however, that is also valuable and powerful, which precedes and supports these other tools, is data visualization. An old piece of advice statisticians still tell their students and clients is: *Look at your data*! This is *data visualization*.

Looking at your data is solid advice, but for databases that are small and easily manageable; that is, *Small Data*. With modern databases, this sage advice is difficult to implement. They are now measured in terabytes and more, so we now have *Big Data*. Paczkowski (2018) and Paczkowski (2016) discuss Big Data and ways to handle it. Any one piece of a Big Data database that you may want to "look" at may be just too large to deal with. So, how do you visualize this data?

In this chapter, I will discuss ways to "look" at data as an important first step in Deep Data Analysis to remove the veil of data hiding important, but latent, Rich Information. Data visualization is often not an end unto itself, but yet it may be

W. R. Paczkowski, *Business Analytics*,
https://doi.org/10.1007/978-3-030-87023-2_4

all that is needed to learn from data. Or, it could precede and support sophisticated statistical and econometric methods so often immediately applied to any data set, Big or Small. In short, it plays an important and powerful role at any stage of the deep analysis of any size and kind of data.

Visualization, is, of course, a wide area with active research in all its aspects such as perception, displays, three-dimensional, rendering, dynamic rendering, color coordination, and eye movement. I cannot possibly provide examples in all these areas or discuss issues specific to any one business problem. This chapter's focus is narrow –business data– but this narrowness does not diminish its importance because its focal areas are so important for everyday commerce, policy, general research, and the functioning and livelihood of the economy.

4.2 Gestalt Principles of Visual Design

There are several principles that have been proposed for creating effective graphs. What is an "effective graph"? It is one that allows the viewer, you or your client, to quickly see and digest a key message. An ineffective graph hides the key message either because of poor design or the use (or abuse) of "chartjunk." Chartjunk are graphing elements that have either nothing to do with the central message of the graph or cloud that message, adding another veil that hides it. Remember, the main focus of data analysis is to penetrate the veil imposed by data itself, a veil that hides the Rich Information buried inside the data. Chartjunk and poor design just compound a problem that already exists: penetrating the veil due to data. See Tufte (1983) on chartjunk.

There are several *Gestalt Principles of Visual Design* that have emerged as guides for effective graphs. There are four in particular I will refer to in this chapter. These are the:

1. Proximity Principle;
2. Similarity Principle;
3. Connectedness Principle; and
4. Common Fate Principle.

There are several others such as *Continuation*, *Closure*, and *Symmetry & Order*. The exact number seems unclear. I have seen reference to five and at times to seven. The four I listed are the most commonly mentioned and the ones I will refer to in this chapter. A Gestalt Principle is a concept regarding how humans perceive a whole based on its parts. If the parts are disorganized and dissimilar, then the whole will be difficult to perceive and understand. If the parts, however, are well organized and similar, then the whole will be well perceived. Regarding graphs, if graph elements are disorganized and dissimilar, then the message, the Rich Information the graph is attempting to convey, will remain hidden.

The *Gestalt Principle of Proximity* says that objects on a graph that are close to each other are interpreted as a group. The goal should be to group like items to make it easier for you and your client to form comparison judgments. If like items are not in close proximity, then it is a challenge, if not impossible to form the judgment.

The *Gestalt Principle of Similarity* says that objects comprising the groups should be similar in nature. Not only should similar items be placed close to each other (i.e., proximity) but those that are placed near each other should be similar. Colors and shapes help to identify similarities in graphs. If all plotting points in a scatter plot, for example, are black dots, then it is impossible to see any pattern in the points aside from the obvious. But if some are black and others red then you can more easily see patterns. The same holds for bars in a bar chart or slices in a pie chart.

The *Gestalt Principle of Connectedness* refers to lines that connect similar units (*The Principle of Similarity*) and help you see chunks of related information. Not only can the chunks be connected by lines, but the lines could also have different colors as well as forms (e.g., dashed, dotted, solid).

The *Gestalt Principle of Common Fate* is concerned with a tendency for objects to move or trend together. This is probably a more common principle for time series charts in which you want to show trends over time. If you have several time series plotted on the same axes, then you want to be able to show or highlight their commonality ("common fate") so that comparisons can be drawn.

There is a rich psychology and data visualization literature on Gestalt Principles. See Vanderplas et al. (2020) for some discussions about the principles and their application to graphic designs. Also see Kosslyn (2006), Pinker (1990) and Peebles and Ali (2015) for additional discussions and insight into the principles.

4.3 Issues Complicating Data Visualization

There are two issues associated with data visualization:

1. what or how much the human visual system can process; and
2. the tools for data visualization.

I will discuss both of these in the following subsections to lay the foundations for the rest of the chapter.

4.3.1 Human Visual Limitations

When we discuss data visualization, we are referring to what the human eye can process and transmit to the brain for further processing into intelligence and understanding. So a gating item is the limitations of the human eye. Wegman (2003) argues that there is a maximal number of *pixels* the human eye can process: 10^6–10^7 points. A pixel, short for "picture element", is a point or graphical element that composes an image such as a graph, picture, video, and so forth.[1] The human eye initially processes these pixels through cones in the retinal area of the eye. Cones are light-sensitive, especially to bright light. See Healey and Sawant (2012). There are also rods adjacent to the cones that are sensitive to lower-level light. There are about 10^7 cones in the eye so you should be able to process at most about the 10^6–10^7 observations I noted above. For small databases, this is not an issue since most are smaller than 10^7 in size. For example, it is probably safe to say that the majority of statisticians, econometricians, and general data analysts work with fewer than 10,000 observations in their normal analytical work where this figure is the total number of data points in their data sets. The 10,000 is 10^3, much less than the maximal number. You can safely say that the data set most commonly used is "Small Data." "Big Data", the order of magnitude referred to above, is a more recent phenomenon with data sizes far exceeding the maximal eye amount. I provide a taxonomy for data set sizes in Table 4.1 based Wegman (2003) and Huber (1994).

Descriptor	Data set size (bytes)	Storage mode
Tiny	10^2	Piece of paper
Small	10^4	A few pieces of paper
Medium	10^6	A floppy disk
Large	10^8	Hard disk
Huge	10^{10}	Multiple hard disks
Ridiculous	10^{12}	Magnetic tape

Table 4.1 Data set sizes currently defined or in use. Source: Wegman (2003) and Paczkowski (2018)

Data type	Data size	
	Small	Big
Discrete	Boxplot, Bar chart,	Same as small
	Pie chart, Mosaic graph	
	Heatmap	
Continuous	Histogram, Scatterplot,	Parallel Plot, Contour plot, Hex Bin plot
	Scatterplot matrix	
	Time Series plots, Contour plot, Hex Bin Plot	

Table 4.2 Visualization tools by data type and data size

The "Ridiculous" size of 10^{12} bytes in Table 4.1, known as a *Terabyte*, is now very common for laptop hard drives. We have gone way past Terabytes and are

[1] Georges Seurat and Paul Signac developed a branch of impressionist painting called Pointillism that used small dots to create images. If you stand at a distance from one of their paintings, the dots blend together so that the image becomes clear.

now in the realm of Petabytes (10^{15} bytes), Exabytes (10^{18} bytes), Zettabytes (10^{21} bytes), and Yottabytes (10^{24} bytes). The descriptor "Ridiculous" has a new meaning. See Paczkowski (2018) for a discussion.

The traditional visualization tools cannot be applied to data of these sizes, at least without modifications. Visualization tools have to be divided into two categories: Small Data applicable and Big Data applicable. Within these two categories, we also have to distinguish between the two types of data that can be visualized: categorical and continuous data and certainly a mix of the two. I summarize some visualization tools by data sizes in Table 4.2. In the following sections, I will describe these available visualization tools for both data size categories. These are not hard-and-fast rules for what can be done with a particular data set size; they are just guiding principles.

4.3.2 Data Visualization Tools

Python has several visualization packages: Pandas itself and *Seaborn* to mention only two. I will introduce a third later for geographic maps. Recall that Pandas is a data management package with the DataFrame as the main data repository. A convenient feature of a DataFrame is that it has a *plot* method attached to it when it is created. This means you can easily create a plot of your data merely by chaining this method to the DataFrame. For example, if a DataFrame contains a categorical variable X, then you can create a pie chart of it by first chaining to it the *value_counts* method to calculate the distribution of the values and then chain the plot method with an argument for a pie chart. The whole command is *df.X.value_counts().plot(kind = 'pie')*. You can change the pie chart to a bar chart by simply changing the *kind* argument from *pie* to *bar*. The key word *kind* specifies the type of plot. I list options for "kind" in Table 4.3.

The kind of plot to produce: Kind	Graph
'line'	Line plot (default)
'bar'	Vertical bar plot
'barh'	Horizontal bar plot
'hist'	Histogram
'box'	Boxplot
'kde'	Kernel Density Estimation plot
'density'	Same as 'kde'
'area'	Area plot
'pie'	Pie plot
'scatter'	Scatter plot
'hexbin'	Hexbin plot

Table 4.3 This is a list of options for the *kind* parameter for the Pandas plot method

Plot family	plotClass	kind
Relational	relplot	scatter, line
Categorical	catplot	boxplot, bar, count
Distribution	distplot	jointplot, pairplot, distplot, kdeplot, rugplot
Line	lineplot	time series

Table 4.4 This is a categorization of Seaborn's plotting families, their *plotClass*, and the *kind* options. See the Seaborn documentation at https://seaborn.pydata.org/ for details

Seaborn is a data visualization package that has a wide range of capabilities. A nice feature of Seaborn is that it reads Pandas DataFrames. A general Seaborn function call is *sns.plotClass(x = Xvar, y = Yvar, data = data, kind = kind, options)* where *sns* is the conventional Seaborn alias. The *plotClass* is a class or family of plots which I list in Table 4.4. The *Xvar* and *Yvar* parameters are the x-variable and y-variable, respectively, which vary by plotting class; options are hue, size, and style.[2] The *kind* options are the same as those in Table 4.3.

Pandas' and Seaborn's graph functionality are built on *Matplotlib*, a graphing package that allows you to control all aspects of creating a graph. As noted in Paczkowski (2021b), although Pandas and Seaborn are interface to this library you still need to know some of the Matplotlib capabilities to effectively create graphs to explore and understand your data. These are extensive capabilities. See the Matplotlib documentation online, VanderPlas (2017, Chapter 4), and Hunt (2019, Chapters 5 and 6). Also see Paczkowski (2021b) from which some of my comments are drawn.

Matplotlib's graphing power is also its shortcoming because of its complexity. An important sub-module, *pyplot*, acts as a gateway to this functionality to shield you from some of the intricacies. Pandas and Seaborn interface with *pyplot*. Consequently, you need to import the *pyplot* sub-module, although you do not need to do this to use their graphing functions *per se*. Importing the sub-module allows you to add features to your graphs such as titles and labels as well as manipulate chart elements such as tick marks. You import the sub-module using *import Matplotlib.pyplot as plt* where the alias is conventional.

All graphs, regardless of the software you use, are created or "painted" on a canvas. In Matplotlib terminology, the canvas is a *figure*. The actual graph is created inside an *axis* which is a region of the figure. If there is only one graph, then the axis occupies the entire figure area; otherwise, the figure area is divided into sub-regions or a grid arrangement each of which holds a graph. You have control over creating and sizing the figure area and specifying the number and arrangement of the axes inside a figure. You create a figure area using *fig = plt.figure( )*. By default, one axis is created. You can define the axes you want either at the time you create the figure, which is most common, or after you create the figure. At the time you create the figure, use *fig, ax = plt.subplots(nrows = 1, ncols = 1)* to create a simple grid that

[2] See the Seaborn documentation at https://seaborn.pydata.org/ for details.

is 1 × 1 axes, which is the default. To add axes, use *fig.add_subplot(2, 2, 1)* for a 2 × 2 grid (based on the first two digits) with each sub-axis placed according to the last digit: 1: upper left; 2: upper right; 3: lower left; 4: lower right.

Each axis is named. In my examples, the name is *ax* which is conventional. Some examples, taken from Paczkowski (2021b), are:

fig, ax = plt.subplots( ) for one axis in the figure (the default)

fig, axs = plt.subplots(2, 2) for 4 axes in a 2 × 2 grid

fig, (ax1, ax2) = plt.subplots(1, 2) for 2 axes in a 1 × 2 grid.

I illustrate some figure structures in Fig. 4.1.

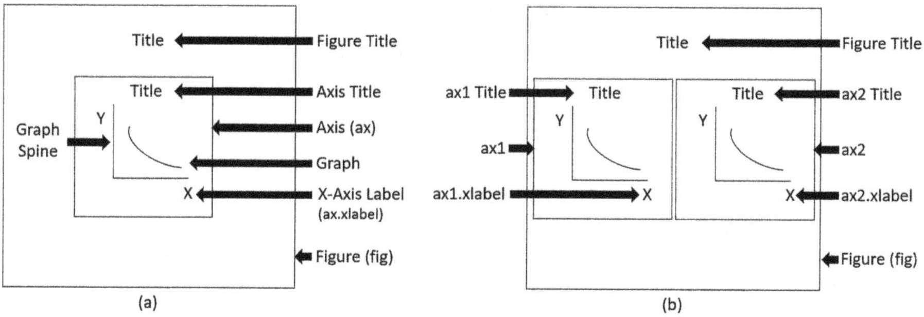

Fig. 4.1 This is the structure for two figures in Matplotlib terminology. Panel (**a**) is a basic structure with one axis (ax) in the figure space. This is created using *fig, ax = plt.subplots( )*. Panel (**b**) is a structure for 2 axes (ax1 and ax2) in a (1 × 2) grid. This is created using *fig, ax = plt.subplots(1, 2)*. Source: Paczkowski (2021b). Permission granted by Springer

As noted by Paczkowski (2021b), the axis name allows you to access parts of a graph such as title, *X* and *Y* axis (i.e., spine) labels and tick-marks, and so forth. You use the figure name to access the figure title. I list options in Table 4.5.

Graph component	Command	Examples
Figure title	suptitle	fig.suptitle(*figureTitle*)
Axis title	set_title	ax.set_title(*axisTitle*); ax1.set_title(*axis1Title*)
X-axis label	xlabel	ax.set_xlabel(*X-label*); ax1.set_xlabel(*X1-label*)
Y-axis label	ylabel	ax.set_ylabel(*Y-label*); ax1.set_xlabel(*Y1-label*)

Table 4.5 These are a few useful Matplotlib annotation commands

4.3.3 Types of Visuals

You most likely think of graphs when you think of visualization, but this thinking is simplistic since there are many types of graphs and for different types of data. A brief listing includes those familiar to most analysts: bar charts, pie charts, histograms, and scatterplots. The list is actually much longer. Not all of these can be applied to any and all types of data. They have their individual uses and limitations that reveal different aspects of the latent information.

Continuity	Number of series	Graph type
Continuous	One	Pie, Boxplot, Histogram, Line
Continuous	Two	Scatter, Contour
Continuous	More than two	Parallel
Discrete	One	Bar, Pie
Discrete	Two	Mosaic, Heatmap
Combination	Two or more	Boxplots, Facets, Contour

Table 4.6 Matching visualization tools to the data

I highlight some possibilities in Table 4.6. Different types of graphs are applicable to different types of data, so the graph must match data features. I show two features here: the *Continuity* and the *Number of Series* or variables to plot.

In the following sections, I will divide the data by their continuity and if they are spatial or temporal.

4.3.4 What to Look for in a Graph

Most people just "look" at a graph and then say something –anything– about what they see. "Looking", however, requires training and experience. "Lookers" can be classified as novices or experts. Novices have either no training in what to look for in a graph, or are just beginning that training. Their tendency is to miss the important messages conveyed by a graph which results in Rich Information remaining hidden. Experts, through training and experience, know what to look for and look for it more rapidly not only to cull the salient messages, the Rich Information, but also translate them into actionable recommendations. This is referred to as *graphicacy*. See Peebles and Ali (2015).

The reason for training in reading and interpreting graphs is that graphs have a symbol system that differs from text. We are all trained from an early age to read and interpret written text. The symbol system for text consists of the alphanumeric characters and punctuation marks, not to overlook a non-symbol system such as paragraphs, indentation, an order from left to right progressing down a page, capitalization, agreement of nouns and verbs, and so on.[3] See Pinker (2014) for an excellent discussion of writing issues. Graphs have a different symbol system consisting of lines, bars, colors, dots, and other marks that tell a story, convey a message, no different than text. You have to be trained to "read" and interpret a graph symbol system just as you were trained to read the text symbol system. There are features, clues if you wish, that help you extract the key hidden, latent messages from text; so also, with graphs. I advocate five guiding features you should look for:

1. Distributions;
2. Relationships;
3. Patterns;
4. Trends; and
5. Anomalies.

4.3.4.1 Feature #1: Distributions

The distribution of your data is their arrangement or shape on a graph axis. Distributions could be symmetric, skewed (left or right), or uni/multi-modal. I show four possible shapes for distributions in Fig. 4.2.

Symmetry means that the shape of the distribution to the left and right of a center point, typically the mean or median, is the same. The normal distribution is the canonical symmetric distribution for a continuous random variable, but certainly it is not the only one. The probability density function (*pdf*) for a normally distributed random variable is

$$f(y) = \frac{1}{\sqrt{2\pi\sigma^2}} e^{-\frac{(Y-\mu)^2}{2\sigma^2}} \tag{4.3.1}$$

where μ and σ^2 are population parameters for the mean and variance, respectively. To show symmetry about the mean, add a small amount, say δ, to the mean of the random variable so that the new value of the random variable is $Y = \mu + \delta$. The *pdf* becomes

$$\begin{aligned} f(\mu + y) &= \frac{1}{\sqrt{2\pi\sigma^2}} e^{-\frac{([\mu+\delta]-\mu)^2}{2\sigma^2}} \\ &= \frac{1}{\sqrt{2\pi\sigma^2}} e^{-\frac{\delta^2}{2\sigma^2}}. \end{aligned}$$

[3] See https://en.wikipedia.org/wiki/Graphicacy. Last accessed January 13, 2021.

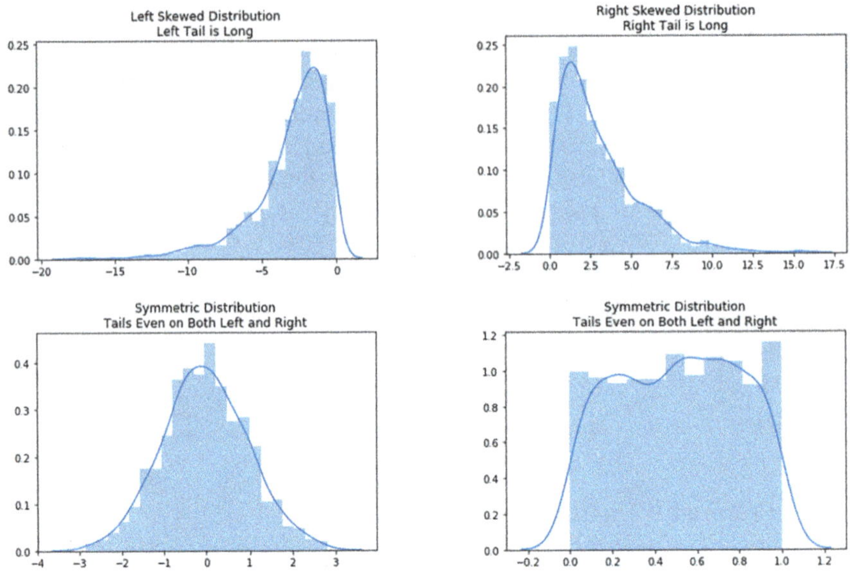

Fig. 4.2 Four typical distributions are illustrated here. The top left is **left** skewed the top right is **right** skewed. The two bottom ones are symmetric. The lower right is **almost** uniform while the lower left is almost normal. The one on the lower left is the most desirable

Now subtract that small amount from the mean to get

$$
\begin{aligned}
f(\mu - y) &= \frac{1}{\sqrt{2\pi\sigma^2}} e^{-\dfrac{([\mu - \delta] - \mu)^2}{2\sigma^2}} \\
&= \frac{1}{\sqrt{2\pi\sigma^2}} e^{-\dfrac{(-\delta)^2}{2\sigma^2}} \\
&= \frac{1}{\sqrt{2\pi\sigma^2}} e^{-\dfrac{\delta^2}{2\sigma^2}}
\end{aligned}
$$

The two *pdf*s are identical so the normal density curve is symmetric about the mean μ.[4] This symmetry is desirable because three key measures of central tendency –mean, median, and mode– are equal under symmetry; they differ under asymmetry.

[4] See https://online.stat.psu.edu/stat414/node/149/ for a similar demonstration. Last accessed March 8, 2010. Also see https://math.stackexchange.com/questions/1506875/proof-that-the-gaussian-distribution-is-symmetric. Last accessed March 8, 2010.

Skewness is defined as the elongation of the tail of the *pdf*, either to the left or right but not both simultaneously. The mean, in particular, deviates from the median when the distribution is skewed. If the distribution is right skewed, then the mean is larger than the median and gives an inaccurate indication of central tendency. The reverse holds for a left skewed distribution. I show the two possibilities in Fig. 4.2

The skewness of the normal *pdf* is used as the benchmark for determining the skewness of other distributions. Since the normal *pdf* is symmetric, its skewness is zero; the areas in tails are equal. A test of skewness can be done using a Z-test to compare a skewness measure for a distribution to zero for a normal distribution. This skewness measure is the *Fisher-Pearson Coefficient of Skewness*:[5]

$$g_1 = \frac{m_3}{m_2^{3/2}} \tag{4.3.2}$$

where $m_i = 1/n \sum_{j=1}^{n}(x_j - \bar{x})^i$. It is a biased sample central moment. An unbiased version is the *Adjusted Fisher-Pearson Standardized Moment Coefficient*:

$$G_1 = \frac{\sqrt{n \times (n-1)}}{n-2} \times g_1 \tag{4.3.3}$$

The Null Hypothesis for the skewness test is that there is no difference between the skewness of a normal distribution and the tested distribution. The test returns a Z-score and associated p-value. If the Z-score is negative and significant, then the tested distribution is left skewed; otherwise, it is right skewed. If the Z-score is insignificant, then the distribution is not skewed. See Joanes and Gill (1998) and Doane and Seward (2011). A test for skewness is based on D'Agostino et al. (1990) which I illustrate in Fig. 4.3.

```
from scipy.stats import skewtest
skewtest([1, 2, 3, 4, 5, 6, 7, 8])
```

```
SkewtestResult(statistic=1.0108048609177787, pvalue=0.3121098361421897)
```

Fig. 4.3 This is an example of the skewness test. This is a Z-test. A Z value less than zero indicates left skewness; greater than zero indicates right skewness. The p-value is used to test the Null Hypothesis skewness that the skewness is zero. Since the p-value is greater than 0.05, the Null Hypothesis of no skewness is not rejected

There are two reasons I mentioned the normal distribution. First, since it is the canonical symmetric distribution, the symmetry of any other portrayal of a distribution is judged relative to the normal via this Z-test. The other portrayal is an empirical distribution such as a histogram derived from data. Second, a normal

[5] See https://docs.scipy.org/doc/scipy/reference/generated/scipy.stats.skew.html for documentation. Last accessed March 26, 2020.

distribution is often overlaid on top of the empirical distribution to show the extent of agreement with or departure from the normal as an aid for visualizing skewness. This is evident in the four panels of Fig. 4.2.

4.3.4.2 Feature #2: Relationships

Relationships are not just associations (i.e., correlations), but more cause-and-effect behavior between or among two or more variables. Products purchased and distribution channels is one example. Customer satisfaction and future purchase intent is another. These relationships could be spatial, temporal, or both.

4.3.4.3 Feature #3: Patterns

Patterns would be groupings of data points such as clusters or segments.

4.3.4.4 Feature #4: Trends

Trends are developments or changes over time (e.g., a same-store sales tracking study or attrition rates for R&D personnel for an *HR* study). These would be mostly temporal.

I treat trend as separate from pattern, even though it is a pattern, because trend is usually associated with a time series. A time series, which I discuss in Chap. 7, consists of measures of an object at different points in time, those points progressing in an orderly manner. The measures are usually plotted against time and are connected by lines. The lines actually convey meaning: that the series is to be interpreted as a single entity with the line binding the plotted points together. So a trend is a movement of that single entity through time. A pattern is a more encompassing concept. It represents grouping and organization. A trend is a pattern, but a pattern is not necessarily a trend. This reflects the *Gestalt Connectedness Principle*.

4.3.4.5 Feature #5: Anomalies

Anomalies or *outliers* are points that differ greatly from the bulk of the data. But not all outliers are created equal: some are innocuous while others are pernicious and must be inspected for their source and effects. The innocuous ones have no effect on analytical results whereas the pernicious ones do. For example, in Fig. 4.4 I show the effect of an outlier on a regression line and then the effect on that line when the outlier is removed. It is clear that the line was pulled by the large outlier. If the goal is to provide Rich Information, then data with a pernicious outlier will not allow you to meet that goal.

Identifying outliers is a daunting task especially when you are dealing with multivariate DataFrames. The Python package *PyOD* has a comprehensive set of data visualization tools for examining data for potential outliers. You can install *PyOD* using *pip install pyodbc* or *conda install -c conda-forge pyod*.

```
##
## Plot data with and without outlier
##
## Set figure space
##
fig, ( ax1, ax2 ) = plt.subplots( 1, 2, figsize = ( 10, 5 ) )
plt.tight_layout()
plt.subplots_adjust( wspace = 0.3 )
ticks = np.linspace(50,200,4)*1000
##
## Import data
##
tmp = pd.read_csv( '../data/outliers.csv' )
##
## Scatter plots
##
ax1.set_xlim( 0, 200000 )
ax1.set_xticks( ticks )
ax1.set_ylim( 0, 120000 )
ax1.set_title( 'Regression with Y Outlier\nSimulated Data', fontsize = 18 )
sns.regplot( x = 'X', y = 'Y', data = tmp, ci = None, ax = ax1 )
ax1.annotate( 'Pernicious\nOutlier', ( 150000, 111000 ) )
##
tmpClean = tmp.query( 'Y < 100000' )
ax2.set_xlim( 0, 200000 )
ax2.set_xticks( ticks )
ax2.set_ylim( 0, 120000 )
ax2.set_title( 'Regression with Y Outlier Removed\nSimulated Data', fontsize = 18 )
sns.regplot( x = 'X', y = 'Y', data = tmpClean, ci = None, ax = ax2 )
ax2.annotate( 'Innocuous\nOutliers', ( 145000, 60000 ) );
```

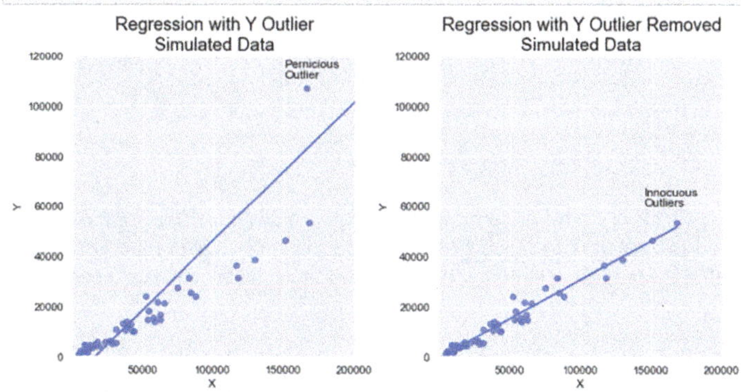

Fig. 4.4 This illustrates the effect of an outlier on a regression line. The left panel shows how the outlier pulls the line away from what appears to be the trend in the data. The right panel shows the effect on the line with the outlier removed

4.4 Visualizing Spatial Data

In this section, I will consider continuous, or floating point, numbers. They can be combined with discrete or integer numbers where the latter are used for categorization purposes. I will use the Perfect Order Index (*POI*) data as a Case Study.

4.4.1 Data Preparation

The *POI* DataFrame is in panel format, meaning that it has a combination of spatial and temporal data. The spatial aspect is baking facilities and/or their geographic locations, either state or marketing region. The temporal dimension is day of delivery. So, the Data Cube is slightly more complex in the spatial than in the temporal dimension. Since I am concerned with the spatial properties of the data, the temporal aspect of the Cube must be collapsed. At first, I aggregate the data for each *FID*. For each aggregated *FID*, the *POI* is calculated through multiplication. I show how to do this in Fig. 4.5. I will discuss other aggregations later.

```
##
## Aggregate POI measures by FIDS, thus collapsing
## the temporal dimension of the cube.
## Aggregate measure is the mean.
##
lst = [ 'fids', 'complete', 'damage', 'document', 'ontime' ]
df_agg = df[ lst ].groupby( [ 'fids' ], as_index = False ).agg( 'mean' )
##
## Calculate POI at the FID Level as the product of the four measures
##
df_agg[ 'poi' ] = df_agg.complete * df_agg.damage * df_agg.document * df_agg.ontime
df_agg.head().style.set_caption( 'Aggregate POI Measures by FIDS').\
    set_table_styles( tbl_styles ).hide_index()
```

Aggregate POI Measures by FIDS

fids	complete	damage	document	ontime	poi
100	0.956712	0.955068	0.934795	0.963288	0.822788
101	0.964932	0.950685	0.936986	0.974247	0.837404
102	0.954521	0.950137	0.936438	0.966575	0.820893
103	0.962740	0.954521	0.939178	0.974795	0.841308
104	0.961096	0.955616	0.938082	0.973699	0.838911

Fig. 4.5 This code shows how the data for the spatial analysis of the *POI* data are aggregated. This aggregation is over time for each *FID*. Aggregation is done using the *groupby* function with the mean function. Means are calculated because they are sensible for this data. The DataFrame is called *df_agg*

This aggregated DataFrame can be merged with a DataFrame that contains data specific to each *FID*: marketing region, state (and state code), location (rural, suburban, or urban), and store type (convenience, grocery, or restaurant). I show this merging in Fig. 4.6.

4.4.2 Visualizing Continuous Spatial Data

Examining distributions of continuous spatial data is common. The *boxplot*, sometimes called a *box-and-whisker plot*, is a powerful tool for continuous data. It was developed by John Tukey in the 1970s. See Tukey (1977) as well as Frigge et al. (1989). I show the parts of the boxplot in Fig. 4.7. The boxplot reflects the statistical fact that simple descriptive statistics based on moments (e.g., the mean or average) are not robust or resistant to outliers. Descriptive statistics based on percentiles,

```
##
## Merge FIDs data with aggregate data
##
df_agg = df_agg.merge( df_fids, on = 'fids' )
##
## Add a warning if POI is <= 0.8
##
df_agg[ 'warning' ] = [ 'Low' if x <= 0.8 else 'High' for x in df_agg.poi ]
df_agg.head( ).style.set_caption( 'Merged Aggregate POI Measures by FIDS').\
    set_table_styles( tbl_styles ).hide_index()
```

Merged Aggregate POI Measures by FIDS

fids	complete	damage	document	ontime	poi	region	State	Code	location	store	warning
100	0.956712	0.955068	0.934795	0.963288	0.822788	MW	Illinois	IL	Urban	Convenience	High
101	0.964932	0.950685	0.936986	0.974247	0.837404	MW	Illinois	IL	Urban	Grocery	High
102	0.954521	0.950137	0.936438	0.966575	0.820893	MW	Illinois	IL	Urban	Grocery	High
103	0.962740	0.954521	0.939178	0.974795	0.841308	MW	Illinois	IL	Urban	Convenience	High
104	0.961096	0.955616	0.938082	0.973699	0.838911	MW	Illinois	IL	Urban	Grocery	High

Fig. 4.6 This code shows how the data are merged. The new DataFrame is called *df_agg*

however, are robust so they are often preferred, or at least highly recommended. Of all the possible percentiles, five are the most common as I show in Table 4.7. These comprise the *Five Number Summary* of the data.

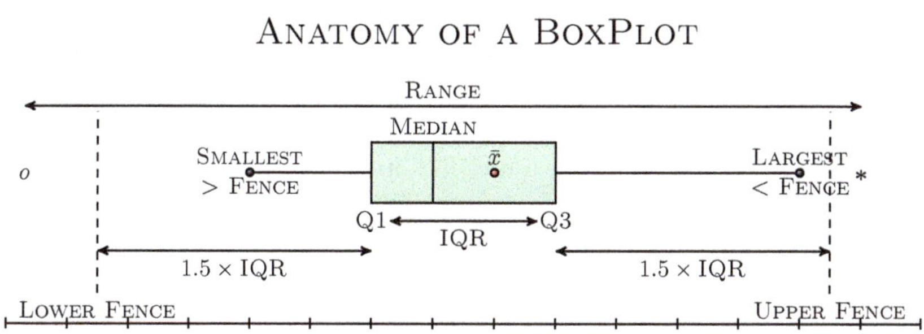

Fig. 4.7 Definitions of parts of a boxplot. Source: Paczkowski (2021b). Permission granted by Springer

Percentile	Definition	Boxplot position
Minimum	Minimum value	Lowest point
Q1	First or lower quartile	Bottom of box
Q2 or median	Second quartile/middle value	Middle line in box
Q3	Third or upper quartile	Top of box
Maximum	Maximum value	Highest point

Table 4.7 The Components of a Five Number Summary. A sixth measure is sometimes added: the arithmetic average or mean. This is shown as another symbol inside the box

Distributional measures can be calculated (or inferred) from the boxplot components in Fig. 4.7. The *Range* ($= Maximum - Minimum$) and the *Interquartile Range* or *IQR* ($= Q3 - Q1$) can be quickly determined. Skewness is also determined

```
##
## Boxplot of aggregated POI
##
ax = sns.boxplot( y = 'poi', data = df_agg )
ax.set_title( 'Perfect Order Index\nAggregated over all Facilities and Months',\
              fontsize = font_title )
ax.set( ylabel = 'Perfect Order Index' );
```

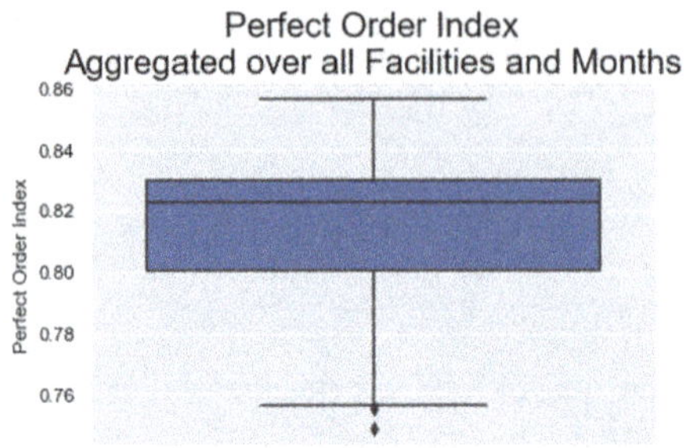

Fig. 4.8 Boxplot for a single continuous variable

by the relationship between $Q3 - Median$ and $Median - Q1$. If $Q3 - Median > Median - Q1$, the distribution is skewed right; if $Q3 - Median < Median - Q1$, then it is skewed left; otherwise, it is symmetric. The whiskers extending from the top and bottom of the box go as far as $1.5 \times IQR$ or to the minimum/maximum value greater/less than $1.5 \times IQR$. The $1.5 \times IQR$ thus defines a fence the whisker cannot pass. Any points beyond the fences are outliers. Almost every software package uses 1.5. Some allow you to specify the multiplier but with 1.5 as the default.[6] The *IQR* can be used to estimate the standard deviation since $s \approx {}^{IQR}/_{1.35}$. See Wan et al. (2014) for some comments on and limitations of this approximation.

I show a boxplot for the aggregated *POI* data in Fig. 4.8. The distribution is indicated by the elongation of the box (i.e., if it is long or squat depending on its orientation) and the length of the whiskers or tails extending from both ends of the box. In general, look for the following in a boxplot:

- Box elongation
  - Small box: small variance/spread
  - Large box: large variance/spread
- Skewness
- Outliers
  - Points outside the whiskers

For the boxplot in Fig. 4.8, there is a narrow elongation of the box implying little variation of 50% of the data. There is also some left skewness in the distribution because the bottom portion of the box, representing 25% of the data, is more

[6] See https://math.stackexchange.com/questions/2140168/statistics-calculating-quartiles-or-box-and-whisker-plots-where-did-the-1-5-co for a discussion of the 1.5 value.

elongated than the top portion, which is also 25% of the data. So, the spread in the data for the bottom 25% is greater than that for the top 25%. In addition, the lower tail is longer than the upper tail, also suggesting left skewness. This can be tested using the *skewtest* I described above. The Null Hypothesis is no skewness. The Z-value is −6.654 and the p-value is 0.0000 so the Null Hypothesis is rejected at any level of significance. Finally, there are a few outliers or extreme values at the lower, left, end of the distribution.

Another, more classic way to visualize a distribution is to use a histogram. In particular, a histogram is a tool for estimating the *probability density function* of the values of a random variable, X. Let $f(x)$ be the density function. As noted by Silverman (1986), knowing this probability density function allows you to calculate probabilities since

$$Pr(a < X < b) = \int_a^b f(d)dX. \tag{4.4.1}$$

Assume a random sample of n observations on a continuous random variable X: $x_1, x_2, \ldots, x_n$. The values of the observations are assigned to *bins* or groups to discretize the continuous data. Let h be the number of units that define the width of a bin on the X scale. If the X scale, for example, is thousands of people, you could have $h = 10,000$ people. You could specify varying widths so you could have h_i for the ith bin. It is typical, however, for the bin widths to be constant so $h_i = h, \forall i$.

A histogram is defined by h; an origin, x_0, for where you want to begin binning your data; and the number of bins, M. The origin does not have to be the first ordered value of the data although it typically is the first value. Any values in the data set before x_0 are simply ignored. The *Scipy* stats histogram function has an argument for the lower and upper values for the range of the histogram. This is specified as a tuple. If a tuple is not given, "a range slightly larger than the range of the values ... is used." This range is given by $(x.min() - s, x.max() + s)$ where $s = 1/2 \times {}^{(x.max()-x.min())}/_{(\#bins-1)}$.

The bins are defined as $B_i = [x_0 + (i-1)h, x_0 + ih)$ where $i = 1, 2, \ldots, M$ is a series of positive integers identifying the bins themselves. For example, for the first bin the interval is $[x_0, x_0 + h)$; it is $[x_0 + h, x_0 + 2h)$ for the second bin; and so forth. These intervals are usually half-open on the right: the left value is included and the right value is excluded. So the interval of integers $[1, 4)$ contains the values 1, 2, 3.

Define an indicator function $\mathbb{I}$ such that

$$\mathbb{I} = \begin{cases} 1 & \text{if } x_j \in B_i \\ 0 & \text{otherwise} \end{cases} \tag{4.4.2}$$

for $j = 1, 2, \ldots, n$, $i = 1, 2, \ldots, M$ and where B_i is the ith bin. The tally of data points in B_i is then $\sum_{j=1}^{n} \mathbb{I}(x_j \in B_i)$ and the total sample is $n = \sum_{i=i}^{M} \sum_{j=1}^{n} \mathbb{I}(x_j \in B_i)$. A bin's tally is referred to as the *frequency* or *count* in B_i. The normalized area for B_i is

$$f_i h = \frac{1}{n} \sum_{j=1}^{n} \mathbb{I}(x_j \in B_i) \tag{4.4.3}$$

where $f_i h$ is the proportion of the total sample in B_i and f_i is a count representing the "height" of the bin. The quantity $f_i h$ is interpreted as the proportion of the total sample per unit of the X scale times the number of units of X in the bin. If $h = 1$, then f_i is merely the proportion of the sample in B_i. This is sometimes referred to as the *relative frequency* in B_i.

You can now write

$$\sum_{i=i}^{M} f_i h = \frac{1}{n} \sum_{i=i}^{M} \sum_{j=1}^{n} \mathbb{I}(x_j \in B_i) \tag{4.4.4}$$

$$= 1. \tag{4.4.5}$$

The *density* of B_i is

$$f_i = \frac{1}{n \times h} \sum_{j=1}^{n} \mathbb{I}(x_j \in B_i) \tag{4.4.6}$$

as in Silverman (1986). This density is composed of two parts:

1. the proportion of the sample in B_i ($= 1/n \sum_{j=1}^{n} \mathbb{I}(x_j \in B_i)$), and
2. the per unit of the X scale.

It is "density" because it is the proportion of the sample per unit of X. This is analogous to population density: 10% of the population per square mile. See Silverman (1986) for a classic discussion of histograms and their problems. Also see the paper by Weglarczyk (2018). Finally, see Freedman et al. (1978) for an intuitive discussion of a density and a histogram.

I show an example histogram for the aggregate *POI* data in Fig. 4.9. The histogram can be enhanced with a smooth density curve to more effectively highlight the distribution pattern. This smooth curve is a *Gaussian Kernel Density Estimate* or *KDE* and is an estimate of the f_i in (4.4.6). More specifically, it is a non-parametric estimate of the normal *pdf* widely used to show how the distribution of one data series compares to the normal distribution. The histogram in Fig. 4.9 shows that the distribution has bimodality rather than a single modality as for a normally distributed random variable. These two distributions could account for the left skewness in Fig. 4.8. Typically, a multimodal distribution is indicative of a mixture of underlying distributions, but not always. You could have a mixture of two distributions that is unimodal as well as a single distribution that is bimodal by its nature. I discuss mixture models in Chap. 12.

```
##
## Histogram of aggregated POI
##
ax = sns.distplot( df_agg.poi )
ax.set_title( 'Perfect Order Index\nAggregated over all Facilities and Months',
             fontsize = font_title )
ax.set( ylabel = 'Probability Density', xlabel = 'Perfect Order Index' );
```

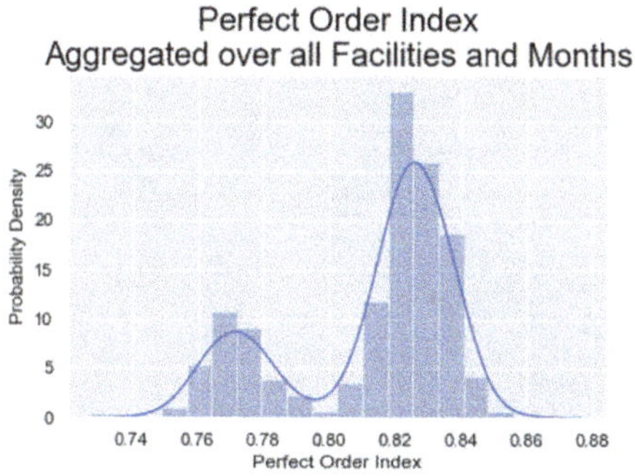

Fig. 4.9 Histogram for a single continuous variable

Two continuous variables could be plotted together is a *scatter plot*. I show an example in Fig. 4.10. You are probably familiar with a simple scatter plot which is relatively easy to interpret. You should look for chunks of data that are close (*Gestalt Proximity Principle*). If three variables are plotted on the same axes, each is given a different color or *hue* indicated in a legend. The *Gestalt Similarity Principle* is used to note chunks of like hues.

There is a problem when the number of data points becomes very large. This is sometimes referred to as the *Large-N* problem. See Carr et al. (1987). With *n* very large, a scatter plot would appear as just a big black ball (or whatever color is used for the points). This is useless for analysis because it is very difficult to discern any relationship between the two variables. So the two Gestalt principles of *Proximity* and *Similarity* breakdown. This is clearly the situation in Fig. 4.10 which is based on the *POI* data aggregated for each week for each year for each *FID*. The aggregated DataFrame has 104,800 records. This scatter plot of *POI* vs *Damage* is not revealing.

There are four ways to handle this situation. You could use a:

1. random sample of the data;
2. *Kernel density estimator* or *contour* plot;
3. *hexagonal binning plot*; or
4. *Lowess* curve, perhaps omitting the plot points.

Random sampling (without replacement) is useful for reducing the *Large-N* problem and it allows you to do cross-validation. Cross-validation means that you could draw several samples using each as a check against the others. You should see the same relationships, more or less, but not dramatic differences. Since you are sampling, you might not have the best view of the data on any one draw simply

```
##
## Scatter plot of monthly aggregated data
##
## Add some random jitter to plotting points
##
np.random.seed( 42 )
mu = 0
sigma = 0.1
n = df_grp.shape[ 0 ]
df_grp[ 'damage' ] = df_grp.damage + np.random.normal( mu, sigma, n )
df_grp[ 'poi' ] = df_grp.poi + np.random.normal( mu, sigma, n )
##
## Plot the data
##
ax = sns.scatterplot( x = 'damage', y = 'poi', data = df_grp );
ax.set_title( 'POI vs Damage\nMonthly Average Data', fontsize = font_title )
ax.set( xlabel = 'Damage', ylabel = 'POI' );
```

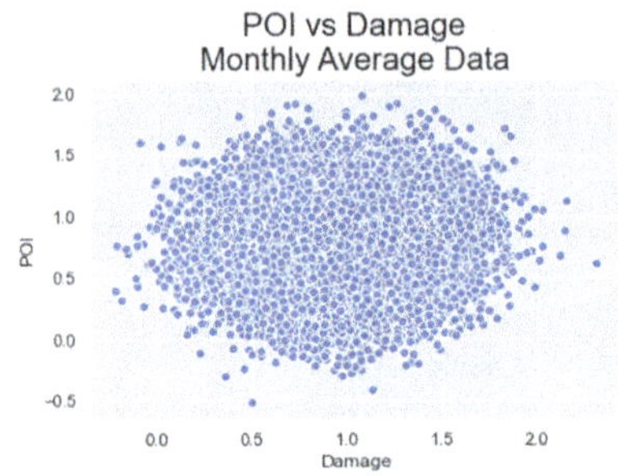

Fig. 4.10 Scatter plot for two continuous variables

because of the "luck of the draw" of the sample. For example, outliers and the mode may be hidden by the sample or you may actually introduce outliers and modes that are not really there, but appear just by the draw of the sample. These are problems with any sampling, whether for graphs, which is our concern, or for surveys, or clinical trials, and so forth. See Carr et al. (1987) for some discussion.

The kernel density plot (*KDE*) is used to show contours of the data. Basically, you can imagine your data plotted in a 3D space: the $X - Y$ two-dimensional (2D) plane which is your scatter plot plane plus a third dimension, perpendicular to this plane, which is the density of the plotting points. The third dimension is a surface rising from the 2D plane which can be sliced horizontally (i.e., parallel to the $X - Y$ plane) at an infinite number of points to reveal contours. These are then projected down to the 2D plane. The density of the data is revealed by the density of the projected contours. Points that are close, that are tightly packed or dense, are shown as dark areas while points further from the dense pack are shown as lighter areas. These dark and light areas reflect the Gestalt principles of *Proximity* and *Similarity*. I show an example in Fig. 4.11 of the data in Fig. 4.10. Two clusters are evident as two black spots while further points have varying degrees of gray. Notice in Fig. 4.11 that *KDE* histograms are on the margins of the graph to emphasize the individual distributions.

An alternative to the contour plot is the *hexagonal binning* plot. This is sometimes better because it involves drawing small hexagonal shapes, the coloring or shading of the shapes indicating the number of data points inside the hexagon. The darker the color or shade, the denser the data. Why hexagons? They have

```
##
## Contour plot of the POI and Damage data.
##
ax = sns.jointplot( y = 'poi', x = 'damage',
                    kind = "kde", data = df_agg )
ax.ax_joint.set_xlabel( 'Damage' )
ax.ax_joint.set_ylabel( 'POI' );
```

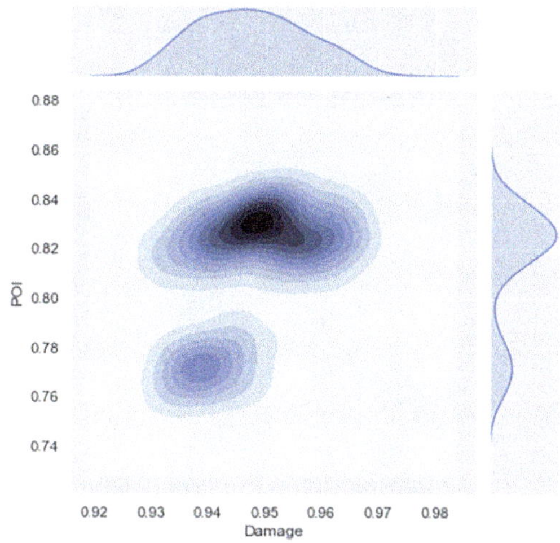

Fig. 4.11 A contour plot of the same data used in Fig. 4.10

better visual properties than squares and better span the plotting space. Hexagons are 13% more efficient for covering the *X*-*Y* plane. See Lewin-Koh (2020) for an extensive discussion and an implementation in R. I provide an example hex bin plot in Fig. 4.12 for the *POI* data in Fig. 4.10. Notice the histograms on the margins of the graph plane.

A final approach is to draw a line through the cloud of data points that shows the general trend of the data which is indiscernible otherwise. This reflects the *Gestalt Common Fate Principle*. The scatter points could be omitted to emphasize the line. The line basically smooths the data to reveal the trend of the relationships between the two variables. The line is sometimes simply called a *smooth*. At one extreme, the smooth is just an *OLS* straight line which I review in Chap. 6; at the other extreme, it connects as many of the data points as possible. A smooth between both extremes is best. Since at one extreme the smooth is an *OLS* line, called a *least squares regression line*, the methodology for determining the smooth should be similar to that used in *OLS*. An *OLS* line is determined by minimizing a loss function specified as the sum of the squared differences between values for an endogenous variable and a prediction of that variable based on an exogenous variable. All the data are used to develop this smooth and they are all given equal weight in determining the smooth.

This approach is modified in two important ways for more flexibility for determining a smooth line that is not necessarily a straight line but yet allows for a straight line as a special case. In short, the modification of the method produces

```
##
## Contour plot of the POI and Damage data.
##
ax = sns.jointplot( y = 'poi', x = 'damage',
                    kind = "hex", data = df_agg )
ax.ax_joint.set_xlabel( 'Damage' )
ax.ax_joint.set_ylabel( 'POI' );
```

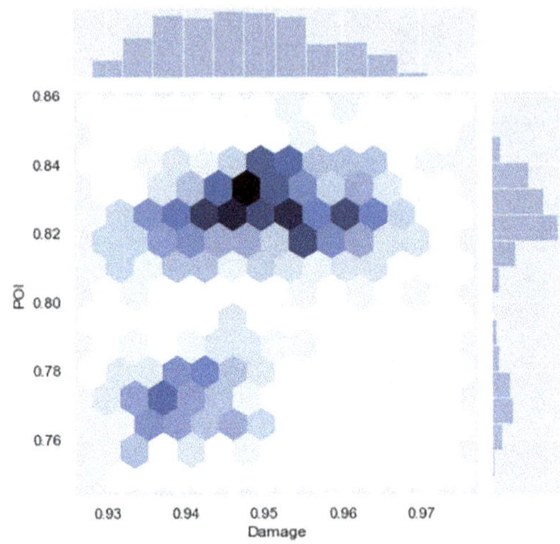

Fig. 4.12 A hex bin plot of the same data used in Fig. 4.10

a more general case for determining a line. The modification uses a small window of data centered around a value for the exogenous variable, say x_0, as opposed to using all the data available. The size of the window is sometimes called the *span* of the window and is symmetric such that half the window is on each side of x_0. The span that is half on each side is called the *half-width*. In addition, the values for the exogenous variable inside the window are weighted so that values close to x_0 have higher weight and those further away have lower weight. A loss function to be minimized is a weighted sum of squared loss based on the data inside the window. The estimation method yields a predicted value for the endogenous variable at the point x_0 that is plotted with x_0 as a point on a line. The window is then moved one data point over for a new set of data and a new weighted loss function is estimated yielding a new predicted value. A second point for a line is determined. This is continued until all the data have been used. Fundamentally, a least squares regression line is fitted for subsets of the data (i.e., the data in a window). The subsets are said to be local around a point, x_0, so the regression is a *local regression*. A series of local regressions is determined each resulting in a point for a line, the line being optimally determined. This line is called a *Locally Weighted Scatterplot Smooth* or *LOWESS*.[7]

[7] An alternative name is *LOESS* for *local regression*.

The smoothness of the *LOWESS* line is determined by the size of the window, and therefore the size of the locality, around the value x_0. The larger the locality, the smoother the line; the smaller the locality the more curvature to the line since it reflects the nature of the data in the smaller localities. There is the potential for a large number of localized regressions for one data set which means the procedure is computationally intensive. This is the cost of using this approach with *Large-N* data. The computational cost can be reduced by increasing the fraction of the data used in the locality but this just makes the line straighter, or by increasing the size of the step from one data point to the next. Instead of moving the locality window over one point, it could be moved, say, 10 or 100 points. This would greatly reduce computational time but may impact the insight from the line. Plotting values are based on linear interpolation between the data points.

In addition to the locality, the size of the weights impacts the smoothness of the line. The weights are sometimes based on a *tricube weight function* defined as

$$w(x) = \begin{cases} (1 - |d|^3)^3 & \text{for } |d| < 1 \\ 0 & \text{for } |d| \geq 1 \end{cases} \tag{4.4.7}$$

where d is the distance of a point in the window, x_i, from the point x_0. This distance is scaled to the range 0 to 1 by dividing by the half-width. That is $d = {}^{(x_i - x_0)}/_{h_0}$. The interval $[x_i - h_0, x_i + h_0]$ equals the span. See Fox (2019).[8] See Cleveland (1979) and Cleveland (1981) for the development of the *LOWESS* approach.[9]

I reproduce Fig. 4.10 in Fig. 4.13 but with a LOWESS smooth added to reveal the general trend and pattern of the data excluding all the noise exhibited by the scatter of the points. Figure 4.14 shows the LOWESS smooths for the same data but without the scatter points and for different values for the span.

Another way to display multiple variables is by drawing a series of lines, one for each row of a data table. The points on the line are the values for the variables. This is called a *parallel chart* and reflects the Gestalt *Common Fate*, *Similarity* and *Connectedness* Principles.[10] See Wegman (1990) for a discussion of this type of graph. In Fig. 4.15, the four marketing regions are compared for the *POI* components. This type of graph is commonly used with high dimensional data most often found in Big Data.

If you have spatial data for geographic areas, you can effectively use a geographic map to display your data. In this case, different levels of the continuous data would be indicated by a temperature gauge that continuously varies from the minimum

---

[8] See https://en.wikipedia.org/wiki/Local_regression on the weight function. Last accessed March 18, 2020. Also see https://www.itl.nist.gov/div898/handbook/pmd/section1/pmd144.htm last accessed March 19, 2020.

[9] Also see the blog posting by Xavier Bourret Sicotte, May 24, 2018 at https://xavierbourretsicotte.github.io/loess.html. Last accessed March 18, 2020.

[10] This is also called a *spaghetti* chart.

```
##
## Scatter plot of monthly aggregated data
## with LOWESS curve added
##
ax = sns.scatterplot( x = 'damage', y = 'poi', data = df_grp )
ax.set_title( 'POI vs Damage\nMonthly Average Data\nLOWESS Smooth Added (10% Span)',\
             fontsize = font_title )
ax.set( xlabel = 'Damage', ylabel = 'POI' )
##lowess = sm.nonparametric.lowess
low = lowess( df_grp.poi, df_grp.damage, frac = 0.10 )
plt.plot( low[ :,0 ], low[ :, 1 ], 'k-', lw = 2 );
```

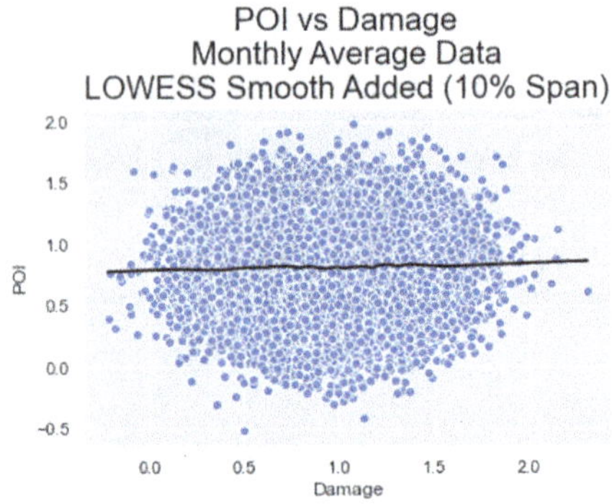

Fig. 4.13 A scatterplot of the same data used in Fig. 4.10 but with a LOWESS smooth overlayed

to the maximum values. The gauge utilizes color variations or pattern differences reflecting the *Gestalt Similarity Principle*. Discrete data can also be used. This type of geographic map is called a *choropleth map*. A map can be developed for any areal unit such as a city, county, state, region, country, or continent.[11] See Peterson (2008) and Buckingham (2010).

There are issues with the type of data used to shade the map. They are less important for qualitative, discrete data since one shade can be used to unambiguously represent a whole category. For example, if states are classified by their US Census region designation (i.e., Midwest, Northeast, South, and West), a discrete categorical designation, then one shade of color would unambiguously represent each region. However, a continuous variable such as population size, population density, percent unemployed and so forth would have a fine gradation and it may be harder to discern levels from one state to another, especially neighboring states. The issue is compounded if these variables are measured at a finer areal level such as counties. See Peterson (2008).

Choropleth maps can be used for classification, but there are also problems with this use. See Andrienko et al. (2001) for uses of these maps and classification issues.

I provide an example choropleth map in Fig. 4.16. This map is based on the *POI* DataFrame at state-level data. Previously, the data were aggregated by marketing region. Now they are aggregated by states.

[11] "Areal" refers to a physical area.

```
##
## Lowess smooth comparison
##
## Instantiate Lowess smoother
##
lowess = sm.nonparametric.lowess
##
## Create figure space
##
fig, (ax1, ax2) = plt.subplots( 1, 2, figsize = ( 10, 4 ) )
##
## Plot curves
##
ax1.set_xlim( 0.4, 1.5 )
low = lowess( df_grp.poi, df_grp.damage, frac = 0.10 )
ax1.plot( low[ :, 0 ], low[ :, 1 ], 'k-', lw = 2 )   ## 'k-'' = black solid line
ax1.set_title( 'POI vs Damage\nMonthly Average Data\nLowess Smooth (10% Span)' )
ax1.set_xlabel( 'Damage' )
ax1.set_ylabel( 'POI' )
##
ax2.set_xlim( 0.0, 1.5 )
low = lowess( df_grp.damage, df_grp.poi, frac = 1 )
ax2.plot( low[ :, 0 ], low[ :, 1 ], 'k-', lw = 2 )   ## 'k-'' = black solid line
ax2.set_title( 'POI vs Damage\nMonthly Average Data\nLowess Smooth (100% Span)' )
ax2.set_xlabel( 'Damage' )
ax2.set_ylabel( 'POI' )
##
plt.tight_layout()
```

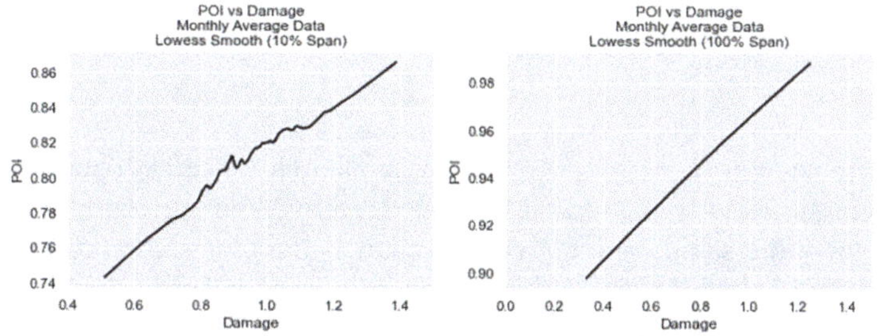

Fig. 4.14 The same data used in Fig. 4.10 is used here to compare different extreme settings for the LOWESS span setting. The scatter points were omitted for clarity

4.4.3 Visualizing Categorical Spatial Data

Discrete data have definite values or levels. They can be numeric or categorical. As numeric, they are whole numbers without decimal values. Counts are a good example. As categorical, they could be numeric values arbitrarily assigned to levels of a concept. In a survey, for example, demographic questions are included to help classify respondents. Gender is such a question. The response is numerically encoded with a number such as "1 = Male" and "2 = Female" although any number can be used. The values are just for classification purposes and are not used in calculations. This is *label encoding*. Using this example, even though gender is encoded as 1 and 2, these values should never be used in a model, such as a regression model. In the modeling context, the variable is encoded as dummy (also called *one-hot encoding* in machine learning) or effects coding. See Paczkowski (2018) for an extensive discussion of dummy and effects encoding in modeling. I will discuss this encoding along with label encoding in Chap. 5.

```
##
## Aggregate POI by Region
##
lst = [ 'region', 'complete', 'damage', 'document', 'ontime' ]
df_region = df_agg[ lst ].groupby( 'region', as_index = False ).agg( 'mean' )
##
df_region[ 'poi' ] = df_region.complete * df_region.damage * df_region.document * df_region.ontime
lst.pop( 0 )
ax = pd.plotting.parallel_coordinates( df_region, 'region', cols = lst )
ax.set_title( 'Parallel-Coordinate Plot\nPOI by Region', fontsize = font_title )
ax.set( ylabel = 'POI Proportions' )
ax.legend( loc = 2, fontsize = 'x-small', ncol = 2 );
```

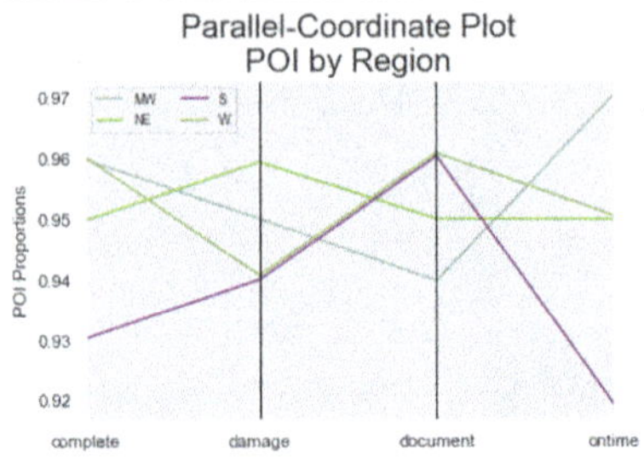

Fig. 4.15 Parallel plot of the *POI* components for each of the four marketing regions. The Southern region stands out

There are several types of graphs you can use with discrete data. Perhaps the most popular, especially in a business context, are pie and bar charts. You can use both to display the same data, but the bar chart is the preferred one. Tufte (1983) once said: "*The only thing worse than a pie chart is several of them.*"

There are several visual issues for not recommending a pie chart. The effectiveness of a pie chart relies on our ability to discern differences in the angles, those created by the slices of the pie. Sometimes, the differences are slight enough that we cannot distinguish between two slices, yet they may be critical for a business decision. For example, consider Fig. 4.17 which shows two pie charts. Both show the market share for a product in three markets simply labeled *A*, *B*, and *C*. Panel *A* does not show the size of the shares. Which market has the smaller share? Panel *B* shows the shares. Clearly Market *C* is smaller and by 5% points. The issue is the angle between the slices. See Paczkowski (2016) for a similar analysis.

Now consider the same data displayed as a bar chart in Fig. 4.18. It should be clear to you that Market B has the smaller share. We are visually better at discerning differences in length so the bar chart, which relies on lengths of bars, is significantly better at conveying the needed information about market shares then the pie chart which relies on angles. Since the data are the same, the bar chart is an *unrolled pie chart*.

Pie and bar charts are good for a single variable such as market share. If there are two variables, then a comparison is needed but the pie chart is inadequate for this. The bar chart could be modified to display several variables. One possibility is shown in Fig. 4.19 for two products sold in the three markets shown above.

A *mosaic graph* is another possible graph. This is an innovative chart that is becoming more popular because it shows the distribution of two categorical variables. The distribution is shown as areas of different sizes based on the

```
##
## Aggregate POI by state Codes
##
lst = [ 'Code', 'poi' ]
tmp = df_agg[ lst ].groupby( 'Code' ).agg( 'mean' )
tmp.reset_index( inplace = True )
##
## Set mapping arguments
##
data = tmp[ 'poi' ].astype( float )
locations = tmp[ 'Code' ]
colorbar_title = "POI"
colorscale = 'Reds'
locationmode = 'USA-states'
##
## Map states
##
map = go.Figure(
    data = go.Choropleth(
        locations = locations,          # Spatial coordinates
        z = data,                       # Data to be color-coded
        locationmode = locationmode,    # set of locations match entries in `locations`
        colorscale = colorscale,
        colorbar_title = colorbar_title,
    )
)
map.update_layout(
    geo_scope = 'usa', # limit map scope to USA
)
```

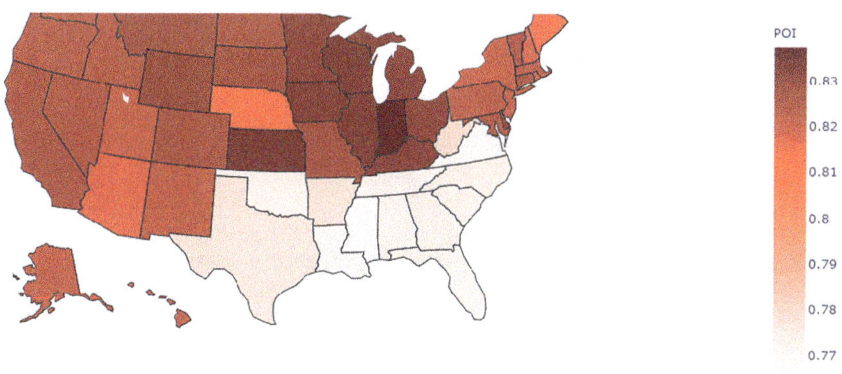

Fig. 4.16 Choropleth map of mean *POI* data by U.S. states

proportion of the data accounted for by the crossing of the two variables. In a sense, a mosaic graph is a display of a cross-tab. Consider the cross-tab in Fig. 4.20 of the *POI* warnings and store type. Now consider the mosaic graph in Fig. 4.21 of the same data, effectively the same table. The mosaic graph is more striking and clearly shows that High/Grocery dominates. Figure 4.20 is actually a small table. It is only 2×3 so it should be easy to spot a pattern. Clearly, a much larger table would be a challenge. The mosaic graph clearly helps you identify patterns. A *correspondence analysis* is better, but is outside this chapter's scope. See Chap. 8 for my discussion of cross-tabs and correspondence analysis.

Heatmaps show the intensity or concentration of discrete values in a visual cross-tab of two discrete variables. You can see an example in Fig. 4.22.

```
##
## Pie chart comparison
##
labels = [ 'C', 'B', 'A' ]
sizes = [ 0.35, 0.30, 0.35 ]
siz = [ '(35%)', '(30%)', '(35%)' ]
lbl_sizes = [i + "\n(" + str( j*100 ) + "%)" for i, j in zip(labels, sizes) ]
##
## Set figure space
##
fig = plt.figure()
##
## Plot pies
##
ax1 = fig.add_axes( [ 0, 0, 0.5, 0.5 ], aspect = 1 )
ax1.pie( sizes, labels = labels, startangle = 90, radius = 1.75 )
ax1.set_title( '(Pie A)', pad = 40, fontsize = font_title )
##
ax2 = fig.add_axes( [ 0.65, 0, 0.5, 0.5 ], aspect = 1 )
ax2.pie( sizes, labels = lbl_sizes, startangle = 90, radius = 1.75 )
ax2.set_title( '(Pie B)', pad = 40, fontsize = font_title );
```

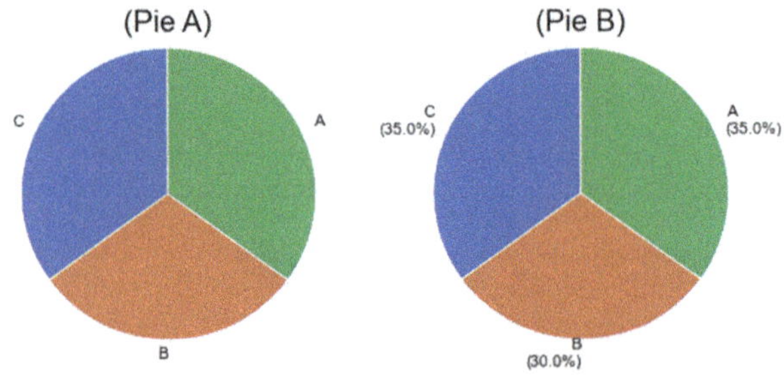

Fig. 4.17 Our inability to easily decipher angles makes it challenging to determine which slice is largest for *Pie A*

4.4.4 Visualizing Continuous and Categorical Spatial Data

Data are not always strictly categorical or continuous. You could have a mix of both. In this case, graph types can be combined to highlight one variable conditioned on another. For example, you could have a continuous variable and you want to compare the distribution of that variable based on the levels of a categorical variable. Figure 4.23 shows the distribution of *POI* by location (Rural and Urban). With the conditioning on location, Fig. 4.23 clearly shows that *POI* is more varied in the Rural areas than the Urban ones. There are also more outliers in the Urban areas and left skewness in both areas.

A second categorical variable can be added to form a *facet*, *trellis*, *lattice*, or *panel* (all interchangeable terms) plot. I show an example in Fig. 4.24 where *POI* is shown relative to locations and store type. It is striking how much more variance there is for *POI* for grocery stores in rural areas.

```
##
data = { 'mkt':[ 'A', 'B', 'C' ], 'share':[ 0.35, 0.30, 0.35 ] }
df_bars = pd.DataFrame( data, index = [ 1,2,3 ] )
ax = sns.barplot( x = 'mkt', y = 'share', data = df_bars )
ax.set( ylabel = 'Share', xlabel = 'Market' )
##
## Add labels over bars
##
for p in ax.patches:
    percentage = '{:,.0%}'.format( p.get_height() )
    width, height = p.get_width(), p.get_height()
    x = p.get_x() + width/2
    y = height + 0.01
    ax.annotate( percentage, ( x, y ) )
```

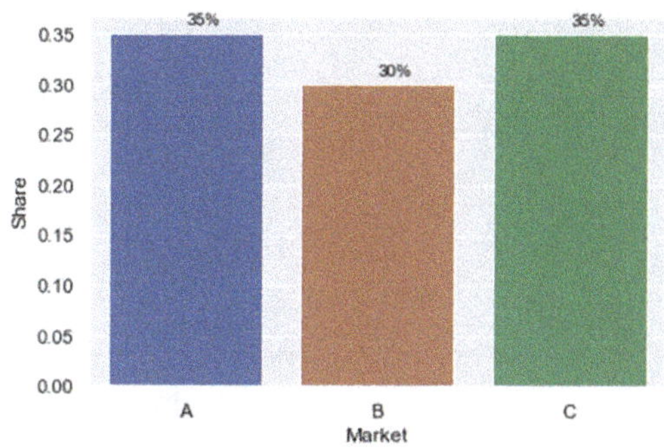

Fig. 4.18 Bar Chart view of *Pie A* of Fig. 4.17. This is easier to read and understand. Market *B* clearly stands out

```
##
## Create crosstab of warnings by store type
## Use pivot_table function
##
df_xtab = pd.pivot_table( data = df_agg, index = 'warning',
                          values = 'fids', columns = [ 'store' ],
                          aggfunc = 'count' )
df_xtab.index.rename( 'Warning', inplace = True )
ax = df_xtab.plot( kind = 'bar', stacked = True, rot=0 )
plt.legend( title = 'Store Type' )
ax.set_title( 'Stacked Bar Chart of Warning\nby Store Type', fontsize = font_title )
ax.set( ylabel = 'Count' );
```

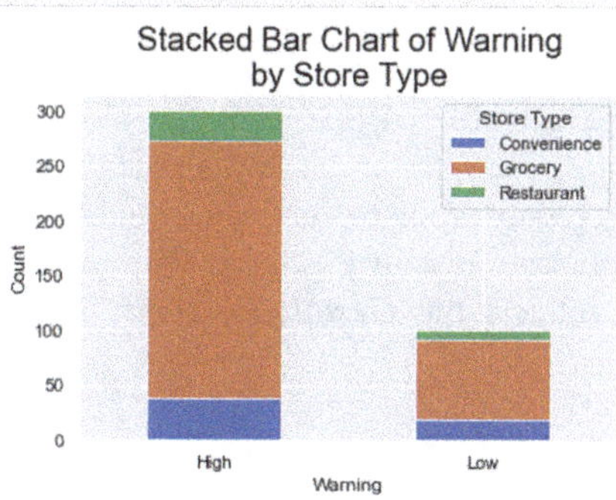

Fig. 4.19 Stacked bar chart

```
##
## Create crosstab of warnings by store type
##
cols = pd.MultiIndex.from_product( [ [ 'Store Type' ], df_agg.store.unique() ] )
xtab = pd.crosstab( df_agg.warning, df_agg.store )
df_xtab = pd.DataFrame( xtab )
df_xtab.columns = cols
df_xtab.index.rename( 'Warning', inplace = True )
df_xtab.style.set_caption( 'Crosstab of Warning by Store Type' ).\
    set_table_styles( tbl_styles )
```

Crosstab of Warning by Store Type

	Store Type		
	Convenience	Grocery	Restaurant
Warning			
High	38	235	27
Low	19	72	9

Fig. 4.20 Cross-tab of *POI* warning and store type

```
##
## Mosaic plot of crosstab
##
props = {}
props[ ( 'Low','Restaurant' ) ] = { 'color':'red' }
props[ ( 'High', 'Restaurant' ) ] = { 'color':'lightblue' }
props[ ( 'Low', 'Grocery' ) ] = { 'color':'lightgreen' }
props[ ( 'High', 'Grocery' ) ] = { 'color':'pink' }
props[ ( 'Low', 'Convenience' ) ] = { 'color':'orange' }
props[ ( 'High', 'Convenience' ) ] = { 'color':'green' }
##
mosaic( df_agg, [ 'warning', 'store' ], properties = props );
```

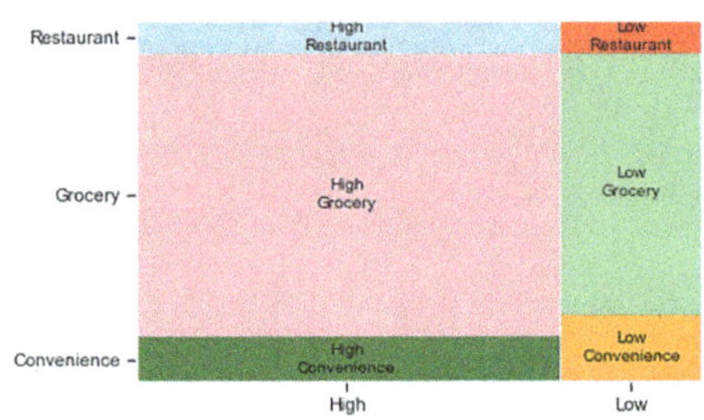

Fig. 4.21 *POI* mosaic graph

A complicated scatter plot can be constructed that sizes the plotted points by another numeric variable and perhaps colors the points by a categorical variable. Because the point sizes can change, they tend to resemble bubbles, hence the graph is called a *bubble graph*. I illustrate one in Fig. 4.25. Patterns, in the form of groupings, are very clear. This reflects the *Gestalt Proximity* and *Similarity Principles*.

```
##
## Heatmap of Correlations
##
lst = [ 'complete', 'damage', 'document', 'ontime' ]
corr = df_agg[ lst ].corr()
ax = sns.heatmap( corr )
ax.set_title( 'Heatmap of Correlations\nPOI Components', fontsize = font_title );
```

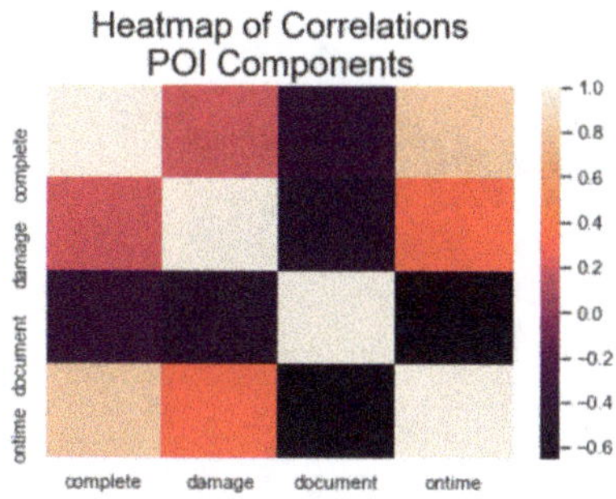

Fig. 4.22 Example of a heatmap

```
##
## Boxplot of a Continuous Variable Conditioned on
## the Levels of a Categorical Variable.
##
ax = sns.catplot( y = "poi",
                  col = "location",
                  data = df_agg, kind = "box",
                  height = 4, aspect = 0.7 )
plt.subplots_adjust( top = 0.85 )
ax.fig.suptitle( 'POI by Location', fontsize = font_title );
```

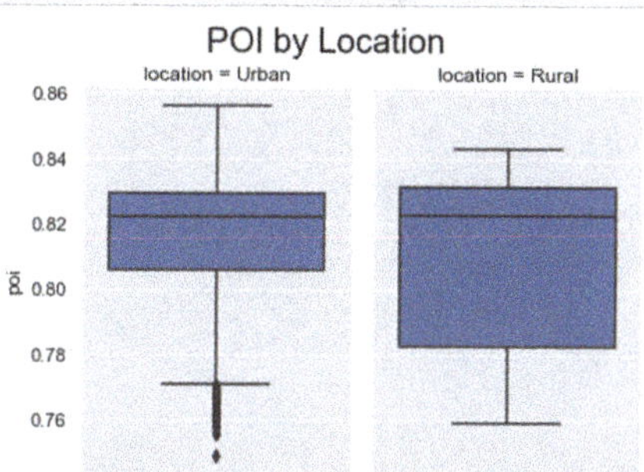

Fig. 4.23 Boxplot of a continuous variable conditioned on the levels of a categorical variable. The conditioning variable is location: Rural and Urban

4.5 Visualizing Temporal (Time Series) Data

I will consider visual displays for temporal or time series data in this section. Some of the displays I previously discussed, such as boxplots, can be used with temporal data. Otherwise, temporal data have their own problems that require variations on some displays.

```
##
## Boxplot of a continuous variable conditioned on
## the levels of a categorical variable.
##
ax = sns.catplot( y = 'poi', row = "location", col = 'store',
                  data = df_agg, kind = 'box', height = 4,
                  aspect = 0.7, margin_titles = True )
plt.subplots_adjust( top = 0.90 )
ax.fig.suptitle( 'POI by Location and Store Type', fontsize = font_title );
```

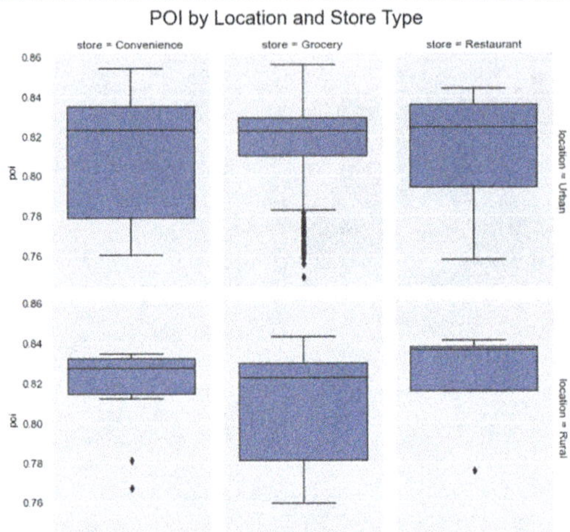

Fig. 4.24 Faceted panel plot of a continuous variable conditioned on two categorical variables

```
##
## Bubble Plot of Ontime Measure and POI
##
ax = sns.scatterplot( x = "ontime", y = "poi", hue = "region",
                      size = "region", data = df_agg )
ax.set_title( 'Bubble Plot of Ontime Measure and POI\nby Region',
              fontsize = font_title )
ax.set( ylabel = 'POI', xlabel = 'Ontime' );
```

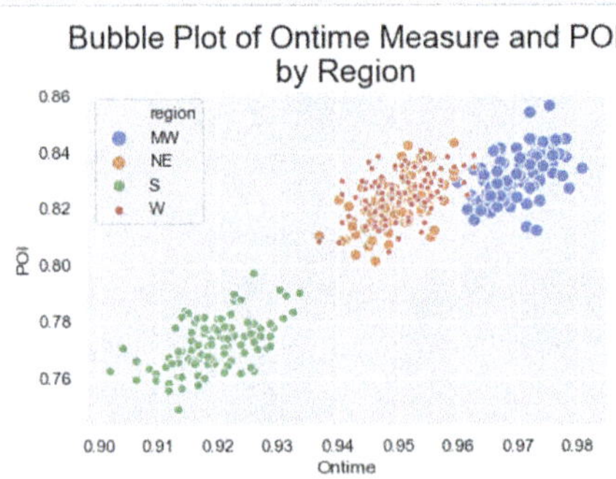

Fig. 4.25 Bubble plot of *POI* by *Ontime* delivery sized by marketing region

4.5.1 Properties of Temporal (Time Series) Data

I discussed the visualization of spatial data in the previous section. Data could also be temporal, which is also known as *time series* data. Time series data are a special breed of data with a host of problems that make visualization more challenging. It is not, however, only the visualization that is complicated; it is the full analysis of time series: data visualization, data handling, and modeling becomes overwhelming. I devote Chap. 7 to times series, albeit at a high level, because of these complications.

A times series, Y_t, can be written (i.e., decomposed) into constituent parts as

$$Y_t = T_t + C_t + S_t + \epsilon_t \tag{4.5.1}$$

$t = 1, 2, \ldots, T$, reflecting trend (T_t), cyclicality (C_t), seasonality (S_t), and random noise (ϵ_t). This is an additive model. A multiplicative version is sometimes used: $Y_t = T_t \times C_t \times S_t \times \epsilon_t$. This could be rewritten, however, as $\ln Y_t = \ln T_t + \ln C_t + \ln S_t + \ln \epsilon_t$ which is an additive model. Subtracting or removing the trend component is *detrending*; subtracting or removing the cyclical component is *decycling*; and subtracting or removing the seasonal component is *deseasonalizing*. The random noise, unfortunately, cannot be removed; it is always present.

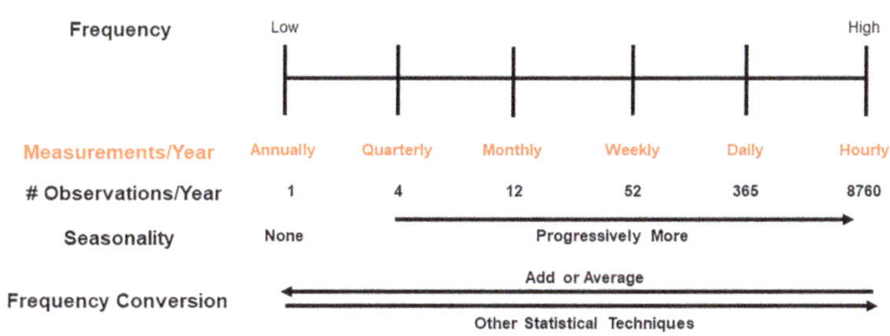

Fig. 4.26 Time series classifications

Time series are measured in terms of frequency, usually with a year as the base (decades are possible but less common). Annual data at one extreme have the lowest frequency at once per year (by definition) and hourly data at another extreme have a high frequency at 8760 (= 24 × 365) times per year. Several time series measured at less than an annual frequency are not always at the same frequency or desired frequency, so a conversion from one level to another may have to be done to put all the series are on the same basis. Averaging or summing are common operations to convert frequencies. For example, you know that sales are a function of real GDP. Real GDP is reported quarterly (at an annual rate) and annually. You want to model sales data which you have on a monthly basis. You can sum sales every three months to a quarterly basis to have sales on the same frequency as quarterly real GDP. See Fig. 4.26 for a guide. I fully discuss aggregating time series in Chap. 7.

Data with frequencies less than a year may exhibit seasonal patterns, and they usually do. *Seasonality* is the repetition of a pattern at more or less the same point each year. Examples are school vacations, Holiday shopping seasons, and, of course, seasons themselves. Annual data have the lowest frequency of occurrence with no seasonal patterns. See Granger (1979) for interesting discussions about seasonality and some reasons for seasonal patterns.

While seasonality is a recurring pattern at intervals less than a year, *cyclicality* is a recurring pattern at intervals more than a year. The business cycle is the prime example. This cycle affects all consumers, businesses, and financial markets in different ways but nonetheless it affects them all as I discussed in Chap. 2.

All time series exhibit a trend pattern, a tendency for the series to rise (or fall or be constant) on average for long periods of time. This reflects the *Gestalt Common Fate Principle*. When the series rises or falls over long periods, the series is said to be "nonstationary". If the series is constant at some level, then it is said to be "stationary." We usually prefer stationary series so a nonstationary series must be transformed to produce a stationary one. A typical transformation is *first differencing*. I will discuss these issues shortly.

Finally, all time series, like any other real-world data, exhibit random variations due to unknown and unknowable causes. These random variations are a noise element we have to recognize and just live with. The noise is sometimes called *white noise*. Statistically, white noise is specified as a normal random variable with zero mean, constant variance, and zero covariance between any two periods: $\epsilon_t \sim \mathcal{N}(0, \sigma^2), cov(\epsilon_t, \epsilon_{t'}) = 0, \forall t$ and $t \neq t'$.

4.5.2 Visualizing Time Series Data

A line chart is probably the simplest time series graph familiar to most analysts. It is just the series plotted against time. You can see an example in Fig. 4.27. You can also plot several series on one set of axes to compare and contrast them. This reflects the *Gestalt Common Fate Principle*.

It may be possible to disaggregate a time series into constituent periods to reveal underlying patterns hidden by the more aggregate presentation. For example, U.S. annual real GDP growth rates from 1960 to 2016 can be divided into six decades to show a cyclical pattern. A boxplot for each decade could then be created and all six boxplots can be plotted next to each other so that the decades play the role of a categorical variable. I show such a graph in Fig. 4.28. This reflects the *Gestalt Similarity Principle*.

```
##
## Subset and Set Index to Dates
##
lst = [ 'dates', 'complete', 'damage', 'document', 'ontime' ]
data = df[ lst ].copy()
data.set_index( 'dates', inplace = True )
##
## Resample the data to shift to lower frequency
##
df_months = data.resample( 'MS', convention = 'end' ).mean()
df_months[ 'poi' ] = df_months.complete * df_months.damage \
* df_months.document * df_months.ontime
df_months.head().style.set_caption( 'POI Data' ).set_table_styles( tbl_styles )
```

POI Data

dates	complete	damage	document	ontime	poi
2016-01-01 00:00:00	0.951129	0.945806	0.954516	0.949032	0.814903
2016-02-01 00:00:00	0.949224	0.946810	0.953276	0.947414	0.811690
2016-03-01 00:00:00	0.951210	0.951129	0.953226	0.947823	0.817407
2016-04-01 00:00:00	0.952083	0.947417	0.955917	0.947083	0.816628
2016-05-01 00:00:00	0.948710	0.950081	0.950403	0.947661	0.811811

```
##
## Monthly Time Series Plot
##
ax = df_months.plot( y = 'poi', legend = False )
ax.set_title( 'POI Time Series\nMonthly Data', fontsize = font_title )
ax.set( xlabel = 'Months', ylabel = 'Monthly Average POI' );
```

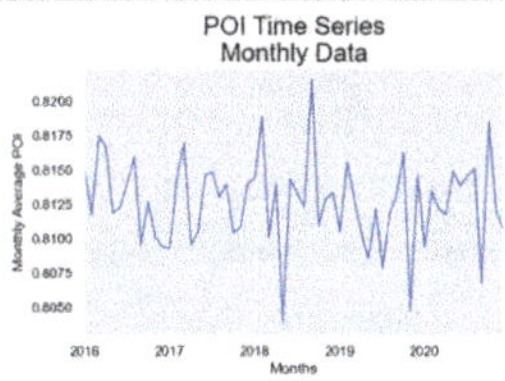

Fig. 4.27 A single, continuous times series of annual data

```
##
## Boxplot of Monthly Data by Year
##
data = df_months.copy()
data.reset_index( inplace = True )
ax = sns.boxplot( y = 'poi', x = data.dates.dt.year, data = data )
ax.set_title( 'Boxplots for Annual Periods', fontsize = font_title )
ax.set( xlabel = '', ylabel = 'POI' );
```

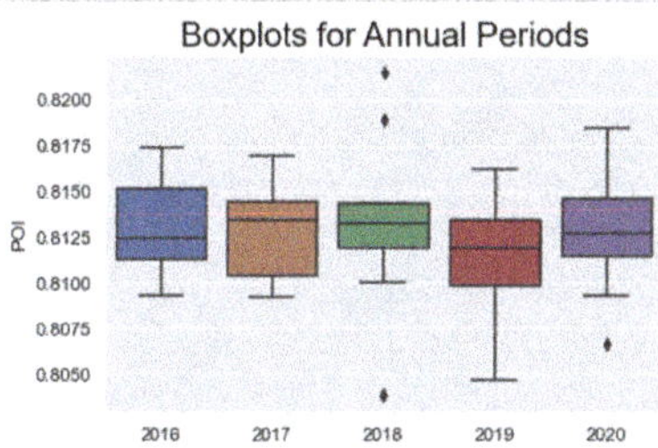

Fig. 4.28 A single, continuous times series of annual data could be split into subperiods with a boxplot created for each subperiod

4.5.3 *Times Series Complications*

Time series data have unique complications which account for why there is so much active academic research in this area. The visualization of time series reflects this work. Some unique problems are:

1. Changing slope through time.
   - This is called *nonstationarity in the mean.*

Solution: Take the difference in the series (usually a first difference will suffice).

2. Changing variance—usually increasing.
   Solution: Plot natural log of series.

   - Straightens curve.
     - Added benefit: slope is average growth rate.
   - Stabilizes variance.

3. Autocorrelation (series correlated with itself).
   Solution: Check correlation with lagged series.

Many time series exhibit a non-zero slope which implies a changing mean. In other words, the series is either rising or falling through time. The changing mean is evident if you take a small section of the data (i.e., a small "window"), calculate the mean within that window, and then slide the window to the left or right one period and calculate the mean for that new section of the data. The two means will differ. For one or two small shifts in the window, the means may not differ much, but for many shifts there will be a noticeable and significant difference. This property of a changing mean is called *nonstationarity*, which is not a desirable property of time series because it greatly complicates any analysis. We want a *stationary* time series: one in which the mean is constant no matter where the window is placed. This is an oversimplification of a complex problem, but the point should be clear.

```
##
## Times Series Plot of Subsetted Data
##
data = df_months.query( "index >= '2019-01-01'" )
ax = data.plot( y = 'ontime', legend = False );
ax.set_title( 'Ontime Time Series\nMonthly Average\n2019 - 2020',
            fontsize = font_title )
ax.set( xlabel = '', ylabel = 'Ontime' );
```

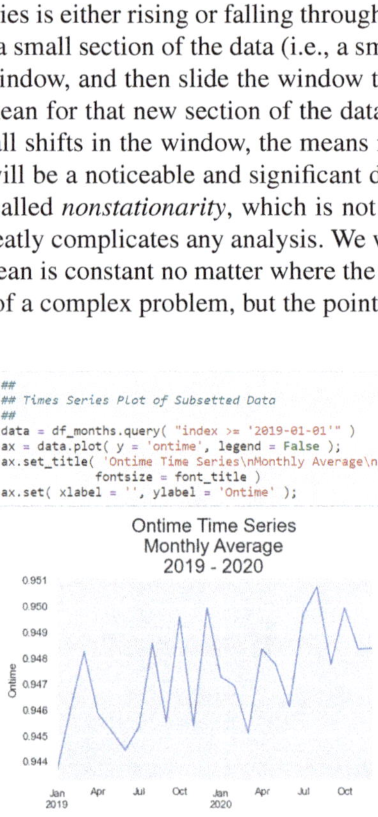

Fig. 4.29 A plot of the *Ontime POI* measure for the 2019–2020 subperiod. This is clearly nonstationary

More formally, stationarity requires the times series generating process to be in a particular state of "statistical equilibrium". A time series, which is a *stochastic process*, is said to be *strictly stationary* if its properties are unaffected by a change

```
##
## Time Series Plot of First Differences
##
data_diff = data.ontime.diff()
ax = data_diff.plot( y = 'ontime', legend = False );
ax.set_title( 'Ontime Time Series\nFirst Differenced Monthly Average\n2019 - 2020',
              fontsize = font_title )
ax.set( xlabel = '', ylabel = 'Ontime First Difference' );
```

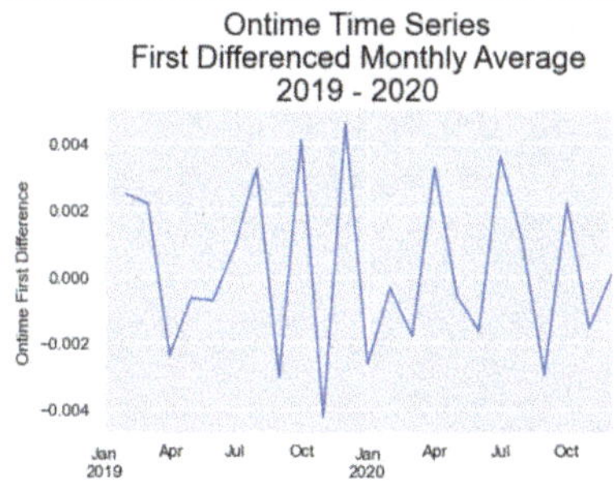

Fig. 4.30 A first differenced plot of the monthly data in Fig. 4.29. This clearly has a constant mean so it is mean stationary as opposed to the series in Fig. 4.29

of time origin. In other words, the joint probability distribution at *any* set of times $t_1, t_2, \ldots, t_m$ must be the same as the joint probability distribution at times $t_1 + k, t_2 + k, \ldots, t_m + k$, where k is an arbitrary shift along the time axis. Basically, the k represents the size of the "window" that is shifted. There are "weak" and "strong" stationarity conditions. These concepts are beyond what we are concerned with here.

You can identify a nonstationary time series by plotting the data against time as I do in Fig. 4.29. This series is clearly nonstationary since the mean constantly increases. Figure 4.30 shows the same data after first differencing. The first difference is a series' value in one period less its value in the previous period: $X_t - X_{t-1}$.

The natural log is a very common transformation used in time series analysis. The natural log is the log to the base e and is written as $\ln(X)$. Always use the natural log in empirical work. Without going into any mathematical details, this log transformation does two things to a times series: it straightens a curve and stabilizes the variance. Suppose $Y = A \times e^{\beta \times X}$. This produces an exponential curve. Taking the natural log of Y yields $\ln(Y) = \ln(A) + \beta \times X$ which is a straight line. Compare the two curves in Fig. 4.31.

You may still find that the log transformation did not completely fix some nonstationarity. This can be remedied by taking the first difference, but always do a log transform first and then the first difference, not the other way around. The reason is that the difference in the logs is the relative change in the data whereas the log of the difference is just the log of the difference. That is, $\ln(X_t) - \ln(X_{t-1}) = \ln(X_t/X_{t-1})$ while $\ln(X_t - X_{t-1})$ is just the log of the difference. The difference in logs is the growth rate. See the Appendix to this chapter for the reason.

A final common graph is the series plotted against itself but lagged one period. That is, plot X_t against X_{t-1}. This graph would show the dynamics from one period

```
##
## Generate fake data
##
x = [ 1, 2, 3, 4, 5, 6, 7 ]
y =  [ 0.5 * np.exp( 0.8 * i ) for i in x ]
log_y = np.log( y )
data = { 'x':x, 'y':y, 'log_y':log_y }
##
## Create DataFrame
##
df = pd.DataFrame( data )
##
## Create two axes
##
fig, ( ax1, ax2 ) = plt.subplots( 1, 2, figsize = ( 10, 5 )  )
## Plot series
##
ax1 = df.y.plot( ax = ax1 )
ax1.set_title( 'Exponential Growth\n$Y = 0.5 \\times e^{0.8 \\times X}$', fontsize = font_title )
ax1.set( xlabel = 'x', ylabel = 'Unlogged y' )
##
ax2 = df.log_y.plot( ax = ax2 )
ax2.set_title( 'Log Growth\n$ln(Y) = 0.5 + 0.8 \\times X$', fontsize = font_title )
ax2.set( xlabel = 'x', ylabel = 'Logged y' );
```

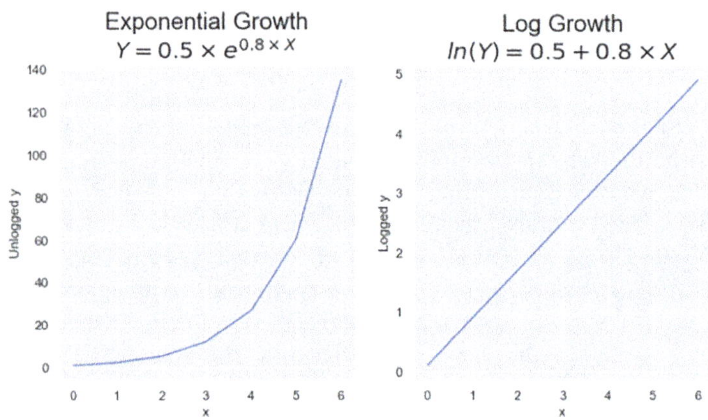

Fig. 4.31 This shows simulated data for an unlogged and logged versions of some data

to the next. Another way to describe this is that the correlation from one period to the next is evident. I illustrate this in Fig. 4.32.

Seasonality is a major problem for time series analysis. See Granger (1979) for a discussion. Boxplots can be quite effective in seasonal visualization. For example, different boxes can be created for each month of a year where the data points behind the box construction extends over several years. I illustrate this for the monthly *POI* damage data in Fig. 4.33.

```
##
## Plot document POI vs lagged document POI
##
## Subset and add one-period lag
##
df_lagged = df_months.copy()
df_lagged = df_lagged[ [ 'document' ] ]
df_lagged[ 'document_lag' ] = df_lagged.document.shift()
##
## Plot document poi vs lagged document poi
##
ax = sns.regplot( y = 'document', x = 'document_lag', ci = False, data = df_lagged )
plt.ylim( 0.945, 0.960 )
plt.xlim( 0.945, 0.960 )
ax.set( title = 'Document Monthly Time Series\nvs\nLagged Document Monthly Time Series' )
ax.set( xlabel = 'Lagged Document Monthly Time Series', ylabel = 'Document Monthly Time Series' );
```

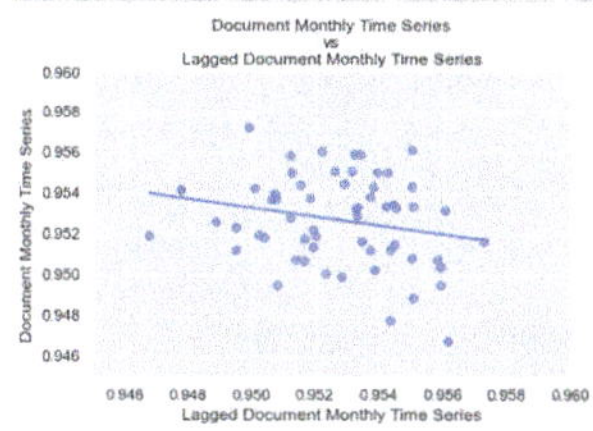

Fig. 4.32 The monthly data for the document component of the *POI* measure plotted against itself lagged one period

```
##
## Boxplot of Monthly Damage Data
##
lst = [ 'dates', 'complete', 'damage', 'document', 'ontime' ]
data = df[ lst ].copy()
data.set_index( 'dates', inplace = True )
##
## Resample the data to shift to lower frequency
##
df_months = data.resample( 'M', convention = 'end' ).mean()
df_months[ 'month' ] = df_months.index.month
##
## Set month abbreviations
##
dt = dict( enumerate( calendar.month_abbr ) )
df_months[ 'month' ] = df_months[ 'month' ].map( dt )
df_months[ 'poi' ] = df_months.complete * df_months.damage \
* df_months.document * df_months.ontime
##
## Create boxplot
##
ax = sns.boxplot( x = 'month', y = 'damage', data = df_months )
ax.set_title( 'Boxplots of Monthly Damage POI Data', fontsize = font_title )
ax.set( xlabel = '', ylabel = 'Damage POI' );
```

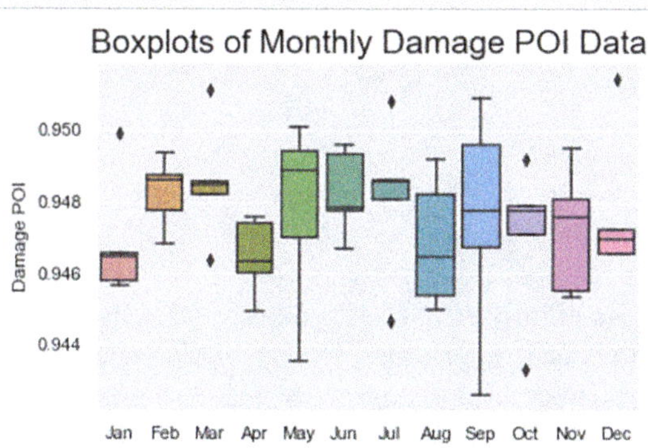

Fig. 4.33 The average monthly damage *POI* data are plotted by months to show seasonality

4.6 Faceted Plots

Sometimes, especially with higher-dimensional data, a scatter plot of each pair of variables helps to reveal relationships. Rather than create each pair separately, they could be created in a square matrix arrangement. The overall graph is called a *scatterplot matrix*. I show an example of the four *POI* measures in Fig. 4.34.

```
##
## Scatterplot Matrix
##
lst = [ 'complete', 'damage', 'document', 'ontime' ]
x = df_agg[ lst ].copy()
sns.pairplot( x )
plt.suptitle( 'Scatterplot Matrix of POI Data', fontsize = font_title, y = 1.03 );
```

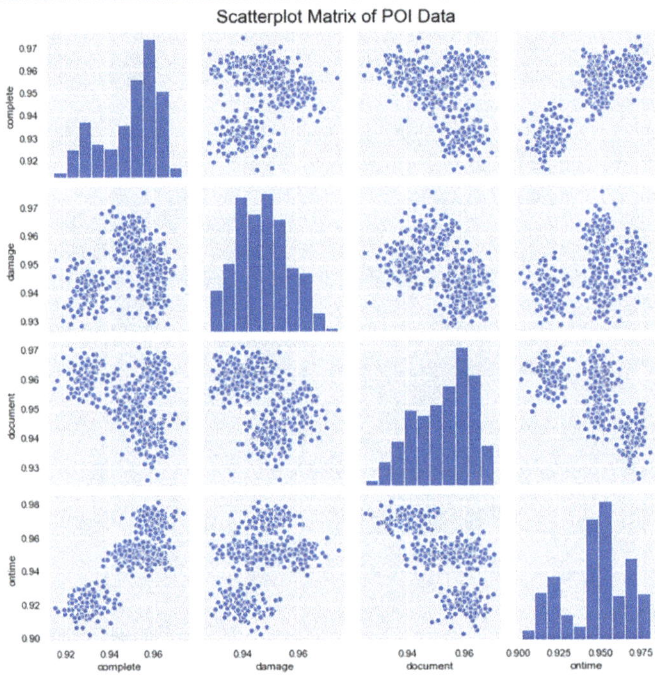

Fig. 4.34 Scatter plot matrix for four continuous variables. Notice that there are $16(= 4 \times 4)$ panels, each presenting a plot of a pair of variables

With four variables, there are 16 cells in the matrix, each panel containing a plot of one pair of variables. With four variables, there are $n \times (n-1)/2 = 4 \times 3/2 = 6$ pairs. The four cells on the main diagonal should contain a plot of a variable against itself which, of course, is uninformative. These diagonal cells are therefore sometimes filled with a variable label; other times, there is the variable label and a single variable distribution graph such as a histogram of that variable. The label in a single diagonal cell identifies the *X* and *Y* axes of the other cells in that column and row, respectively. The *Y* axis labels are all the same in a row while the *X* axis labels are all the same in a column. For example, all the cells in the first row (at the bottom) have the same Y label: "ontime" as indicated in the first cell of that row while all the cells in the first column have that label for the X axis. Notice in Fig. 4.34 that each of the six cells above the main diagonal is a mirror image of the corresponding cell below the main diagonal. The matrix is symmetric around this diagonal. Also notice that the six cells above the main diagonal form a triangle; so do the six cells below. The upper six cells are called the *upper triangle* while the lower six are the *lower triangle*. Both triangles convey the same information so one is redundant. This implies that only one triangle is needed and the other could be dropped. Which is dropped is arbitrary. Figure 4.35 repeats Fig. 4.34 but shows only the lower triangle. This is easier to read.

```
##
## Lower Triangle of Scatterplot Matrix
##
lst = [ 'complete', 'damage', 'document', 'ontime' ]
x = df_agg[ lst ].copy()
sns.pairplot( x, corner = True )
plt.suptitle( 'Scatterplot Matrix of POI Data\nLower Triangle', fontsize = font_title, y = 1.05 );
```

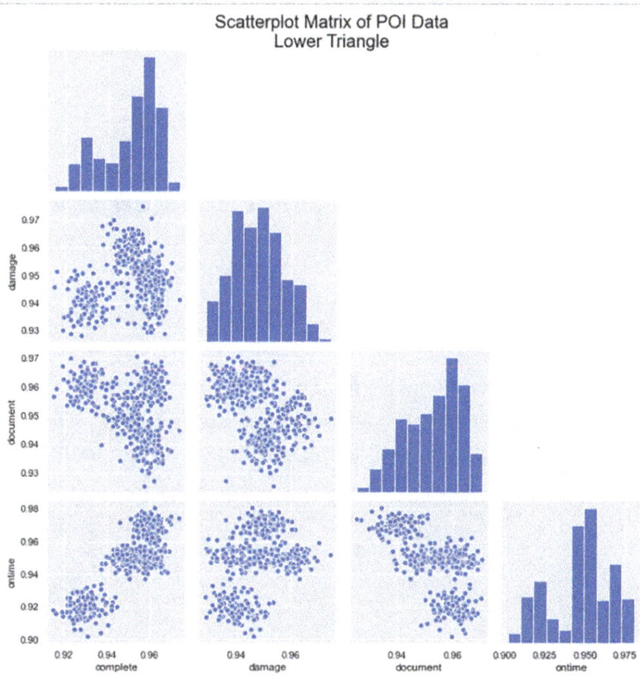

Fig. 4.35 Scatter plot matrix lower triangle of Fig. 4.34

4.7 Appendix

In this Appendix, I will describe how growth rates can be derived using a Taylor Series Expansion.

4.7.1 *Taylor Series Expansion for Growth Rates*

The Taylor Series Expansion of a function, $f(x)$, is

$$f(x) = \sum_{i=0}^{\infty} f^i(a) \times \frac{(x-a)^i}{i!}$$

where a is a point near x, $f^i(a)$ is the ith derivative of $f(x)$ evaluated at the pint a, and $i!$ is the factorial function with $0! \equiv 1$. Let $x = X_t$ and $a = X_{t-1}$ and use the natural log function. Then

$$\ln(X_t) = \ln(X_{t-1}) + \frac{X_t - X_{t-1}}{X_{t-1}} + R$$

where R is all the remaining terms as an infinite sum. You can assume that $R = 0$. Therefore,

$$\ln(X_t) \approx \ln(X_{t-1}) + \frac{X_t - X_{t-1}}{X_{t-1}}$$

with the approximation being "close enough." The second term on the right-hand side is the growth from period $t-1$ to period t. Let his be g. Then clearly $g = \ln(X_t) - \ln(X_{t-1})$ so the growth rate is the difference in the natural logs of X for two successive periods.

Chapter 5
Advanced Data Handling: Preprocessing Methods

A problem faced by those new to Data Science is getting past the only data paradigm they know: textbook data which are always clean and orderly with no or very few issues as I noted at the beginning of Chap. 3. Unfortunately, real world data do not agree with this paradigm. They are, to say the least, messy. They have missing values, are disorganized relative to what you need to do, and are just, well, a mess. Before any meaningful work is done, you have to *process* or, better yet, *preprocess* your messy data. I will discuss four preprocessing tasks:

1. transformation;
2. encoding;
3. dimensionality reduction; and
4. missing data identification and handling.

Note that I list four tasks. As a heads-up, not all are necessary. Which one you do depends on your data. Preprocessing should not be taken literally because even midway through your data analysis you may decide that a new transformation is needed or a new encoding is warranted. So, you may have to restart. Business Data Analytics is an iterative process, not a linear one.

Transformation and encoding are the same in the sense that both turn a variable into something else. Transformation does this by a formula, encoding by a scheme or rule. Transformation produces a new scale; encoding is a code that requires a key to understand and unravel if necessary. Transformations are mostly on interval and ratio data and produce ratio data while encoding are for categorical data and produce nominal or ordinal codes for the categories. This last point is not always strictly true since you could encode interval and ratio data into categories, a process called *binning*. For example, you could bin age into age categories. This means that encoding also changes continuous data to categorical data, but this implies that encoding a continuous variable hides the original values. The same applies to encoding categorical data. One form of encoding of a categorical variable I will discuss is dummy encoding which changes the categories of the variable into a set

W. R. Paczkowski, *Business Analytics*,
https://doi.org/10.1007/978-3-030-87023-2_5

of new variables but you cannot tell the values of the original variable without a key for the encoding.

Dimensionality reduction collapses a number of variables into either one or a few new ones that capture or reflect some aspect of the original ones. This allows you to replace the original ones with the new ones. The aspect usually focused on is the variance of the original variables so that the new, but smaller, set captures most of that variance. This becomes important with high-dimensional data as I will discuss below. High-dimensional data introduce potential problems when you estimate (what I will later call "train") a linear model.

Missing values are always a problem with any form of empirical research. The literature on this topic is vast, to say the least. The reason for this vastness is that missing values might upset the patterns and relationships I discussed in Chap. 4; they might prevent you from estimating (i.e., training) a model, especially a time series model. The impact on time series analysis is especially important because a time sequence must be complete. If there are "holes" in the series, then you cannot tell what the pattern is. This may not be too onerous if there are small holes, but if you have a large number of missing values than you cannot be sure at all of the pattern. And, as you will see in Chap. 7, patterns, primarily lagged patterns, are important for time series analysis.

5.1 Transformations

Data transformation is an old, well-developed topic in statistics and its subdisciplines, such as econometrics. See, as a partial listing of the literature: Tukey (1957), Tukey (1977), Morgenthaler (1997), Emerson and Stoto (1983), Mosteller and Tukey (1977), Carroll and Ruppert (1988), and Box and Cox (1964). For the application of transformations in econometrics, see Zarembka (1974). The Box and Cox (1964) reference is particularly important because it introduced the widely used *Box-Cox Transformation*. I will discuss this transformation below.

There are three reasons for transforming your data. These are to eliminate or reduce:

1. skewness in the data's distribution;
2. outliers and their impact; and
3. scale differences among predictor or explanatory variables.

Transformation is divided into two classes: linear and nonlinear. Linear transformations preserve the distributional properties of the data, such as normality. Nonlinear transformations do not preserve distributional properties.

5.1.1 Linear Transformations

The first linear transformation I will discuss is one you know from your basic statistics course: the Z-transform that *standardizes* data. This is taught in conjunction with the *standard normal distribution*. Basically, you are taught that a normally distributed random variable, $X \sim \mathcal{N}(\mu, \sigma^2)$, is standardized as a $\mathcal{N}(0, 1)$ random variable to solve simple probability problems. The standardization uses the formula:

$$Z_i = \frac{X_i - \bar{X}}{SD_X} \tag{5.1.1}$$

where Z_i is the standardized score for observation X_i, $i = 1, 2, \ldots, n$, $\bar{X}$ is the sample mean of X, and SD_X is the sample standard deviation of X. Subtracting the mean centers the data so that all data values have the mean removed. The mean of the transformed data is zero. This is called *centering*. It is easy to verify that $\bar{Z} = 0$. Dividing by SD_X scales the data so that the variance of Z_i is adjusted to 1.0. This is called *scaling*. See the Appendix for a simple demonstration of each.

The standard deviation in (5.1.1) is the square root of the variance of the random variable, X. The variance is the average of the sum of the squared deviations of the random variable's values from its arithmetic mean, $\bar{X}$. The sum of the squared deviations is $\sum (X_i - \bar{X})^2$. The divisor for the average could be n or $n - 1$. The latter is typically used in a basic statistics course because it produces an unbiased estimator of the true population variance, σ^2, although this is typically not mentioned. There is no difference between n and $n - 1$ as the sample size becomes large. See the Appendix for a demonstration of this.

The Z-transformation in (5.1.1) is a linear transformation that preserves the distribution of X (except for the mean and variance) and relationships. The linearity follows by rewriting (5.1.1) as

$$Z_i = \frac{1}{SD_X} \times X_i - \frac{\bar{X}}{SD_X} \tag{5.1.2}$$

$$= \beta_0 + \beta_1 \times X_i \tag{5.1.3}$$

where $\beta_0 = -\frac{\bar{X}}{SD_X}$ and $\beta_1 = \frac{1}{SD_X}$. Since $X \sim \mathcal{N}(\mu, \sigma^2)$, then $Z \sim \mathcal{N}$ by the *Reproductive Property of Normals*. See Paczkowski (2018) and Dudewicz and Mishra (1988).

The use of the standardized score outside of probability problems is little discussed, if at all, in a basic statistics course. Standardization goes beyond simple probability problems which are typically expressed in terms of one random variable. It puts several variables on the same basis so comparisons can be made. Standardization, therefore, is used for two types of problems: probability problems with one variable and comparison problems involving several variables. The advantage for probability problems is that they are easier to solve with standardized variables. The advantage for comparison problems is that with all variables in a data set on the same scale, one does not overwhelm another and thus distort results. If this is not done so that all variables have an equal chance at explaining something, then one variable with a scale inconsistent with the others could dominate simply because of scale. In fact, if you look at (5.1.1) you will see that the unit of measure cancels out so Z is unitless. For example, if X is measured in dollars, then the dollars cancel in (5.1.1) because both the numerator and denominator are measured in dollars.

Although Z-Score standardization centers data at mean 0 with variance 1, you may want values with a mean of 100 because they could then be interpreted as index values with 100 as the base. Many find it easier to interpret this data. You can change the mean and variance, although changing only the mean is more common. A general transformation statement of each value of the random variable X is:

$$Z_i = \frac{X_i - \bar{X}}{SD_X} \times SD_X^{New} + \bar{X}^{New} \tag{5.1.4}$$

where SD_X^{New} is the new standard deviation and $\bar{X}^{New}$ is the new mean. For example, setting $SD_X^{New} = 2$ and $\bar{X}^{New} = 100$ in (5.1.4) sets the standard deviation at 2 and centers the data at 100:

$$Z_i = \frac{X_i - \bar{X}}{SD_X} \times 2 + 100 \tag{5.1.5}$$

(5.1.4) is a linear transformation that preserves relationships. The original distribution of X is preserved so that the information in the original data is unchanged. It does not, however, reduce the impact of outliers.

The standardized score with all variables in an analysis on the same basis is used with *Principal Components Analysis*, *Support Vector Machines*, and *Cluster Analysis* as three examples. It is not needed for *Decision Trees*. I will discuss these methods in later chapters and use standardized scores where appropriate.

I show several possibilities for the standardization of a random variable, X, in Fig. 5.1 and Fig. 5.2. Note that I calculate the mean and standard deviation of X using the Numpy *mean( )* and *std( )* functions, each using X as an argument. Especially note that the *std( )* function has an extra argument, $ddof = 1$. The $ddof$ stands for *delta degrees-of-freedom* which determines the divisor for the sample variance calculation. The degrees-of-freedom are then $dof = n - ddof$ where n is the sample size. The default is $ddof = 0$ so the divisor is n. As I show in the Appendix, this leads to a biased estimator of the population variance, although the bias disappears for large n. Using $ddof = 1$ sets the divisor to $n - 1$ for an unbiased estimator in small samples. Why does Numpy use n? It is used because this results in the *maximum likelihood* estimator of the population variance for a normally distributed random variable. It turns out that the standard deviation as the square root of the estimated variance, even with $ddof = 1$, is a biased estimator of the population standard deviation. Just the variance is unbiased; not the standard deviation.

I calculated the Z-scores "by hand" (i.e., programmatically) in Fig. 5.1. I could have used scalers in *sklearn*'s preprocessing package. There are two, *scale* and *StandardScaler*, which fundamentally perform the same operation, but yet there are differences. The first, *scale*, accepts a one-dimensional array or a multi-dimensional array; i.e., a DataFrame. It then returns the standardized values using the biased standard deviation as a divisor. The second, *StandardScaler*, only accepts a multi-dimensional array, i.e., a DataFrame. It also uses the biased standard deviation. The *scale* function does not allow you to reuse the scaling operation while the *StandardScaler* does. This is important because as you will learn in Chap. 9, your data set should be divided into two mutually exclusive and completely exhaustive data sets called the *training data set* and the *testing data set*. The former is used in model estimation (more formally, it is used to *train* a model) while the latter is used to test the trained model's predictive ability. *StandardScaler* standardizes the training data set based on the mean and standard deviation for each variable independently, but then stores them and uses them to standardize the testing data. Operationally, you can use *StandardScaler* to *fit* or calculate the mean and standard deviation and then *transform* or standardize the data or perform both operations at once. You can also reverse the standardization to go back to the original data using the simple result $X_i = Z_i \times SD_X + \bar{X}$. The method is *inverse_transform*. I illustrate the use of the preprocessing package's two scalers in Fig. 5.3.

```
##
## Generate random data
##
np.random.seed( seed = 1234 )
x = np.random.random( 25 )
##
## Calculate statistics; Unbiased std. deviation
##
xBar = np.mean( x )
xStd = np.std( x, ddof = 1 )
##
## Create DataFrame and add Z-Score
##
df = pd.DataFrame( { 'X':x } )
df[ 'Z' ] =  df.X.apply( lambda x: ( x - xBar )/ xStd )  ## Base
df[ 'Z_2' ] = df.Z + 2                                   ## New Mean Only
df[ 'Z_3' ] = df.Z*1.5                                   ## New Standard Deviation Only
df[ 'Z_4' ] = df.Z*1.5 + 2                               ## New Mean and Std. Deviation
xmean = df.X.mean()
ymean = df.Z.mean()
##
df.describe().T.round( 2 ).style.set_caption( 'Z-Scores: Descriptive Statistics' ).\
    set_table_styles( tbl_styles )
```

Z-Scores: Descriptive Statistics

	count	mean	std	min	25%	50%	75%	max
X	25.000000	0.550000	0.270000	0.010000	0.360000	0.560000	0.780000	0.960000
Z	25.000000	0.000000	1.000000	-1.980000	-0.680000	0.050000	0.860000	1.520000
Z_2	25.000000	2.000000	1.000000	0.020000	1.320000	2.050000	2.860000	3.520000
Z_3	25.000000	0.000000	1.500000	-2.970000	-1.020000	0.070000	1.280000	2.270000
Z_4	25.000000	2.000000	1.500000	-0.970000	0.980000	2.070000	3.280000	4.270000

Fig. 5.1 A randomly generated data set is standardized using (5.1.1) and (5.1.4). The means and standard deviations are calculated using Numpy functions

```
##
## Plot the random X data against Z-standardizations
##
base = 'Randomly generated X.  Standard deviation is unbiased estimator.'
ax = sns.lineplot( data = df, legend = False, markers = True )
ax.axvline( x = xmean )
ax.axhline( y = ymean )
ax.set_title( 'Z-Score Linear Transformations', fontsize = font_title )
ax.set( xlabel = 'Random X', ylabel = 'Z Standardizations' )
##
new_labels = [ 'Original Z', 'Z + New Mean', 'Z * New Std. Dev.',
               'Z * New Std. Dev.\n+ New Mean' ]
plt.legend( title = 'Key', loc = 'upper left', labels = new_labels,
            fontsize = 'small' )
plt.annotate( 'Lines Shifted by New\nMean for X', xy = (0.48, 3.5 ) )
plt.annotate( 'Lines with Original\nMean for X', xy = (0.25, -2.80 ) )
plt.annotate( 'Mean Z', xy = (0.70, -0.50 ) )
plt.annotate( 'Mean X', xy = (0.57, -2.40 ), rotation = 90 )
##
footer();
```

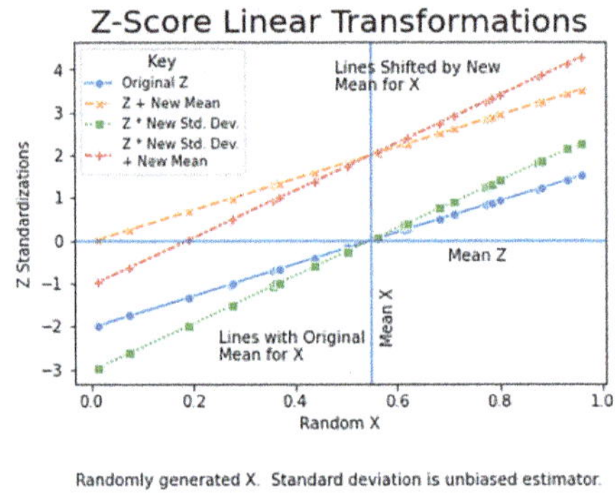

Fig. 5.2 This chart illustrates the Z-transformations in Fig. 5.1. Note the linear relationship between X and Z

Another form of standardization, *MinMax standardization*, recodes data to be uniformly distributed in the range [0, 1]. In this case, the new values can be multiplied by 100 and so they can also be interpreted as index numbers ranging from 0 to 100. The formula is:

$$X_i^{New} = \frac{X_i - min(X)}{max(X) - min(X)} \tag{5.1.6}$$

$$= \frac{X_i - min(X)}{range(X)} \tag{5.1.7}$$

where $range(X)$ is the range of the X values.

This is also a linear transformation that preserves relationships, especially the shape of the original distribution and outliers. A generalization is:

$$X_i^{New} = \frac{X_i - X_{Min}}{X_{Max} - X_{Min}} \times (X_{Max}^{New} - X_{Min}^{New}) + X_{Min}^{New} \tag{5.1.8}$$

```
##
## Use sklearn's preprocessing scale function
##
from sklearn.preprocessing import MinMaxScaler
##
## Generate random data
##
np.random.seed( seed = 1234 )
x = np.random.random( 25 )
##
## Create DataFrame and add Z-Score
##
df = pd.DataFrame( { 'X':x } )
##
## Scale the data
##
## ===> Step 1: Instantiate the scaler <===
##
scaler = MinMaxScaler( ( 1, 5 ) )   ## Notice the tuple: ( 1, 5 ) for a new min and max
##
## ===> Step 2: Scale the data <===
##
tmp = scaler.fit_transform( df )
df_scaled = pd.DataFrame( tmp, columns = df.columns + '_scaled' )
df_scaled.describe().T.style.set_caption( 'MinMax Scaler' ).set_table_styles( tbl_styles )
```

MinMax Scaler

	count	mean	std	min	25%	50%	75%	max
X_scaled	25.000000	3.265864	1.144439	1.000000	2.487202	3.318698	4.245366	5.000000

Fig. 5.3 A randomly generated data set is standardized using the *sklearn* preprocessing package *StandardScaler*. Notice how the package is imported and the steps for the standardization. In this example, the data are first fit (i.e., the mean and standard deviation are first calculated) and then transformed by (5.1.1) using the single method *fit_transform* with the argument *df*, the DataFrame

The linearity follows by rewriting (5.1.7) as

$$X_i^{New} = \frac{X_i}{range(X)} - \frac{min(X)}{range(X)} \tag{5.1.9}$$

$$= \beta_0 + \beta_1 \times X_i \tag{5.1.10}$$

with $\beta_0 = -{Min}/{Range}$ and $\beta_1 = {1}/{Range}$. The generalization in (5.1.8) allows you to set the minimum and maximum for the new variable to any number. You might want to use this in a conjoint study to rescale estimated utilities (called *part-worths*) to be more interpretable. See Paczkowski (2018) and Paczkowski (2020) for background on conjoint studies and part-worth utilities. I show how to scale with (5.1.7) and (5.1.8) in Fig. 5.4. I also show a graph of the standardization Fig. 5.5.

Similar to the *sklearn* Z-transformation, there is a *sklearn MinMaxScaler* preprocessing function. It handles the general transformation in (5.1.8) by specifying a tuple for a new minimum and maximum. The default is (0, 1) which is (5.1.7). It will also do a fit and transformation simultaneously as well as a data reversal in the same manner as the *StandardScaler* described above. I illustrate this in Fig. 5.6.

These two standardization methods suffer from a common problem: the effect of outliers. The StandardScaler relies on the sample mean and standard deviation which are distorted by outliers, the amount of distortion depending on their extent

```
##
## Generate random data
##
np.random.seed( seed = 1234 )
x = np.random.random( 25 )
##
## Create DataFrame and add Z-Score
##
df = pd.DataFrame( { 'X':x } )
##
## Scale the data
##
df[ 'minMax' ] = ( df.X - df.X.min() )/( df.X.max() - df.X.min() )                         ## Base
df[ 'minMax_2' ] = ( df.X - df.X.min() )/( df.X.max() - df.X.min() ) + 1.5                 ## New Min Only (1.5)
df[ 'minMax_3' ] = ( df.X - df.X.min() )/( df.X.max() - df.X.min() )*( 4.0 )               ## New Max Only (4.0)
df[ 'minMax_4' ] = ( df.X - df.X.min() )/( df.X.max() - df.X.min() )*( 5.0 - 1.5 ) + 1.5 ## New Max & Min
##
df.describe().T.round( 4 ).style.set_caption( 'MinMax Data' ).set_table_styles( tbl_styles )
```

MinMax Data

	count	mean	std	min	25%	50%	75%	max
X	25.000000	0.548700	0.270200	0.013800	0.364900	0.561200	0.780000	0.958100
minMax	25.000000	0.566500	0.286100	0.000000	0.371800	0.579700	0.811300	1.000000
minMax_2	25.000000	2.066500	0.286100	1.500000	1.871800	2.079700	2.311300	2.500000
minMax_3	25.000000	2.265900	1.144400	0.000000	1.487200	2.318700	3.245400	4.000000
minMax_4	25.000000	3.482600	1.001400	1.500000	2.801300	3.528900	4.339700	5.000000

Fig. 5.4 A randomly generated data set is standardized using (5.1.7) and (5.1.8)

```
##
## Plot the random X data against Min-Max Transformations
##
base = 'Randomly generated X.'
##
df_tmp = df[ [ 'X', 'minMax', 'minMax_2', 'minMax_3', 'minMax_4' ] ].copy()
df_tmp.set_index( 'X', inplace = True )
ax = sns.lineplot( data = df_tmp )
ax.set_title( 'MinMax Linear Transformations', fontsize = font_title )
ax.axhline( y = 1 )
ax.set( xlabel = 'Random X', ylabel = 'Min-Max Standardizations' )
##
new_labels = [ 'Base', 'New Min Only', 'New Max Only', 'New Min + New Max' ]
plt.legend( title = 'Key', loc = 'upper left', labels = new_labels,
            fontsize = 'small' )
##
plt.show( ax );
```

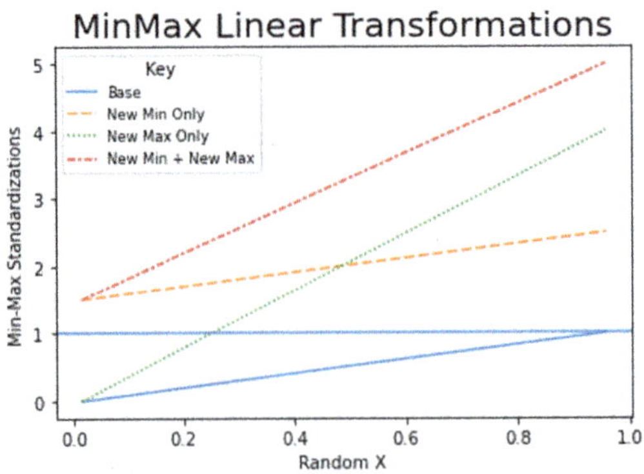

Fig. 5.5 This chart illustrates the MinMax standardization in Fig. 5.4

```
##
## Use sklearn's preprocessing scale function
##
from sklearn.preprocessing import MinMaxScaler
##
## Generate random data
##
np.random.seed( seed = 1234 )
x = np.random.random( 25 )
##
## Create DataFrame and add Z-Score
##
df = pd.DataFrame( { 'X':x } )
##
## Scale the data
##
## ===> Step 1: Instantiate the scaler <===
##
scaler = MinMaxScaler( ( 1, 5 ) )   ## Notice the tuple: ( 1, 5 ) for a new min and max
##
## ===> Step 2: Scale the data <===
##
tmp = scaler.fit_transform( df )
df_scaled = pd.DataFrame( tmp, columns = df.columns + '_scaled' )
df_scaled.describe().T.style.set_caption( 'MinMax Scaler' ).set_table_styles( tbl_styles )
```

MinMax Scaler

	count	mean	std	min	25%	50%	75%	max
X_scaled	25.000000	3.265864	1.144439	1.000000	2.487202	3.318698	4.245366	5.000000

Fig. 5.6 This illustrates the *sklearn* preprocessing function *MinMaxScaler*

and magnitude. This effect is well-known. Outliers skew a distribution: a right skewed distribution pulls the mean to the high end of the scale; a left skewed distribution does the opposite. Consequently, the Z-transform independently applied to each variable in a data set could yield distorted data, contrary to what you might expect. See the *sklearn* documentation, "*Compare the effect of different scalers on data with outliers*", for an interesting discussion. That documentation concludes that "*StandardScaler … cannot guarantee balanced feature scales in the presence of outliers*."[1] The same holds for the *MinMaxScaler*.

There is an alternative to *StandardScaler* called *RobustScaler*. The principle behind this scaler is the same as for *StandardScaler* except that robust percentile measures are used. The median is substituted for the mean and the Interquartile Range (*IQR*) is substituted for the standard deviation. This has the same fit, transform, and reversal functions as *StandardScaler* and *MinMaxScaler*. The equation is

$$Z_i^{Robust} = \frac{X_i - Median(X)}{IQR(X)} \tag{5.1.11}$$

[1] Available at https://scikit-learn.org/stable/auto_examples/preprocessing/plot_all_scaling.html#sphx-glr-auto-examples-preprocessing-plot-all-scaling-py. Last accessed: April 23, 2020.

$Median(X)$ is the median of X and $IQR(X)$ is the Interquartile Range of X. The Interquartile Range is the difference between the third and first quartiles and covers 50% of the data.

A data set could have outliers and not be skewed, but symmetric. There are outliers in both tails of the distribution. But also, a distribution could have outliers in one tail which would make the distribution skewed to that side. This means that the relationship between outliers and skewness is complex. In a multivariate context, which is typical of *BDA* problems, some of the available outlier detection methods are based on symmetrically distributed data. The multivariate Gaussian distribution is the most popular for analytical work for this situation. See Hubert and der Veeken (2008). Skewness is also a problem for low, not high, dimensional data sets which is another reason to consider the dimension reduction techniques I will discuss below. See Hubert and der Veeken (2008) for comments about this issue.

5.1.2 Nonlinear Transformations

A transformation could be nonlinear. One possibility is to rescale a variable so that its values lie in the [0, 1] range but sum to 1.0 using:

$$X_i^{New} = \frac{e^{X_i}}{\sum_{j=1}^{n} e^{X_j}}, i = 1, \ldots, n \tag{5.1.12}$$

$$\sum_i X_i^{New} = 1 \tag{5.1.13}$$

This is nonlinear. The new values, X_i^{New}, can be interpreted as probabilities or shares (i.e., proportions) of the market, preference, or wallet (i.e., how much people will spend). In a MaxDiff choice study, rescaled estimated utilities are more interpretable after transforming them using this transformation. See Paczkowski (2018) for a detailed discussion of MaxDiff choice studies. I provide an example of this nonlinear transformation in Fig. 5.7.

Other nonlinear transformations are available. For example, you could transform a continuous variable to a (natural) log scale. This is an advantage in a linear regression model because the estimated coefficient for the (natural) log variable is interpreted as an *elasticity*.

You could transform a proportion by dividing the proportion (interpreted as the probability of an event happening) by one minus that proportion. This is interpreted as the *odds* of an event happening. The odds are

$$O = \frac{p}{1-p} \tag{5.1.14}$$

```
##
## Generate random data
##
np.random.seed( seed = 1234 )
x = np.random.uniform( -1, 1, 25 )
df = pd.DataFrame( { 'X':x } )
##
num = np.exp( x )
denom = num.sum()
logit = num/denom
df[ 'logit' ] = logit
##
## Graph Logit transformation
##
ax = sns.lineplot( y = 'logit', x = 'X', data = df )
ax.set_title( 'Logit Nonlinear Transformation', fontsize = font_title )
ax.set( xlabel = 'Random Values for X', ylabel = 'Logit Transformed X' );
```

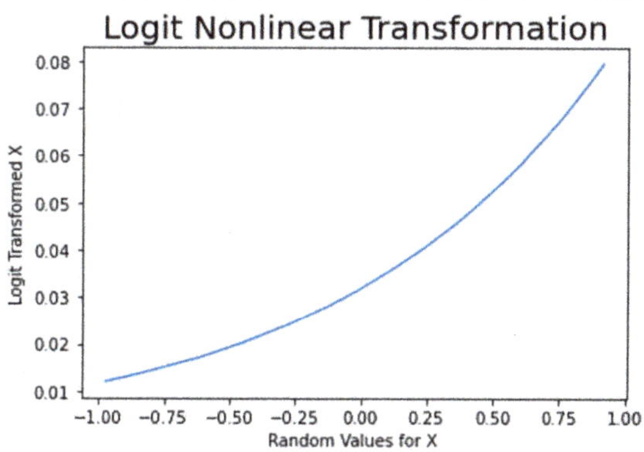

Fig. 5.7 This is an example of the nonlinear transformation using (5.1.13)

Probability of event	Odds of event
0.0–0.5	0.0–1.0
0.5–1.0	1.0–∞

Table 5.1 When the probability of an event is 0.5, then the odds of the event happening is 1.0. This is usually expressed as "odds of 1:1"

where $p = Pr(Event)$ is the probability of an event happening. Since odds are a function of a probability, the equation can be reversed to express the probability in terms of the odds of an event happening:

$$p = \frac{O}{1 + O} \tag{5.1.15}$$

I show the relationships between odds and probabilities in Table 5.1 and the nonlinearity in Fig. 5.8. The natural log of the odds, called the *log-odds* or *logit*, is used as the dependent variable in a logistic regression model, which I will discuss in Chap. 11.

```
##
## Generate random data
##
np.random.seed( seed = 1234 )
prob = np.random.uniform( size = 25 )
odds = prob/(1 - prob)
df = pd.DataFrame( { 'prob':prob, 'odds':odds } )
##
## Graph odds transformation
##
ax = sns.lineplot( y = 'odds', x = 'prob', data = df )
ax.set_title( 'Odds Transformation', fontsize = font_title )
ax.set( xlabel = 'Probability of Event', ylabel = 'Odds of Event');
```

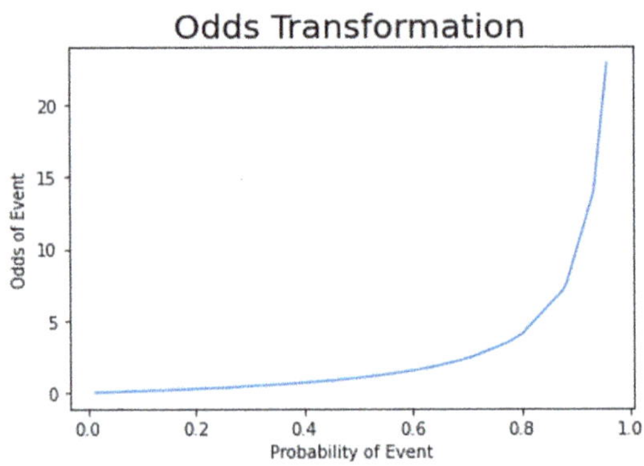

Fig. 5.8 This is an example of the nonlinear odds transformation using (5.1.14)

Sometimes you have a Likert Scale variable which you need to recode. In customer satisfaction studies, for example, satisfaction is typically measured on a 5-point Likert Scale with 1 = Very Dissatisfied and 5 = Very Satisfied. One way to transformation these variables is by converting them to *Top-Two Box* (*T2B*) or *Top-Three Box* (*T3B*) if a 5- or 10-point scale, respectively, is used. The *T2B* is the lowest two points on the scale; these are called *boxes*. For a satisfaction study, they represent someone being satisfied; the remaining boxes, the bottom-three box (*B3B*), collectively represent dissatisfaction. Similarly, $T3B$ is the highest three points. I will demonstrate this transformation in a later chapter.

5.1.3 A Family of Transformations

There is a family of transformations introduced by Box and Cox (1964) called the *power family of transformations*, or simply the *Box-Cox Transformation*, that is applicable for positive data. It is defined as

$$Y^{(\lambda)} = \begin{cases} \dfrac{Y^{\lambda} - 1}{\lambda} & \lambda \neq 0 \\ \ln Y & \lambda = 0 \end{cases} \tag{5.1.16}$$

The λ is the transformation power that must be estimated. See Zarembka (1974) for some discussion on estimating λ using maximum likelihood methods. The natural log portion of (5.1.16) results from

$$\lim_{\lambda \to 0} {}^{(Y^\lambda - 1)}/_{\lambda} = \lim_{\lambda \to 0} {}^{(e^{\lambda \times \ln Y} - 1)}/_{\lambda}.$$

Take the first derivative of the numerator with respect to λ and the first derivative of the denominator with respect to λ and apply *L'Hopital's Rule*: $\lim_{\lambda \to 0} \ln Y \times e^{\lambda \times \ln Y} = \ln Y$. See Fig. 5.9 for an example using simulated data and Fig. 5.10 for a before-after comparison of the transformation. This transformation is used to convert any distribution to a more normal distribution which allows you to use conventional hypothesis testing procedures that rely on normality. This explains why the natural log transformation is used so often in demand analysis. See Coad (2009) and Paczkowski (2018).

```
##
## Based on example in skLearn
##
from sklearn import preprocessing
##
## Generate a Log-normal distribution
##
X_lognormal = np.random.RandomState( 1234 ).lognormal( size = ( 1000, 1 ) ).flatten()
##
## Do Box-Cox on the Log-normal data
##
pt = preprocessing.PowerTransformer( method = 'box-cox', standardize = False )
bc = pt.fit_transform( X_lognormal.reshape(-1, 1) ).flatten()
##
df = pd.DataFrame( { 'logNormal':X_lognormal, 'bc':bc } )
df.head().style.set_caption( 'Box-Cox Transformation' ).set_table_styles( tbl_styles )
```

Box-Cox Transformation

	logNormal	bc
0	1.602292	0.472330
1	0.303925	-1.185288
2	4.190026	1.440996
3	0.731505	-0.312259
4	0.486466	-0.718504

Fig. 5.9 This illustrates the Box-Cox transformation on randomly simulated log-normal data

A restriction on using the Box-Cox transformation is that the Y values must all be greater than zero. In must *BDA* studies involving sales, prices, production, shipments, and so forth, this condition is not an issue. In some cases, however, it may be a restriction. For example, you may be asked to analyze sales net of returns which are sales less returned units. Reasons for returned products include: manufacturing defects; not meeting customers' expectations (e.g., quality, size, color, shape); incorrect product shipped; delivered too late to satisfy customers' needs; or the customer changed his/her mind to mention a few. If the returns are sufficiently high, then net sales might be negative. Revenue, often used as a *Key Performance Indicator* (*KPI*), is another example. Gross or total revenue is simply price times sales quantity: $TR = P \times Q$. This is a positive number. This gross revenue, however,

```
##
## Plot simulated data before and after Box-Cox
##
fig, axes = plt.subplots( 1, 2, figsize = ( 10, 5 ), sharex = False, sharey = False )
##
ax1 = sns.distplot( df.logNormal, ax = axes[ 0 ] )
ax1.set_title( 'Log-Normal\nDistribution', fontsize = font_title )
ax1.set( ylabel = 'Density' )
##
fig.tight_layout( pad = 4.0 )
##
ax2 = sns.distplot( df.bc, ax = axes[ 1 ] )
ax2.set_title( 'Box-Cox\nTransformation',  fontsize = font_title )
ax2.set( ylabel = 'Density' );
```

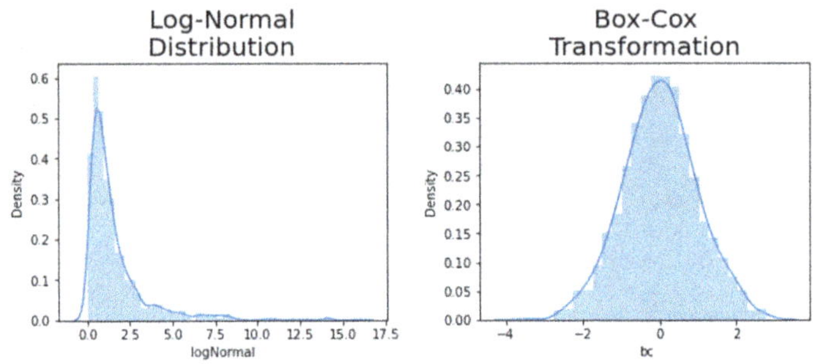

Fig. 5.10 This compares the histograms for the log-normal distribution and the Box-Cox transformation of that data

is not the only revenue. There is gross revenue, billed revenue (revenue that has been billed but not yet booked so it is pending), booked revenue (i.e., revenue received and booked), revenue not booked (i.e., received but not booked yet), and revenue net of returns. And do not forget revenue before and after taxes! These may all fall under the umbrella term of *operating revenue*: revenue due to sales. There is also non-operating revenue: revenue due to non-sales activity (e.g., royalties, interest earned, property rents). The Box-Cox transformation is inappropriate for the net revenue because it might be negative.

Yeo and Johnson (2000) proposed a modification that is implemented in the *sklearn* preprocessing module. This transformation handles negative values for the variable to be transformed. It is defined as

$$\Psi(Y, \lambda) = \begin{cases} \dfrac{(Y+1)^{\lambda} - 1}{\lambda} & \lambda \neq 0, Y \geq 0 \\ \ln(Y+1) & \lambda = 0, Y \geq 0 \\ \dfrac{-\left[(-Y+1)^{2-\lambda} - 1)\right]}{2-\lambda} & \lambda \neq 2, Y < 0 \\ -\ln(-Y+1) & \lambda = 2, Y < 0 \end{cases} \tag{5.1.17}$$

The λ is estimated from the data using a maximum likelihood method as for the Box-Cox case. See Yeo and Johnson (2000). I illustrate this transformation in Fig. 5.11. This is the default in *sklearn*. Also see Fig. 5.12.

```
##
## Based on example in sklearn
##
from sklearn import preprocessing
##
## Generate a log-normal distribution
##
X_lognormal = np.random.RandomState( 1234 ).lognormal( size = ( 1000, 1 ) ).flatten()
##
## Do Box-Cox on the log-normal data
##
pt = preprocessing.PowerTransformer( method = 'box-cox', standardize = False )
bc = pt.fit_transform( X_lognormal.reshape(-1, 1) ).flatten()
##
## Do Yeo-Johnson on the log-normal data.  This is the default method.
##
pt = preprocessing.PowerTransformer( standardize = False )
yj = pt.fit_transform( X_lognormal.reshape(-1, 1) ).flatten()
##
df = pd.DataFrame( { 'logNormal':X_lognormal, 'bc':bc, 'yj':yj } )
df.head().style.set_caption( 'Yeo-Johnson Transformation').set_table_styles( tbl_styles )
```

Yeo-Johnson Transformation

	logNormal	bc	yj
0	1.602292	0.472330	0.652509
1	0.303925	-1.185288	0.237338
2	4.190026	1.440996	0.881842
3	0.731505	-0.312259	0.437826
4	0.486466	-0.718504	0.336033

Fig. 5.11 This illustrates the Yeo-Johnson transformation alternative to the Box-Cox transformation. The same log-normally distributed data are used here as in Fig. 5.9

5.2 Encoding

There are several variable encoding schemes depending on the nature of the variable. If it is categorical with categories recorded as text, then the text must be converted to numeric values before they are used. Statistical, econometric, and machine learning methods do not operate on text *per se* but on numbers. One way to convert the textual categories to numerics is through *dummy coding*, which is also called *one-hot encoding* in machine learning. You could also label encode the categories which means you just assign ordinal values to the alphanumerically sorted category labels. Even though the assigned values are ordinal, this does not mean they represent an ordinal nature of the categories. It is just a convenient mapping of categories to numerics. The categories may or may not have an ordinal interpretation.

If a variable is not categorical but numeric to begin with, you may want to nonetheless categorize it for easier analysis because the categories may be more informative and useful than the original continuous values. Age and income are examples: categorizing people by teen, young adults, middle aged, and senior is more informative that just having their age in years; the same holds for categorizing people by low, middle, and high income.

```
##
## Plot simulated data before and after Box-Cox and Yeo-Johnson
##
fig, axes = plt.subplots( 1, 3, figsize = ( 10, 5 ), sharex = False, sharey = False )
##
ax1 = sns.distplot( df.logNormal, ax = axes[ 0 ] )
ax1.set_title( 'Log-Normal\nDistribution', fontsize = font_title )
ax1.set( ylabel = 'Density' )
##
fig.tight_layout( pad = 4.0 )
##
ax2 = sns.distplot( df.bc, ax = axes[ 1 ] )
ax2.set_title( 'Box-Cox\nTransformation', fontsize = font_title )
ax2.set( ylabel = 'Density' )
##
fig.tight_layout( pad = 4.0 )
##
ax3 = sns.distplot( df.yj, ax = axes[ 2 ] )
ax3.set_title( 'Yeo-Johnson\nTransformation', fontsize = font_title )
ax3.set( ylabel = 'Density' );
```

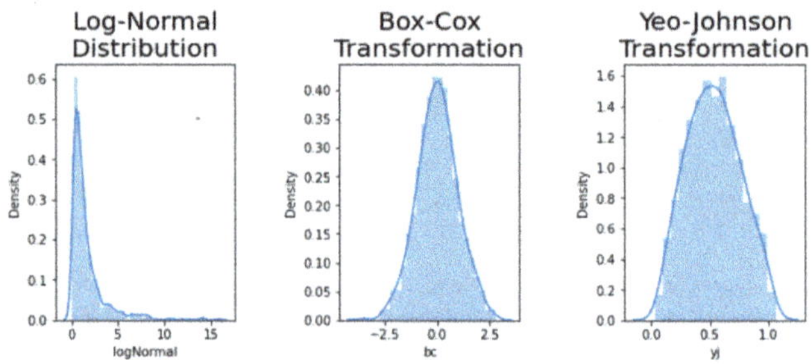

Fig. 5.12 This compares the histograms for the log-normal distribution, the Box-Cox transformation, and the Yeo-Johnson transformation of that data

5.2.1 Dummy or One-Hot Encoding

You may be familiar with *dummy variables* from a statistics course or some work you may have done with econometric modeling. Dummy variables are a form of *encoding*: the assignment of numeric values to the levels of a categorical variable. A categorical variable is a concept, something that is not measured by numbers but is often described by words such as:

- Gender;
- Buy/No Buy;
- Favor/Oppose;
- Regions (e.g., U.S. Census, marketing, world); and
- Marketing segments.

Formally, a categorical variable has categories that are the discrete categories of the categorical concept. These are mutually exclusive and completely exhaustive. For example, a company's marketing department would divide the country into regions and the concept variable "region" would have levels corresponding to each of these (completely exhaustive) and any customer would be located in only one of these regions (mutually exclusive). I list some categorical variables and their levels in Table 5.2. Notice that these have discrete, mutually exclusive levels and the collection of levels is completely exhaustive.

Variable/concept	Levels or categories	#Levels
Gender	Male/Female	2
Purchase	Yes/No	2
Opinion	Favor/Oppose	2
Product returned	Yes/No	2
Credit classification	Good/Medium/Rejected	3
Customer classification	Excellent/Good/Poor	3
Marketing regions: U.S. Census	MW, NE, S, W	4
Marketing regions: world	Continents	7

Table 5.2 These are some categorical variables that might be encountered in Business Analytic Problems

Since categorical variables represent *concepts*, they are, by their nature, non-numeric. All statistical, econometric, and machine learning methods require numeric data. A categorical variable must be encoded to a numeric variable. This is not transformation *per se*, but the creation of a whole new set of variables. The encoding can be done many ways. The most common is variously called:

- *one-of-K*;
- *one-hot*; or
- *dummy encoding.*

"Dummy coding" is a popular label in econometrics while "one-hot encoding" is popular in machine learning for the same encoding.[2] I use the two terms interchangeably. As an example of this encoding, assume that the four U.S. Census Regions—Midwest (*MW*), Northeast (*NE*), South (*S*), and West (*W*)—correspond to your marketing regions. The one-hot encoding of the concept "marketing region" is

$$D_{i1} = \begin{cases} 1, & \text{if } i \in \text{MW} \\ 0, & \text{otherwise} \end{cases} \tag{5.2.1}$$

$$D_{i2} = \begin{cases} 1, & \text{if } i \in \text{NE} \\ 0, & \text{otherwise} \end{cases} \tag{5.2.2}$$

$$D_{i3} = \begin{cases} 1, & \text{if } i \in \text{S} \\ 0, & \text{otherwise} \end{cases} \tag{5.2.3}$$

[2] A fourth encoding, *effect coding*, is popular in market research and statistical *DOE* work. It is dummy encoding with one column subtracted. See Paczkowski (2018) for a discussion of effects coding and its relationship to dummy coding.

$$D_{i4} = \begin{cases} 1, & \text{if } i \in \text{W} \\ 0, & \text{otherwise} \end{cases} \tag{5.2.4}$$

Using the *indicator function*, these are more compactly written as: $\mathbb{I}(MW)$, $\mathbb{I}(NE)$, $\mathbb{I}(S)$, $\mathbb{I}(W)$. Recall that an indicator function returns a binary result (0 or 1) depending on the evaluation of its argument. It is defined as:

$$\mathbb{I}(x) = \begin{cases} 1 & \text{if } x \text{ is true} \\ 0 & \text{otherwise} \end{cases} \tag{5.2.5}$$

For example, the indicator function $\mathbb{I}(MW)$ returns 1 if the region is the Midwest, 0 otherwise. This is compact notation that I use quite often.

As noted by Paczkowski (2021b), the indicator function can be written using set notation:

$$\mathbb{I}_A(x) = \begin{cases} 1 & \text{if } x \in A \\ 0 & \text{if } x \notin A \end{cases} \tag{5.2.6}$$

where A is a set of elements. Then:

$$\mathbb{I}_{A \cap B}(x) = min(\mathbb{I}_A(x), \mathbb{I}_B(x)) \tag{5.2.7}$$

$$= \mathbb{I}_A(x) \cdot \mathbb{I}_B(x) \tag{5.2.8}$$

and

$$\mathbb{I}_{A \cup B}(x) = max(\mathbb{I}_A(x), \mathbb{I}_B(x)) \tag{5.2.9}$$

$$= \mathbb{I}_A(x) \cdot \mathbb{I}_B(x) - \mathbb{I}_A(x) \cdot \mathbb{I}_B(x) \tag{5.2.10}$$

based on the definitions of intersection and union in set theory.

Also, as noted by Paczkowski (2021b), a list comprehension can emulate an indicator function. For example, $\mathbb{I}$ = [1 if $x > 3$ else 0 for x in range(5)] matches

$$\mathbb{I}(x) = \begin{cases} 1 & \text{if } x > 3 \text{ for } x = 0, 2, \ldots, 4 \\ 0 & \text{otherwise} \end{cases}$$

Incidentally, the *range* function returns the values 0, 1, 2, 3, 4 since Python is zero-based.

Notice that four dummy variables are defined in (5.2.1)–(5.2.4). Not all of these are needed or can be used in a linear model for two reasons. First, it is just impractical since the information contained in all the dummies is the same as the information in a subset. If there are J levels so J dummy variables are created, the subset is $J - 1$ of the dummies. There is redundancy in what all the dummies convey. Using the dummy coding above, if an observation is in the Midwest, then the dummy value for the Midwest region, D_{i1}, is 1 while the value for the other three is 0. So, knowing $D_{i1} = 1$ immediately tells you that the other three are 0. Similarly, if the observation is for the South region, then $D_{i3} = 1$ and you know immediately that the other three are all 0. Knowing the value of one dummy tells you the value of the other three. You can then safely drop one dummy and have the exact same information as if you kept all them. Hence the subset of $J - 1$ dummy variables. Including all the dummies might lead you into the *dummy variable trap*. See Gujarati (2003).

The second reason is more important. By using all the dummies in a linear model, one that contains an intercept,[3] then a linear combination of all the dummies exactly equals the value for the intercept variable. The intercept variable equals 1.0 for each observation and multiples the intercept parameter. The linear combination of the dummies, each dummy variable multiplied by a coefficient equal to 1, equals this intercept variable exactly. As a result, you will have *perfect multicollinearity*, a condition that will prevent you from estimating the parameters of the linear model. In simple terms, you have another case of redundancy, this time between all the dummies and the intercept variable; they all have the same value: 1.0.

The remedy for both problems is to select one dummy to drop. This is called the *base* or *reference* dummy. The redundancy among the dummies and with the intercept variable is eliminated. For the region example, the Midwest is first in alphanumeric order, so it could be the base. This is the choice automatically made by *Statsmodels* which I will demonstrate in Chap. 6. Other software automatically use the last in alphanumeric order. You could select any of the dummies as the base, but this is problem specific. Multicollinearity is an important topic, especially in *BDA* when large, *high dimensional* data sets are used. A data set is high dimensional when it has a large number of variables or features. The likelihood that some of these features are related is, therefore, high. I will discuss this in Chap. 10. See Gujarati (2003) and Greene (2003) for discussions about multicollinearity. Also see Belsley et al. (1980) for a technical discussion and diagnostic checks for multicollinearity.

Variables are one-hot encoded using different methods depending on how your data are stored, either in Pandas DataFrames or Numpy Arrays, and how you will use them. I recommend that you always use Pandas for managing your data, so I will focus on encoding categorical variables using a DataFrame with Pandas functions as well as *sklearn* functions.

5.2.1.1 Pandas Dummy Encoding

For a categorical variable in Pandas, you can use the Pandas function *get_dummies*. This is the most convenient approach. It takes several arguments, but the most important are:

[3] Your linear models should always contain an intercept.

- the DataFrame name (required first positional argument);
- the columns, as a list, to encode (default is to encode all columns that have a data type of *object* or *category*);
- the base (an indicator whether or not to drop the first level in alphanumeric order; default is False); and
- how to handle missing (NaN) values (encode them as well or not; default: do not encode)

See the Pandas *get_dummies* documentation for more arguments.

5.2.1.2 sklearn Dummy Encoding

The *sklearn* preprocessing module has the *OneHotEncoder* function which dummifies a variable or set of variables. There is a two-step process for using this encoder, unlike Pandas' *get_dummies* function which just takes a DataFrame name as an argument. The *OneHotEncoder* first requires *instantiation*. This means you must create an instance of the function. The function is a class of code that just sits in computer memory. You have to activate it before you can use it; this activation is the instantiation. This involves calling the function and assigning it a name. For the *OneHotEncoder*, a typical name is *ohe*. In the process of calling it, you could assign values to its parameters or just use the defaults, if any. Once the function is instantiated, you can use it by chaining a method to the assigned name. For *OneHotEncoder*, the most commonly used methods are

- fit;
- transform;
- fit_transform; and
- inverse_transform.

The *fit* method calculates whatever statistics or values are needed for the transformation and stores them, but it does not do the transformation. The *transform* method uses the stored values and does the transformation. The *fit_transform* method does both in one call and is typically used rather than the other two. I introduced these three before. The *inverse_transform* method reverses the transformation and restores the original data. The input for the *OneHotEncoder* is an array-like object, usually a Numpy array, of integers or strings such as the names of regions (Fig. 5.12).

5.2.2 *Patsy Encoding*

Patsy is a statistical formula system that allows you to specify a modeling formula, such as an *OLS* linear model, as a text string. That string can then be used in a *statsmodels* modeling function. An important component of the Patsy system is the

$C(\cdot)$ function which takes a categorical variable as an argument and appropriately dummifies it. I will show you an example of its use in Chap. 6.

5.2.3 Label Encoding

You may just need to assign nominal or ordinal values to the levels of a categorical variable. This may occur when dummy variables are not appropriate. The U.S. Census regions, for example, could be nominally encoded in alphanumeric order as: Midwest = 1, Northeast = 2, South = 3, West = 4. Management levels could be ordinally encoded: Entry Level = 1, Mid Level = 2, Executive = 3. This is called *label encoding*. You will see an application of this when I discuss *decision trees* in Chap. 11.

5.2.4 Binarizing Data

You may have numeric data (i.e., floats or ints) that you want to convert to dummy values based on a threshold. Any value less than the threshold is encoded as 0; otherwise, 1. For example, you may have data on years of service for employees. Anyone with less than 5 years are to be encoded as 0; otherwise, they are encoded as 1. The sklearn *Binarizer* function does this. Its parameters are an array-like object and a threshold. The default threshold is 0.0. This function also has a *fit*, *transform*, and *fit_transform* method, but the *fit* method really does nothing. There is no inverse_transform method. I show an example in Fig. 5.13.

This type of encoding is sometimes used to make it easier to interpret a continuous variable. But there is a cost to doing this: a loss of information. Extracting information from data, especially Rich Information, is the goal of *BDA*, but this encoding may do the opposite. The information lost is the distribution of the continuous data, including any indication of skewness, location, and spread, as well as trends, patterns, relationships, and anomalies. These may be important and insightful but are hidden by this encoder.

Another form of bin encoding using the sklearn's *KBinDiscretizer* allows you to encode by allocating the values of each feature into one of $b > 2$ mutually exclusive and completely exhaustive bins. The default is $b = 5$ bins. The encoding is based on three parameters:

1. the number of bins (default is 5);
2. the type of encoding (default: *onehot*); and
3. an encoding strategy (default: *quantile*).

The types of encoding that are used are *one-hot* (the default), *one-hot dense*, and *ordinal*. The difference between the one-hot and one-hot dense encoding is that the former returns a *sparse array* and the latter returns a *dense array*. A sparse array

```
##
## Randomly generate data
##
np.random.seed( 42 )
x = np.random.randn( 25 )*10
x = x.reshape( 5, 5 )
print( f'Random Data:\n{x}' )
##
## Instantiate the binarizer using threshold of 5
##
binarizer = preprocessing.KBinsDiscretizer( n_bins = 3,
                                            encode = 'ordinal',
                                            strategy = 'uniform' )
##
## Bin the data and put in a DataFrame
##
binned = binarizer.fit_transform( x )
cols = [ 'X' + str( i ) for i in range( 0, 5 ) ]
df_bin = pd.DataFrame( binned, columns = cols )
df_bin.style.set_caption( 'Binned Data' ).format( '{:0.2}' ).\
    set_table_styles( tbl_styles )
```

```
Random Data:
[[  4.96714153  -1.38264301   6.47688538  15.23029856  -2.34153375]
 [ -2.34136957  15.79212816   7.67434729  -4.69474386   5.42560044]
 [ -4.63417693  -4.65729754   2.41962272 -19.13280245 -17.24917833]
 [ -5.62287529 -10.1283112    3.14247333  -9.08024076 -14.12303701]
 [ 14.65648769  -2.257763     0.67528205 -14.24748186  -5.44382725]]
```

Binned Data

	X0	X1	X2	X3	X4
0	1.0	1.0	2.0	2.0	1.0
1	0.0	2.0	2.0	1.0	2.0
2	0.0	0.0	0.0	0.0	0.0
3	0.0	0.0	1.0	0.0	0.0
4	2.0	0.0	0.0	0.0	1.0

Fig. 5.13 Several continuous or floating point number variables or features can be nominally encoded based on a threshold value. Values greater than the threshold are encoded as 1; 0 otherwise. In this example, the threshold is 5

has many zeros that, if exploited or handled efficiently, can lead to computational efficiencies. This handling involves not storing the zeros and not using them in computations. A dense array has zeros but none of them are treated differently from non-zero values. Sparse arrays are common in *BDA*, for example in text processing where an array, called a *Document Term Matrix* (*DTM*), is created that contains the frequency of occurrence of words or *tokens* in documents, but not all words occur in each document. Those non-occurrences are represented by zeros. For a detailed discussion of spare and dense matrices, see Duff et al. (2017).

The *KBinDiscretizer* also has the four methods: *fit*, *transform*, *fit-transform*, and *inverse_transform*.

The ordinal encoding is an ordered set of integers with base zero. This means they have values 0, 1, 2, ... The strategy is how the encoding is developed. It could be *uniform*, *quantile*, or *kmeans*. The uniform strategy returns bins of equal width; quantile returns bins with the same number of data points; and kmeans returns bins with the same nearest center. This latter strategy is based on *K-Means clustering* which I discuss in Chap. 12.

This form of encoding has the same information issue I described above. I illustrate this type of encoding in Fig. 5.14. There is a method available to reverse this encoding. See the *sklearn* documentation for this.

```
##
## Randomly generate data
##
np.random.seed( 42 )
x = np.random.randn( 25 )*10
x = x.reshape( 5, 5 )
print( f'Random Data:\n{x}' )
##
## Instantiate the binarizer using threshold of 5
##
binarizer = preprocessing.Binarizer( threshold = 5 )
##
## Bin the data and put in a DataFrame
##
binned = binarizer.transform( x )
cols = [ 'X' + str( i ) for i in range( 0, 5 ) ]
df_bin = pd.DataFrame( binned, columns = cols )
df_bin.style.set_caption( 'Binned Data' ).format( '{:0.2}' ).\
    set_table_styles( tbl_styles )
```

```
Random Data:
[[  4.96714153  -1.38264301   6.47688538  15.23029856  -2.34153375]
 [ -2.34136957  15.79212816   7.67434729  -4.69474386   5.42560044]
 [ -4.63417693  -4.65729754   2.41962272 -19.13280245 -17.24917833]
 [ -5.62287529 -10.1283112    3.14247333  -9.08024076 -14.12303701]
 [ 14.65648769  -2.257763     0.67528205 -14.24748186  -5.44382725]]
```

Binned Data

	X0	X1	X2	X3	X4
0	0.0	0.0	1.0	1.0	0.0
1	0.0	1.0	1.0	0.0	1.0
2	0.0	0.0	0.0	0.0	0.0
3	0.0	0.0	0.0	0.0	0.0
4	1.0	0.0	0.0	0.0	0.0

Fig. 5.14 Several continuous or floating point number variables or features are ordinally encoded. Notice that the *fit_transform* method is used

A final two, somewhat related functions for creating bins are the Pandas *cut* and qcut functions. The *cut* function "assigns values of the variable to bins you define or to an equal number of bins." Furthermore, "if you specify bins as a list [18, 40, 60, 110], then the bins are interpreted by Pandas as (18, 40], (40, 60] and (40, 100]. Note the brace notation. This is standard math notation for half-open intervals, in this case open on the left. This means the left value in not included but the right value is included; including the right value is the default. So, the interval (18, 40] is interpreted as $18 < age \leq 40$. You can change the inclusion of the right value using the argument $right = False$ and the left value using $include\_lowest = True$ in the *cut* function. The function *qcut* does the same thing as *cut* but uses quantile-based binning. The quantiles could be the number of quantiles (e.g., 10 for deciles, 4 for quartiles) or an array of quantiles (e.g. [0, 0.25, 0.50, 0.75, 1.0])." From Paczkowski (2021b). Also see McKinney (2018) for discussions and examples.

5.3 Dimension Reduction

Another common problem, especially with high dimensional data sets as I have mentioned before, is that there are too many variables that contain almost the same information. This is the multicollinearity problem I discussed above. Technically, the implication of this high degree of linear relationship is that the parameters of a linear model cannot be estimated when the collinearity is perfect. One solution is to collapse the dimensionality of the data, meaning that the variables are collapsed to a few new ones that are the most important. The measure of importance is the variance of the data. Specifically, new variables are created such that the first, called the *first principal component*, accounts for most of the variance. The next variable created, called the *second principal component*, accounts for the next largest amount of the variance not captured by the first principal component AND is independent of that first component. The independence is important because, if you recall, the issue is the multicollinearity or linear relationship among the original variables in high-dimensional data. You want to remove this dependency which is what this restriction does. Each succeeding new variable accounts for the next amount of the variance and is independent of the preceding new variables.

If there are p original variables, then there are at most p new ones. Usually, only the first few (maybe two) principal components are retained since they account for most of the variance of the original data. These new, mutually orthogonal principal components can be used in a linear model, usually called a *principal components regression*.

Principal components are extracted using a matrix decomposition method called *Singular Value Decomposition* (*SVD*).[4] This is a very important method used in many data analytic procedures as you will see elsewhere in this book. Fundamentally, *SVD* decomposes a $n \times p$ matrix into three parts which are themselves matrices. I will refer to these three matrices as the left, center, and right matrices. If $\mathbf{X}$ is the $n \times p$ data matrix, then the *SVD* of $\mathbf{X}$ is

$$\mathbf{X} = \mathbf{U}\Sigma\mathbf{V}^{\top} \tag{5.3.1}$$

where $\mathbf{U}$ is an $n \times n$ orthogonal matrix such that $\mathbf{U}^{\top}\mathbf{U} = \mathbf{U}\mathbf{U}^{\top} = \mathbf{I}_n$ where $\mathbf{I}_n$ is the $n \times n$ identity matrix; $\mathbf{\Sigma}$ is an $n \times r$ diagonal matrix; and $\mathbf{V}^{\top}$ is an $r \times p$ orthogonal matrix such that $\mathbf{V}^{\top}\mathbf{V} = \mathbf{V}\mathbf{V}^{\top} = \mathbf{I}_p$ where $\mathbf{I}_p$ is $p \times p$.

As noted by Paczkowski (2020), "two vectors are orthogonal if their inner or dot product is zero. The inner product of two vectors, $\mathbf{a}$ and $\mathbf{b}$, is $\mathbf{a} \cdot \mathbf{b} = \sum_{i=1}^{n} a_i \times b_i$ and the two vectors are orthogonal if $\mathbf{a} \cdot \mathbf{b} = 0$. This is the same as saying they are perpendicular to each other, or the cosine of the angle between the two vectors is zero." See Lay (2012), Strang (2006), and Jobson (1992) on matrices and *SVD*. Also

[4] This section is based on Paczkowski (2020).

see Paczkowski (2020) for a discussion of the use of *SVD* in text analysis and new product development.

5.4 Handling Missing Data

Missing data are always an issue, and a big one. You have to identify:

1. which features have missing values;
2. the extent of the missingness;
3. the reasons for the missingness; and
4. what to do about them.

The first and second are handled using Pandas' *info( )* method. This is called by chaining *info( )* to the DataFrame name. A report is returned that lists each feature in the DataFrame, the feature's data type (int64, object, float64, and datetime64[ns]),[5] the non-null count (i.e., number of records without missing values) as well as information about the index and memory usage.

Another approach is to create a missing value report function that has one parameter: the DataFrame name. I created such a function which I display in Fig. 5.15. This function relies on the package *sidetable* which supplies an accessor, *stb*, to the DataFrame. The accessor has a function called *missing* which calculates some missing value information. These are put into a DataFrame and the displayed. I show a typical display from this function in Fig. 5.16. The package *sidetable* is installed using *pip install sidetable* or *conda install -c conda-forge sidetable*.

The third problem is difficult. There are many reasons for missing data. For very large, high-dimensional data sets, the reasons could become overwhelming. More importantly, however, the data typically used in *BDA* problems is secondary which suggests that you may never know the cause of the missingness. Data collection was out of your hands; you had no control. You could inquire about the reasons for the missingness from your *IT* department and, hopefully, those reasons would be documented in the Data Dictionary. Unfortunately, this is often not the case. See Enders (2010) for a very good treatment of missing data.

The final problem is also a challenge. There are two options:

1. delete records with missing values; and
2. impute the missing values.

Pandas has a simple method, *dropna*, that deletes either rows (axis = 0, the default) or columns (axis = 1) that has any missing values. Missing values are recorded as *NaN*s. Since this is a method, it is chained to the DataFrame name. That could be subsetted on specific columns.

[5] There is also a count of each data type.

```
def mvReport( df ):
    """
    Purpose: Calculate and display missing value report.
    Packages: Use sidetable package accessor stb.
    Parameters:
        df: dataFrame handle.
    """
    x = df.stb.missing( )
    ##
    ## Reorder columns and capitalize
    ##
    cols = [ 'total', 'missing', 'percent' ]
    x = x[ cols ]
    x.columns = x.columns.str.capitalize()
    ##
    ## Display
    ##
    base = 'Base: n = ' + str( df.shape[ 0 ] )
    fmt_dict = { 'Missing':'{:,.0f}', 'Total':'{:,.0f}', 'Percent':'{:.3}%' }
    display( x.style.set_caption( 'Missing Value Report for DataFrame ' + get_df_name( df ) ).\
        format( fmt_dict ).set_table_styles( tbl_styles ) )
    print( base )

def get_df_name( df ):
    """
    Purpose: Get DataFrame name
    Source: https://stackoverflow.com/questions/31727333/get-the-name-of-a-pandas-dataframe
    Parameters: DataFrame handle
    Return: DataFrame name as string
    """
    name = [ x for x in globals() if globals()[ x ] is df ][ 0 ]
    return name
```

Fig. 5.15 A missing value report function using the package *sidetable*. This function also relies on another function, *get_df_name* to retrieve the DataFrame name. An example report is in Fig. 5.16

There are problems with using this method. If your data are a time series, then dropping a record introduces a break in the time sequence. A central characteristic of time series is that there are no breaks, although in many practical situations, they cannot be avoided. For example, there are weather events, strikes, production failures, pandemics, civil unrest, and so forth that cause markets, production, deliveries and other activities to stop. Consequently, data are not collected for the affected time period. But these are natural events (not all, of course) which differ from you dropping records because of missing values. Also, if you have a small data set, then dropping records, aside from the breaks that are introduced, reduce your data set even further. If too many records are dropped, then estimation (i.e., "training") is jeopardized.

An alternative to dropping records is to impute or fill-in missing values. Pandas has a *fillna* method that allows you to specify the way to impute the missing values: with a scalar (e.g., 0), using a dictionary of values, a Series, or another DataFrame. You could specify the value explicitly or you could calculate it from other values in your DataFrame either in the feature that has the missing value or a combination of features. For example, you could calculate the sample mean for a feature and use that mean as the scalar. To do this, you could use *x_mean = df.X.mean( )* followed by *df.X.fillna(x_mean, inplace = True)* or *df.X.fillna(df.X.mean( ), inplace = True)* where "X" is the feature in *df* with the missing values. You also specify the axis (axis = 0 for rows, axis = 1 for columns) to fill by. Another possibility is the *interpolate* method which, by default, does a linear interpolation of the missing values.

```
mvReport( df_agg )
```

Missing Value Report for DataFrame df_agg

	Total	Missing	Percent
Unnamed: 0	779	0	0.0%
CID	779	0	0.0%
Region	779	0	0.0%
loyaltyProgram	779	0	0.0%
buyerRating	779	0	0.0%
buyerSatisfaction	779	0	0.0%
totalUsales	779	0	0.0%
meanPprice	779	0	0.0%
meanDdisc	779	0	0.0%
meanOdisc	779	0	0.0%
meanCdisc	779	0	0.0%
meanPdisc	779	0	0.0%
sat_t2b	779	0	0.0%

```
Base: n = 779
```

Fig. 5.16 A missing value report function using the function in Fig. 5.15

The sklearn package has a function in its *impute* module called *SimpleImputer* that replaces missing values. This function does not operate on a DataFrame, instead it operates on a Numpy array. This function is imported using *from sklearn.impute import SimpleImputer*. As for the encoders, you first instantiate the imputer, in this case using a parameter to specify the imputing strategy (i.e., the way to impute) and the missing value code that identifies the missing values. The codes are specified as *int*, *float*, *str*, *np.nan* or *None* with default equal *np.nan*. The strategy is a string such as "mean", "median", "most_frequent" (for strings or numeric data), or "constant." If you use "constant", then you need to specify another parameter, *fill_value* as a string or numeric value; the default is None, which equates to 0. After instantiation, you can use any of the four methods I mentioned in other contexts: *fit*, *transform*, *fit_transform*, and *inverse_transform*. The parameter for them is the array to impute on.

5.5 Appendix

In this Appendix, I summarize and demonstrate a few statistical concepts I mentioned in this chapter.

5.5.1 Mean and Variance of Standardized Variable

To show that the mean of the standardized variable in (5.1.1) is zero, let

$$Z_i = \frac{X_i - \bar{X}}{SD_X}$$

for the random variable X_i, $i = 1, 2, \ldots, n$. Then, the mean is

$$\begin{aligned}\bar{Z} &= \frac{1}{n \times SD_X} \times \sum_{i=1}^{n} (X_i - \bar{X}) \\ &= 0\end{aligned}$$

since $\sum_{i=1}^{n} (X_i - \bar{X}) = 0$.

To show that the variance of the standardized variable in (5.1.1) is 1, note that

$$\begin{aligned}V(Z) &= \frac{\sum_{i=1}^{n} (Z_i - \bar{Z})^2}{n-1} \\ &= \frac{1}{n-1} \times \sum_{i=1}^{n} Z_i^2 \\ &= \frac{1}{n-1} \times \sum_{i=1}^{n} \frac{(X_i - \bar{X})^2}{SD_X^2} \\ &= \frac{n-1}{(n-1) \times SD_X^2} \times \frac{\sum_{i=1}^{n} (X_i - \bar{X})^2}{n-1} \\ &= 1\end{aligned}$$

5.5.2 Mean and Variance of Adjusted Standardized Variable

The mean of (5.1.4) is found using

$$\begin{aligned}\bar{Z} &= \frac{1}{n} \times \sum_{i=1}^{n} Z_i \\ &= \frac{1}{n} \times \sum_{i=1}^{n} \left(\frac{X_i - \bar{X}}{SD_X} \times SD_X^{New} + \bar{X}^{New} \right) \\ &= \frac{SD_X^{New}}{n \times SD_X} \times \sum_{i=1}^{n} (X_i - \bar{X}) + \frac{1}{n} \sum_{i=1}^{n} \bar{X}^{New} \\ &= \bar{X}^{New}\end{aligned}$$

The variance of (5.1.4) is found using

$$
\begin{aligned}
V(Z) &= \frac{1}{n-1} \times \sum_{i=1}^{n} (Z_i - \bar{Z})^2 \\
&= \frac{1}{n-1} \times \sum_{i=1}^{n} \left(\frac{X_i - \bar{X}}{SD_X} \times SD_X^{New} + \bar{X}^{New} - \bar{X}^{New} \right)^2 \\
&= \frac{\left(SD_X^{New}\right)^2}{SD_X^2} \times \frac{\sum_{i=1}^{n} (X_i - \bar{X})^2}{n-1} \\
&= \left(SD_X^{New}\right)^2.
\end{aligned}
$$

5.5.3 *Unbiased Estimators of μ and σ^2*

An estimator of a population parameter, θ, is *unbiased* when

$$E\left(\widehat{\theta}\right) = \theta$$

That is, an estimator is unbiased if it equals the true value of the parameter "in the long run, on the average." In essence, the amount by which you overestimate and underestimate the true value of the parameter just balances so that the estimated value is correct on average. You have a biased measure when $E\left(\widehat{\theta}\right) \neq \theta$.

If X_i, $i = 1, 2, \ldots, n$, are *independent and identically distributed random variables (iid)* with mean $E(X) = \mu$ and variance $E(X - \mu)^2 = \sigma^2$, then $E(\bar{X}) = \mu$. This is easy to show:

$$
\begin{aligned}
E(\bar{X}) &= \frac{1}{n} \times E(\sum_{i=1}^{n} X_i) \\
&= \frac{1}{n} \times \sum_{i=1}^{n} E(X_i) \\
&= \frac{1}{n} \times n \times \mu \\
&= \mu
\end{aligned}
$$

Now consider the sample estimator, s^2, for the population variance, σ^2. You want $E(s^2) = \sigma^2$ so that s^2 is an unbiased estimator of σ^2. Before showing unbiasedness, note that:

$$\begin{aligned} X_i &= X_i + 0 + 0 \\ &= X_i - \bar{X} + \bar{X} - \mu + \mu \\ X_i - \mu &= X_i - \bar{X} + \bar{X} - \mu \\ &= (X_i - \bar{X}) + (\bar{X} - \mu). \end{aligned}$$

Squaring both sides, you get:

$$(X_i - \mu)^2 = (X_i - \bar{X})^2 + (\bar{X} - \mu)^2 + 2 \times (X_i - \bar{X})(\bar{X} - \mu)$$

Summing and rearranging, you get

$$\sum (X_i - \bar{X})^2 = \sum (X_i - \mu)^2 - \sum (\bar{X} - \mu)^2$$

It is left as an exercise to show that the sum of the cross products equals zero.

Now define another estimator $\tilde{s}^2 = \dfrac{\sum(X - \bar{X})^2}{n}$. Notice that the divisor is n, the sample size. Then

$$\begin{aligned} E(\tilde{s}^2) &= E\left[\frac{\sum(X - \bar{X})^2}{n}\right] \\ &= E\left[\frac{\sum(X - \mu)^2}{n}\right] - E\left[\frac{\sum(\bar{X} - \mu)^2}{n}\right] \\ &= \frac{n\sigma^2}{n} - \frac{n\frac{\sigma^2}{n}}{n} \\ &= \sigma^2 - \frac{\sigma^2}{n} \\ &= \sigma^2\left(\frac{n-1}{n}\right) < \sigma^2 \end{aligned}$$

So, $\tilde{s}^2$ is a biased estimator. For an unbiased estimator, define s^2 as

$$\begin{aligned} s^2 &= \left(\frac{n}{n-1}\right)\tilde{s}^2 \\ &= \frac{\sum(X - \bar{X})^2}{n-1} \end{aligned}$$

This is the estimator taught in a basic statistics course. Note that $\lim_{n\to\infty}\left(\frac{n}{n-1}\right) = 1$ so there is no difference between s^2 and $\tilde{s}^2$ for large n. Using the divisor $n - 1$ for the sample variance yields an unbiased estimator for the population variance.

Part II
Intermediate Analytics

I begin the development of statistical/econometric/machine learning methodologies in this second part of the book. The theory part of the methods are introduced so you will gain a perspective on strengths, weaknesses, issues, and limitations. The methods I will discuss will allow you to begin a more formal and detailed analysis of business data but without getting overly involved in high-powered methods. After reading the first two parts of this book, you will be able to do sophisticated analyses of business data.

Chapter 6
OLS Regression: The Basics

A major part of Business Data Analytics focuses on understanding relationships and then using them to predict most likely outcomes of business decisions. An example of the former is a price elasticity used for repricing an existing product or setting the initial price for a new one. In either case, the elasticity is calculated based on the relationship between price and quantity, allowing for other factors such as income levels, seasons of the year, geographic locations, and so forth. The exact relationship between price and quantity, other than being negative all economic theory and experience, can only be determined by examining data, lots of it. For the latter use of relationships, predicting, it is almost always the case that you will need to know the outcome of a decision for a key performance measure (*KPM*) such as revenue, sales, shipments, and so forth. Predictions are very important and complicated to make, but they are, nonetheless, like the key measures, based on relationships among variables or features in a data set.

I will set the stage for determining relationships among variables in this chapter by restricting my discussion to linear relationships. I will cover other relationships in succeeding chapters. The main tool for determining a linear relationship is *ordinary least squares regression analysis* (*OLS*). This is a well-developed statistical and econometric topic with many excellent books covering its essentials. I will only highlight some of these. For more detailed coverage, see Goldberger (1964) for a classical approach, (Greene, 2003; Gujarati, 2003) for advanced theoretical developments, and (Draper and Smith, 1966; Neter et al., 1989) for more applied development. I will assume that you have some familiarity with basic *OLS*, although I will review basic concepts to introduce my notation and terminology.

W. R. Paczkowski, *Business Analytics*,
https://doi.org/10.1007/978-3-030-87023-2_6

6.1 Basic *OLS* Concept

An underlying concept for regression analysis is the existence of a population line relating a minimum of two variables, simply called X and Y. The X is an *independent* or *exogenous variable* or *feature* and the Y is a *dependent* or *target* variable. If p is the number of features, then $p = 1$. Both X and Y are assumed to be floating point numbers although more advanced frameworks consider other possibilities. See Chap. 11 for these. The population relationship between the two variables is given by a *population regression line* (*PRL*) expressed as the linear relationship

$$E(Y) = \beta_0 + \beta_1 \times X \tag{6.1.1}$$

where β_0 and β_1 are unknown population parameters that have to be estimated from data and $E(Y)$ is the expected value of Y. An *expected value* is the population mean weighted over all possible values in the population, the weights being the probabilities of seeing the values. The reason for the expected value is that there is random variation in the observations that cause the observed data to deviate, or be *disturbed*, from the population line. There is no particular reason for the disturbance except for pure random noise. The implication is that an actual observation deviates from the *PRL* so any observation is written as

$$Y_i = E(Y) + \epsilon_i \tag{6.1.2}$$

$$= \beta_0 + \beta_1 \times X_i + \epsilon_i \tag{6.1.3}$$

for $i = 1, 2, \ldots, n$ where n is the number of sample observations and ϵ_i is a *disturbance term.*

6.1.1 *The Disturbance Term and the Residual*

The ϵ_i is the random noise associated with the ith observation. This term is assumed to be drawn from a normal distribution with mean 0 and variance σ^2: $\epsilon_i \sim \mathcal{N}(0, \sigma^2)$. It is also assumed that $cov(\epsilon_i, \epsilon_j) = 0, \forall i, j, i \neq j$. These are the *Classical Assumptions* for *OLS*:

- $\epsilon_i \sim \mathcal{N}(0, \sigma^2)$
- $cov(\epsilon_i, \epsilon_j) = 0, \forall i, j, i \neq j$.

The case with a constant σ^2 is called *homoskedasticity*; the case with a non-constant σ^2 is called *heteroskedasticity*.

A central feature of the *PRL* is that it is unknown so you have to estimate it using data. In particular, you have to estimate the two parameters. A parameter is a constant, unknown numeric characteristic of the population. The unknown feature is the reason for estimation. Let $\hat{\beta}_0$ and $\hat{\beta}_1$ be the estimators for these two parameters. Once these are known, then you can calculate the estimated *sample regression line* (*SRL*) as

$$\hat{Y}_i = \hat{\beta}_0 + \hat{\beta}_1 \times X_i. \tag{6.1.4}$$

Notice that the *SRL* does not have a disturbance term because this line is known with certainty unlike the *PRL*. There is, however, a value comparable to the disturbance term called the *residual* that is the difference between the actual observation, Y_i, and the value estimated by the *SRL*, $\hat{Y}_i$. The residual is the vertical difference between the actual observation and the estimated line: $e_i = Y_i - \hat{Y}_i$ where e_i is the residual. The residual is interpreted as an error because it is the difference between the actual and predicted observations. It is easy to note that

$$\begin{aligned} e_i &= Y_i - \hat{Y}_i & (6.1.5)\\ &= \beta_0 + \beta_1 \times X_i + \epsilon_i - \hat{\beta}_0 - \hat{\beta}_1 \times X_i & (6.1.6)\\ &= (\beta_0 - \hat{\beta}_0) + (\beta_1 - \hat{\beta}_1) \times X_i + \epsilon_i. & (6.1.7) \end{aligned}$$

If $\hat{\beta}_0$ is a very good estimator of β_0, then their difference should be close to zero. Similarly for $\hat{\beta}_1$. This means that $e_i \approx \epsilon_i$. The residual, e_i, is observable and measurable, while the disturbance is not, but the residual approaches the disturbance. The implication is that the residuals should have the same properties as the disturbance terms: zero mean, constant variance, zero covariance, and be normally distributed. If the residuals have a zero mean, then $\sum_{i=1}^{n} e_i = 0$.

6.1.2 OLS *Estimation*

A goal is to minimize the residuals, actually a function of them since there are so many residuals, one for each of the *n* observations. The reason for the minimization is the interpretation of a residual as a loss or cost: intuitively, you should want to minimize your losses. Since the residual approximates the disturbance, and the disturbance is the deviation of the actual observation from the *PRL*, then minimizing a function of the residuals will simultaneously minimize the same function of the disturbances. This should bring the *SRL* close to the *PRL*, which is still hidden from you.

There are two ways to define a loss: as a squared loss or absolute value loss:

$$\begin{aligned} L(e) &= (Y - \hat{Y})^2 \quad \text{Squared Loss} & (6.1.8)\\ L(e) &= \mid Y - \hat{Y} \mid \quad \text{Absolute Value Loss} & (6.1.9) \end{aligned}$$

The mean squared loss function is the mean of the squared losses, or

$$MSE = \frac{1}{n} \times \sum_{i=1}^{n} (Y_i - \hat{Y}_i)^2 \tag{6.1.10}$$

and is referred to as the *Mean Squared Error* (*MSE*). The absolute value loss function is the mean of the absolute losses, or

$$MAE = \frac{1}{n} \times \sum_{i=1}^{n} \mid Y_i - \hat{Y}_i \mid . \tag{6.1.11}$$

The *MAE*, referred to as the *L1 Loss* or *L1-norm Loss*, is not differentiable at $e_i = 0$ so it is not mathematically easy to use. The problem with the *MSE*, referred to as the *L2 Loss* or *L2-norm Loss*, is that it is subject to the effect of outliers, but it is mathematically tractable and, so, it is the one most often used. It is the basis for ordinary least squares. Since the $1/n$ factor in (6.1.10) is a constant, the focus is on the numerator which is the sum of the squared residuals (*SSE*). This is

$$SSE = \sum_{i=1}^{n} e_i^2 \tag{6.1.12}$$

$$= \sum_{i=1}^{n} (Y_i - \hat{Y}_i)^2 \tag{6.1.13}$$

$$= \sum_{i=1}^{n} (Y_i - \hat{\beta}_0 - \hat{\beta}_1 \times X_i)^2. \tag{6.1.14}$$

I provide a graphical comparison between the squared and absolute value residuals in Fig. 6.1. Notice that the residuals vary between a negative and positive number implying that there are situations where you underestimate and overestimate the average value. Also notice that for the absolute value, there is a point where the derivative is undefined since the slope at that point is infinite.

The reason for squaring the residuals is that the sum of the unsquared residuals is always zero (as long as the model has a constant, which it always should). Then minimizing $\sum_{i=1}^{n} e_i$ does not make sense. Also note that a squared residual defines the area of a geometric square with sides e_i. So, the SSE is the total area of all these geometric squares.

The minimization of (6.1.14) with respect to the estimators of the unknown parameters, $\hat{\beta}_0$ and $\hat{\beta}_1$, is a basic calculus exercise. Since there are two unknowns, two equations, called the *normal equations*, result which are solved simultaneously using algebra. The solutions are the intercept and slope estimators as:

$$\hat{\beta}_0 = \bar{Y} - \hat{\beta}_1 \times \bar{X} \tag{6.1.15}$$

```
##
## Generate some data
##
## Use Numpy's linspace to generate 1000 evenly spaced numbers including
##   the end points.
##
e = np.linspace( -5, 5, 1000 )
e_sq = np.square( e )
e_ab = np.abs( e )
##
data = { 'residual':e, 'residual_squared':e_sq, 'residual_absolute':e_ab }
df = pd.DataFrame( data )
df.to_csv( 'c:/temp/residuals.csv' )
##
## Print sums
##
print( 'Sums of:' )
print( f'\tResiduals: {np.sum( e )}' )
print( f'\tSquared residuals: {round( np.sum( e_sq ), 1)}' )
print( f'\tAbsolute value of residuals: {round( np.sum( e_ab ), 1)}' )
##
## Plot data
##
fig, ( ax1, ax2 ) = plt.subplots( 1, 2, figsize = ( 12, 5 ) )
##
ax1 = df.plot( x = 'residual', y = 'residual_squared', legend = False, ax = ax1 )
ax1.set_title( 'Squared Residuals', fontsize = 18 )
ax1.set( xlabel = 'Residuals', ylabel = 'Squared Residuals' )
##
ax2 = df.plot( x = 'residual', y = 'residual_absolute', legend = False, ax = ax2 )
ax2.set_title( 'Absolute Value of Residuals', fontsize = 18 )
ax2.set( xlabel = 'Residuals', ylabel = 'Absolute Value of Residuals' )
circle1 = plt.Circle( ( 0, 0 ), 0.2, color = 'k', fill = False)
ax2.add_patch( circle1 )
ax2.text( 0, 1.5, 'Not\nDifferentiable', horizontalalignment = 'center' )
ax2.arrow( 0, 1.4, 0, -0.70, width = 0.1 );
```

```
Sums of:
        Residuals: 0.0
        Squared residuals: 8350.0
        Absolute value of residuals: 2502.5
```

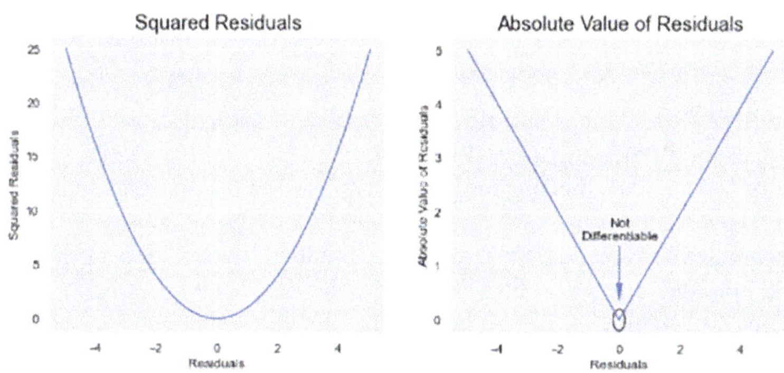

Fig. 6.1 This is a comparison of the squared and absolute value of the residuals which are simulated. I used the Numpy *linspace* function to generate 1000 evenly spaced points between −5 and +5 with the end points included. Notice that the sum of the residuals is 0.0

and

$$\hat{\beta}_1 = \frac{\sum \left(X_i - \bar{X}\right) \times \left(Y_i - \bar{Y}\right)}{\sum \left(X_i - \bar{X}\right)^2}. \qquad (6.1.16)$$

Notice from (6.1.15) that if you do not have an X at all, just a batch of Y data, then $\hat{\beta}_0 = \bar{Y}$.

Anything that is estimated has a variance. This holds for the regression line. The estimated *variance of the regression*, s^2, is

$$s^2 = \frac{SSE}{n-2} \tag{6.1.17}$$

with standard error

$$s = \sqrt{s^2}. \tag{6.1.18}$$

The standard error of $\hat{\beta}_1$ is:

$$s_{\hat{\beta}_1} = \sqrt{\frac{s^2}{SXX}} \tag{6.1.19}$$

$$= \frac{s}{\sqrt{SXX}} \tag{6.1.20}$$

where $SXX = \sum (X_i - \bar{X})^2$. The estimate of the standard error of the intercept estimate, $\hat{\beta}_0$, is:

$$s_{\hat{\beta}_0} = s \times \sqrt{\frac{1}{n} + \frac{(\bar{X})^2}{SXX}}. \tag{6.1.21}$$

Notice from (6.1.21) that if your model does not have an X so all you have is a batch of data for Y, then the second term in the radical does not exist (it is not $0/0$; it just is not there). Then the standard error is $s/\sqrt{n}$ which is the standard error of the mean from basic statistics.

These standard errors are used to calculate t-statistics used for hypothesis testing of the individual parameters. The t-statistic for $\hat{\beta}_1$ is defined as

$$t_{C,\hat{\beta}_1} = \frac{\widehat{\beta_1} - \beta_1}{s_{\widehat{\beta_1}}}$$

$$= \frac{\widehat{\beta_1}}{s_{\widehat{\beta_1}}}$$

under the Null Hypothesis. The "C" in the subscript indicates that this is a calculated value. The Null and Alternative Hypotheses are

$$H_0 : \beta_1 = 0$$

$$H_A : \beta_1 \neq 0$$

although the Alternative Hypothesis could also be $\beta_1 > 0$ or $\beta_1 < 0$ depending on the problem. A *p-value* is associated with the t-statistic and this value is compared to a prestated level of significance, or α value; usually $\alpha = 0.05$. The Null Hypothesis is rejected if p-value < 0.05 and not rejected otherwise. The t-statistic for $\hat{\beta}_0$ is similarly defined.

6.1.3 *The Gauss-Markov Theorem*

Just because you can derive formulas for calculating the parameters using data does not mean those formulas, the estimators, are good formulas useful for applications. In order to be acceptable, they must meet some criteria. Several have been proposed and are well accepted in the statistical area. These are that the estimators must be:

- linear;
- unbiased;
- have the minimum valiance in the class of linear unbiased estimators; and
- be consistent.

Linear means they can be written as a linear function of the variables; unbiased means they give the right answer on average; minimum variance means they have the smallest variance; and consistency means the estimators will equal the true parameter as the sample size becomes large. A very important theorem, called the *Gauss-Markov Theorem*, shows that the *OLS* estimators satisfy these criteria. It is because of this theorem that we have confidence in using them. See Hill et al. (2008) and Greene (2003) for the Gauss-Markov Theorem. See Kmenta (1971) for derivations showing the *OLS* estimators satisfy these criteria.

6.2 Analysis of Variance

A useful step in the analysis of a regression model is the decomposition of the variance of the dependent variable into two parts. This decomposition is called *Analysis of Variance* (*ANOVA*), an important, well-developed part of statistics in general. See Draper and Smith (1966), Neter et al. (1989), and Hocking (1996) for discussions of *ANOVA* and its use in regression analysis.

To motivate the *ANOVA*, note that you can trivially write $Y_i = Y_i$ which you can then rewrite as $Y_i = Y_i + \hat{Y}_i - \hat{Y}_i + \bar{Y} - \bar{Y}$ so nothing changed. Now rearrange terms, square both sides, and sum over all observations to get

$$\sum(Y_i - \bar{Y})^2 = \sum(\hat{Y}_i - \bar{Y})^2 + \sum(Y_i - \hat{Y}_i)^2 + 2\sum(\hat{Y}_i - \bar{Y}) \times (Y_i - \hat{Y}_i). \tag{6.2.1}$$

The last cross-product term equals zero, so you get:

$$\underbrace{\sum(Y_i - \bar{Y})^2}_{SST} = \underbrace{\sum(\hat{Y}_i - \bar{Y})^2}_{SSR} + \underbrace{\sum(Y_i - \hat{Y}_i)^2}_{SSE}. \tag{6.2.2}$$

Recall from basic statistics that the formula for the sample variance of a batch of data, Y, is

$$s_Y^2 = \frac{\sum_{i=1}^{n}(Y_i - \bar{Y})^2}{n - 1}. \tag{6.2.3}$$

The numerator in (6.2.3) is really the variance; the denominator just scales it. This numerator is the *total sum of squares* (SST): $SST = \sum_{i=1}^{n}(Y_i - \bar{Y})^2$. Let SSR be the *regression sum of squares*: $SSR = \sum_{i=1}^{n}(\hat{Y}_i - \bar{Y})^2$. Then you can write:

$$SST = SSR + SSE. \tag{6.2.4}$$

This is a fundamental identity in statistics for the analysis of variance. Each component of it has associated degrees-of-freedom. The SST has $n - 1$ degrees-of-freedom from basic statistics. The SSE has $n - 2$: you lose 1 degree-of-freedom for each parameter, hence the 2. The SSR has 1 degree-of-freedom reflecting the one feature. I summarize the component parts and their degrees-of-freedom in a conventional table, called the *ANOVA Table*, in Table 6.1.

Source of variation	DOF	Sum of squares	Mean squares	F_C
Regression	1	SSR	$MSR = SSR/1$	MSR/MSE
Error	n - 2	SSE	$MSE = s\frac{SSE}{n-2}$	
Total	n - 1	SST	$MST = SST/n-1$	

Table 6.1 This is the general *ANOVA* table structure. The mean squares are just the average or scaled sum of squares. The statistic, F_C, is the calculated F-statistic used to test the fitted model against a subset model. The simplest subset model has only an intercept. I refer to this as the restricted model. Note the sum of the degrees-of-freedom. Their sum is equivalent to the sum of squares summation by (6.2.4)

The F-statistic, F_C, in Table 6.1 is used to test the hypothesis that the linear model is better than a subset model. The simplest subset is one with only a constant term. I refer to this as the *restricted model*. See Weisberg (1980). The estimated model is called an *unrestricted model*. The hypotheses are

$$H_O : \text{Restricted Model is Better; i.e., } \beta_1 = 0 \tag{6.2.5}$$

$$H_A : \text{Unrestricted Model is Better; i.e., } \beta_1 \neq 0. \tag{6.2.6}$$

Notice that H_A is not concerned with whether or not the parameter is > 0 or < 0, but only that it is not 0. This is different from the t-test which can have > 0, < 0, or $\neq 0$.

The F-statistic has a p-value since it is a statistic calculated from data. The decision rule is as before:

- Reject H_0 if p-value < 0.05
- Do not reject otherwise.

A basic statistic that is overworked and abused is the R^2 defined as

$$R^2 = \frac{\sum_{i=1}^{n}(\hat{Y}_i - \bar{Y})^2}{\sum_{i=1}^{n}(Y_i - \bar{Y})^2} \tag{6.2.7}$$

$$= \frac{SSR}{SST}. \tag{6.2.8}$$

This shows the proportion of the variation in the dependent variable explained or accounted for by the model. As a proportion, $0 \leq R^2 \leq 1$. There is an issue with this value that I will explain later. Note that you can also write

$$R^2 = \frac{SSR}{SST} \tag{6.2.9}$$

$$= \frac{SST - SSE}{SST} \tag{6.2.10}$$

$$= 1 - \frac{SSE}{SST}. \tag{6.2.11}$$

Since R^2 is a function of SSR, you should suspect that R^2 and the F-statistic are related since they both have a common factor. In fact, they are related as can be seen from (6.2.12).

$$F = \frac{(n-2) \times SSR}{SSE} \tag{6.2.12}$$

$$= \frac{(n-2) \times {}^{SSR}/_{SST}}{{}^{SSE}/_{SST}} \tag{6.2.13}$$

$$= \frac{(n-2) \times R^2}{1 - R^2}. \tag{6.2.14}$$

Other statistics associated with *OLS* are the *AIC*, *BIC*, Durbin-Watson, Jarque-Bera and many others available to assess the regression model fit and help you interpret results. See any of the references I cited above for background on these. I will discuss *AIC* and *BIC* below and the Durbin-Watson statistic in Chap. 7.

6.3 Case Study

The furniture transactions data will be used to illustrate a simple *OLS* regression. The transactions are sales of living room blinds to local boutique retailers. The manufacture's sales force offers discounts, or none at all, at their discretion to these retailers: a dealer discount (i.e., a reward for prior business with the company), a competitive discount (to meet local market competition), an order size discount (i.e., a volume discount), and a pick-up discount (i.e., an incentive to avoid shipping). The discounts reduce the list price for the blinds to a *pocket price* which is the amount the manufacturer actually receives per blind sold. The objective is to estimate a price elasticity for the effect of pocket price on unit sales. See Paczkowski (2018) for a thorough discussion of price elasticities and their estimation.

6.3.1 Basic OLS *Regression*

The first step is to examine the distribution of unit sales. I show a histogram in Panel (a) of Fig. 6.2. Notice that the distribution is heavily right skewed so there are outliers in the far-right tail. These outliers impact estimations so they must be corrected. A (natural) log transformation is used for this purpose. The log transformation I use, however, has a slight twist: it is the log of 1 plus the unit sales: $\ln(1 + USales)$. The reason for the addition is simple. If any sales are zero, then the log is undefined at that point. That is, $\ln(0) = \infty$. Adding 1 to each observation avoids this issue since $\ln(1) = 0$. I show the distribution of the log transformed unit sales in Panel (b) of Fig. 6.2. The log transform has clearly normalized the skewed distribution. I use the same transformation on the pocket price.

All statistical software packages ask you to follow four steps for estimating a model. How you do this varies, but, nonetheless, you are asked to do all four. These steps are:

1. specify a model;
2. specify or *instantiate* the estimation procedure and the data you will use with it;
3. fit the model; and
4. print summary results.

6.3.2 The Log-Log Model

Since I applied a log transformation to both Y and X for the Case Study, the resulting model is called a *log-log model*. This is written as:

$$\ln(Y) = \beta_0 + \beta_1 \times \ln(X). \tag{6.3.1}$$

```
##
## Unit sales histogram
##
ax = sns.distplot( df.Usales )
ax.set_title( 'Unit Sales Distribution', fontsize = font_title )
ax.set( xlabel = 'Unit Sales', ylabel = 'Proportion' );
```

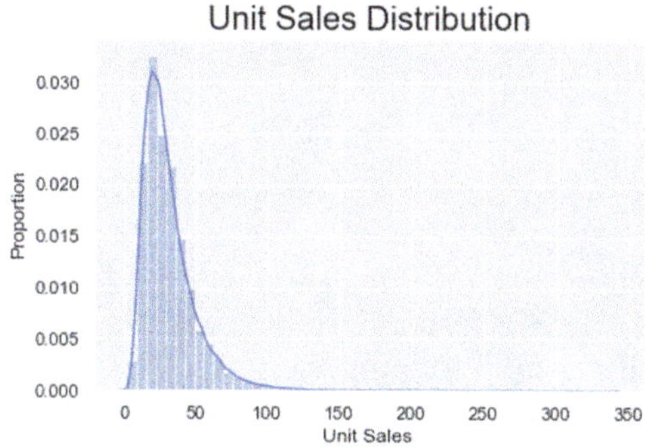

(a) Unit Sales

OLS Regression Results

Dep. Variable:	log_Usales	R-squared:	0.197
Model:	OLS	Adj. R-squared:	0.197
Method:	Least Squares	F-statistic:	2151.
Date:	Mon, 25 May 2020	Prob (F-statistic):	0.00
Time:	15:53:41	Log-Likelihood:	-48263.
No. Observations:	70257	AIC:	9.654e+04
Df Residuals:	70248	BIC:	9.663e+04
Df Model:	8		
Covariance Type:	nonrobust		

	coef	std err	t	P>|t|	[0.025	0.975]
Intercept	6.4614	0.034	190.144	0.000	6.395	6.528
C(Region)[T.Northeast]	0.0026	0.006	0.425	0.671	-0.010	0.015
C(Region)[T.South]	-0.0064	0.006	-1.074	0.283	-0.018	0.005
C(Region)[T.West]	-0.0037	0.005	-0.810	0.418	-0.013	0.005
log_Pprice	-1.7302	0.015	-118.762	0.000	-1.759	-1.702
Ddisc	-0.0188	0.056	-0.335	0.737	-0.129	0.091
Odisc	0.0184	0.127	0.145	0.884	-0.230	0.267
Cdisc	-0.2431	0.091	-2.657	0.008	-0.422	-0.064
Pdisc	-0.1093	0.158	-0.691	0.489	-0.419	0.201

Omnibus:	47.728	Durbin-Watson:	1.992
Prob(Omnibus):	0.000	Jarque-Bera (JB):	47.713
Skew:	0.061	Prob(JB):	4.36e-11
Kurtosis:	2.964	Cond. No.	185.

(b) Log Unit Sales

Fig. 6.2 Panel (**a**) shows the raw data for unit sales of the living room blinds while Panel (**b**) shows the log transformed unit sales. The log transform is $log(1 + Usales)$ to avoid any problems with zero sales. I use the Numpy log function: $log1p$. This function is the natural log by default

Inverting this model gives

$$Y = e^{\beta_0} \times X^{\beta_1}. \tag{6.3.2}$$

Notice that if you take the total differential of (6.3.1), you get

$$\frac{1}{Y} \times dY = \beta_1 \times \frac{1}{Y} \times dY \tag{6.3.3}$$

so,

$$\beta_1 = \frac{X}{Y} \times \frac{dY}{dX} \tag{6.3.4}$$

which is the elasticity of Y with respect to X, η_X^Y.[1]

6.3.3 Model Set-up

I provide the regression set-up with these four steps in Fig. 6.3 Panel (a) so you can see the general structure for doing a regression estimation. The regression summary from step #4 is in Fig. 6.3 Panel(b). The first step, specify a model, is done with a *Patsy* formula character string. A Patsy formula is a succinct and efficient way to write a formula for use in a wide variety of modeling situations. The formula uses a "$\sim$" to separate the left-hand side from the right-hand side of a model and a "+" to add features to the right-hand side. A "−" sign is used to remove columns from the right-hand side (e.g., remove or omit the constant term which is always included by default). A Patsy formula is succinct because it omits obvious pieces of a model statement. For example, it is understood that the unknown parameters (i.e., β_0 and β_1) are part of the model and, in fact, are the targets for estimation, so they do not have to be specified. Only the dependent and independent variables are needed. To omit the constant, use $Y \sim -1 + X$; to include it by default just use $Y \sim X$. Patsy notation is efficient because complex combinations of variables (i.e., interactions) are simply expressed. I will not include interactions so only the simplest Patsy statement will be used.

Model instantiation means you have to specify the estimation procedure, the formula, and the DataFrame to create an instance of the model. In this example, the estimation procedure is *OLS* so the *ols* function in the *Statsmodels* package is accessed using dot notation. The Patsy formula and the DataFrame name are arguments to the function. The fully instantiated model is stored in a variable object, which I call *mod* in Fig. 6.3, Panel (a). This object just holds the model specification; it does not do anything. The model is estimated or fitted using the *fit* method associated with the *mod* object. The estimation results are stored in the variable object *reg01* in Fig. 6.3, Panel (a). My recommended naming convention is to use the procedure name (e.g., "reg") followed by a sequence number which increments for new models: *reg01*, *reg02*, etc. I could have combined the model instantiation and fit using one statement by chaining: *smf.ols(formula, data = df).fit()*.

[1] If a log-log model is not used, then the elasticity is evaluated at the means as $\eta_X^Y = \bar{X}/\bar{Y} \times \beta_1$. See Paczkowski (2018) for a detailed discussion.

6.3.4 Estimation Summary

The estimation results are displayed in a conventional regression model layout in Fig. 6.3, Panel (b) using the *summary()* method for the *reg01* object. There are three sections to this layout:

1. a housekeeping section in the top left with estimation date and time, estimation type, etc.;
2. a diagnostic section in the top right with other supporting information in the bottom; and
3. the parameter estimates in the middle section.

Notice from the diagnostics section that the R^2 is 0.197 so only about 20% of the variation in log sales is explained by log pocket price; this is not very good. There must be other variables that account for sales, perhaps the discounts and the marketing region of the local retailers. I will explore this in the next section. See Paczkowski (2021b) for details.

I noted above that there is an R^2 version of the F-Statistic. See (6.2.12). I summarize the calculations in Fig. 6.5 to verify this relationship. Notice that the values are as they should.

6.3.5 ANOVA *for Basic Regression*

The F-Statistic in this section has a value of 17192.714 as shown in the regression summary, Fig. 6.3, Panel (b), and in the *ANOVA* table, Fig. 6.4. This tests the estimated unrestricted model against the restricted model with no explanatory variable, which is just $Y_i = \beta_0 + \epsilon_i$. The Null Hypothesis asserts that the restricted model is better. The p-value for this F-Statistic is 0.0 which indicates that the Null Hypothesis is rejected. See Weisberg (1980) for the use of the F-test for comparing these two models. I show the *ANOVA* table in Fig. 6.4.

6.3.6 Elasticities

From the parameter estimate section, you can see that the coefficient for log pocket price is -1.7249, which indicates that (log) sales is highly elastic with a p-value equal to 0.0. The estimated coefficient is the price elasticity of demand and indicates that sales of living room blinds are highly price elastic: a 1% decrease in price results in a 1.7% increase in sales. This should be expected since this product has many good substitutes such as other manufactures' living room blinds as well as drapes, shades, and, of course, nothing at all for windows.

```
##
## OLS
##
## There are four steps for estimatng a model:
##
##   1. define a formula (i.e., the specific model to estimate)
##   2. instantiate the model (i.e., specify it)
##   3. fit the model
##   4. summarize the fitted model
##
## ===> Step 1: Define a formula <===
##
## The formula uses a "~" to separate the left-hand side from the right-hand side
## of a model and a "+" to add columns to the right-hand side.  A "-" sign (not
## used here) can be used to remove columns from the right-hand side (e.g.,
## remove or omit the constant term which is always included by default).
##
formula = 'log_Usales ~ log_Pprice'
##
## ===> Step 2: Instantiate the OLS model <===
##
mod = smf.ols( formula, data = df )
##
## ===> Step 3: Fit the instantiated model <===
##        Recommendation: number your fitted models
##
reg01 = mod.fit()
##
## ===> Step 4: Summarize the fitted model <===
##
reg01.summary()
```

(a)

OLS Regression Results

Dep. Variable:	log_Usales	**R-squared:**	0.197
Model:	OLS	**Adj. R-squared:**	0.197
Method:	Least Squares	**F-statistic:**	1.719e+04
Date:	Mon, 25 May 2020	**Prob (F-statistic):**	0.00
Time:	12:17:21	**Log-Likelihood:**	-48289.
No. Observations:	70270	**AIC:**	9.658e+04
Df Residuals:	70268	**BIC:**	9.660e+04
Df Model:	1		
Covariance Type:	nonrobust		

	coef	**std err**	**t**	**P>|t|**	**[0.025**	**0.975]**
Intercept	6.4267	0.024	270.400	0.000	6.380	6.473
log_Pprice	-1.7249	0.013	-131.121	0.000	-1.751	-1.699

Omnibus:	48.641	**Durbin-Watson:**	1.990
Prob(Omnibus):	0.000	**Jarque-Bera (JB):**	48.654
Skew:	0.062	**Prob(JB):**	2.72e-11
Kurtosis:	2.966	**Cond. No.**	30.9

(b)

Fig. 6.3 A single variable regression is shown here. (**a**) Regression setup. (**b**) Regression results

```
##
## ANOVA table
##
aov = anova_lm( reg01 )
aov.style.set_caption( 'ANOVA Table' ).set_table_styles( tbl_styles )
```

ANOVA Table

	df	sum_sq	mean_sq	F	PR(>F)
log_Pprice	1.000000	3978.929871	3978.929871	17192.714901	0.000000
Residual	70268.000000	16262.204417	0.231431	nan	nan

Fig. 6.4 *ANOVA* table for the unit sales regression

```
##
## R-Squared version of F-Statistic
##
rsquared = reg01.rsquared
dof = reg01.df_resid
##
F = (dof * rsquared)/(1 - rsquared )
##
print( f'Calculated R-Squared version of F-statistic: {F}' )
print( f'Reported F-Statistic: {reg01.fvalue}' )
```

```
Calculated R-Squared version of F-statistic: 17192.714900897543
Reported F-Statistic: 17192.714900897547
```

Fig. 6.5 There calculations verify the relationship between the R^2 and the F-Statistic. I retrieved the needed values from the *reg01* object I created for the regression in Fig. 6.3

You might want to know the impact of a price change on *KPMs*, especially revenue. As shown in Paczkowski (2018), if η_P^Q is the price elasticity for unit sales, then the price elasticity for revenue is $\eta_P^{TR} = 1 + \eta_P^Q$. For this problem, $\eta_P^Q = -1.7$, so $\eta_P^{TR} = -0.7$. If price is decreased 1%, revenue will increase 0.7%. See Paczkowski (2018) for a discussion of price and revenue elasticities and the use of a log model to estimate and interpret them.

6.4 Basic Multiple Regression

The basic *OLS* model can be extended to include multiple independent variables ($p > 1$) and specialized independent variables. The specialized independent variables are lagged variables (lagged dependent and/or independent) to capture dynamic time patterns; time trend to capture an underlying time dynamic; and dummy or one-hot encoded concept variables. I introduced dummy encoding in Chap. 5.

The use of $p > 1$ independent variables requires a modification of how the estimates are expressed. Matrix algebra notation is used, although the results are the same, regardless of the form of expression. See Lay (2012) and Strang (2006) for a review of matrix algebra. For its use in multiple regression, see any econometrics textbook such as Goldberger (1964) or Greene (2003). Using matrix notation, the dependent variable is expressed as a $n \times 1$ vector. Similarly for the disturbance term, ϵ. The p independent variables plus the constant are collected into a $n \times (p + 1)$ matrix, $\mathbf{X}$. The linear model is now written in matrix notation as

$$\mathbf{Y} = \mathbf{X}\boldsymbol{\beta} + \boldsymbol{\epsilon}. \tag{6.4.1}$$

The problem is exactly the same as before despite this notation change: you minimize a loss function, *SSE*, and solve the resulting $p+1$ normal equations to get (in matrix notation):

$$\hat{\boldsymbol{\beta}} = \left(\mathbf{X}^{\top}\mathbf{X}\right)^{-1}\mathbf{X}^{\top}\mathbf{Y}. \tag{6.4.2}$$

The vector $\hat{\boldsymbol{\beta}}$ is $(p+1) \times 1$ and includes the constant term. The term $\left(\mathbf{X}^{\top}\mathbf{X}\right)^{-1}$ is the inverse of the sum of squares and cross-products matrix formed from the $\mathbf{X}$ matrix. The predicted value of $\mathbf{Y}$ is

$$\hat{Y} = \mathbf{X}\hat{\boldsymbol{\beta}}. \tag{6.4.3}$$

This is the *SRL* but as a hypersurface rather than a straight line. The variance of $\hat{\boldsymbol{\beta}}$, which is needed for hypothesis testing, is $\sigma^2_{\hat{\boldsymbol{\beta}}} = \sigma^2 \left(\mathbf{X}^{\top}\mathbf{X}\right)^{-1}$. Since multiple regression is just a generalization of what I covered earlier, the Gauss-Markov Theorem still holds.

6.4.1 ANOVA *for Multiple Regression*

The *ANOVA* table is modified to handle the multiple independent variables but the interpretation is the same. I provide a modified table in Table 6.2. You can see that the structure is unchanged.

Source of Variation	DOF	Sum of Squares	Mean Squares	F_C
Regression	p	SSR	$MSR = SSR/p$	MSR/MSE
Error	n - p - 1	SSE	$MSE = s\frac{SSE}{n-p-1}$	
Total	n - 1	SST	$MST = SST/n-1$	

Table 6.2 This is the modified *ANOVA* table structure when there are $p > 1$ independent variables. Notice the change in the degrees-of-freedom, but that the degrees-of-freedom for the dependent variable is unchanged. The p degrees-of-freedom for the *Regression* source accounts for the p independent variables which are also reflected in the *Error* source

The F-test still tests the restricted model against the unrestricted one, but the unrestricted model has $p > 1$ explanatory variables unlike before when it had $p = 1$. The hypotheses are now

$$H_0 : \beta_1 = \beta_2 = \ldots = \beta_p = 0 \tag{6.4.4}$$

$$H_A : \text{At Least 1 } \beta_i \text{ Not Zero.} \tag{6.4.5}$$

I compare the F-test for the simple and multiple regression cases in Table 6.3. Notice that in the simple model, the Null Hypothesis for the F-test is the same as for the t-test. See Neter et al. (1989, p. 97) and Draper and Smith (1966, p. 25).

Model	Null Hypothesis	Alternative Hypothesis
Simple OLS ($p = 1$)	$\beta_1 = 0$	$\beta_1 \neq 0$
Multiple OLS ($p > 1$)	$\beta_1 = \beta_2 = \ldots = \beta_p = 0$	At least 1 $\beta_i \neq 0$

Table 6.3 The F-test for the multiple regression case is compared for the simple and multiple regression cases

The F-statistic in the multiple regression case is defined as

$$F_C = \frac{(SSR_U - SSR_R)/p'}{SSE_U/(n-p-1)} \sim F_{p',n-p-1} \tag{6.4.6}$$

where SSR_U and SSE_U are from the unrestricted model and SSR_R is from the restricted model. The degrees-of-freedom for the numerator are the difference between those for the unrestricted and restricted models. So, if $p_U + 1$ is the number of variables in the unrestricted model plus a constant and $p_R + 1$ is the number for the restricted model plus a constant, then the difference is just $p' = p_U - p_R > 0$. Note that if $p_U + 1 = 2$ for a simple one-variable model and $p_R = 1$ for a constant-only model, then the difference is $p' = 1$ which is the degrees-of-freedom for SSR in Table 6.1.

The definition of the R^2 is unchanged, but there is an adjustment you have to make. As you add more explanatory variables to your model, the SSR automatically increases, even if the added variables have little or no explanatory power. See Neter et al. (1989). Consequently, R^2 will automatically increase as you add more variables. As a penalty for this inflation, an *adjusted-*R^2 (symbolized as $\bar{R}^2$) is used in a multiple regression context. This is defined as

$$\bar{R}^2 = 1 - \left[\frac{n-1}{n-p-1}\right] \times \left(1 - R^2\right) \tag{6.4.7}$$

$$= 1 - \left[\frac{n-1}{n-p-1}\right] \times \left(\frac{SSE}{SST}\right) \tag{6.4.8}$$

$$= 1 - \frac{SSE/_{n-p-1}}{SST/_{n-1}} \tag{6.4.9}$$

$$= 1 - \frac{s^2}{s_Y^2} \tag{6.4.10}$$

The adjusted-R^2 has certain properties:

1. $\bar{R}^2 \leq R^2$
2. Unlike R^2 which will only increase as more variables are added, $\bar{R}^2$ can decline if an added independent variable has no or little explanatory power. The reason is the increase in p decreases $n - p - 1$ and so $SSE/n-p-1$ increases.

6.4.2 Alternative Measures of Fit: AIC and BIC

A modeling objective is often just a good fit for a linear model to the data. This translates to having a large $\bar{R}^2$ (or R^2 for a 1-variable model). A problem with the $\bar{R}^2$ is that it is only applicable for comparing nested models. A *nested model* is a subset model of a larger linear model in which the dependent variable is the same and all the independent variables in a subset are also in the larger model. So, the model $Y = \beta_0 + \beta_1 \times X_1 + \epsilon$ is nested under $Y = \beta_0 + \beta_1 \times X_1 + \beta_2 \times X_2 + \epsilon$. You can use the adjusted R^2 to compare these two models. You cannot, however, compare $Y = \beta_0 + \beta_1 \times X_1 + \epsilon$ and $\ln(Y) = \beta_0 + \beta_1 \times X_1 + \beta_2 \times X_2 + \epsilon$ because the first is not nested under the second; the dependent variables are different. Consequently, the SSTs are different so a comparison is not possible. See Kennedy (2003, p. 73 and p.74) for discussions about R^2. You can pick the "best" model in your portfolio using *Akaike's Information Criterion* (*AIC*) defined as

$$AIC = 2 \times (p + 1) - 2 \times \ln(L) \tag{6.4.11}$$

for a model with a constant term where $\ln(L)$ is the *log-likelihood* value. For a model without a constant, the first term is $2 \times p$.

The *AIC* measures the "badness-of-fit" of a model rather that the "goodness-of-fit" as measured by the adjusted-R^2. This means it measures the amount of variation in the dependent variable left unexplained by the model; it measures the amount of information left unaccounted for by the variables. The adjusted-R^2 measures the amount of variation accounted by them. The goal for using *AIC* is to select the model with the smallest *AIC*. An alternative to *AIC* is the *Bayesian Information Criterion* (*BIC*), which is also a function of the log-likelihood. This is defined as

$$BIC = \ln(n) \times (p + 1) - 2 \times \ln(L) \tag{6.4.12}$$

for a model with a constant. For a model without a constant, the first term is $\ln(n) \times p$.

The likelihood, L, is sometimes loosely interpreted as the probability of occurrence of an event. This is incorrect. A *probability density function*, $f(X;\theta)$, is a function of the random variable X with fixed parameters, θ, such as

$$f(X_i;\theta) = \frac{1}{\sqrt{2\pi\sigma^2}} e^{-\frac{(X_i-\mu)^2}{2\sigma^2}} \tag{6.4.13}$$

where μ and σ^2 are given parameters. Therefore, $\theta = [\mu, \sigma^2]$. In this example, X_i is a random variable. For a discrete distribution this is called a *probability mass function* since there is mass (i.e., area) under the curve for a fixed value of X. A density function is not a probability because it has no mass under the curve for a fixed value of X. The area under the curve for X in the range a to b (i.e., $a < X < b$) is a probability. But, as this range narrows so that eventually $a = b$, then the associated area goes to zero, so the probability goes to zero. We typically refer to the densities as probabilities, but they are not. These densities could be greater than 1.0 if the mass of the distribution is concentrated in a narrow range.

The likelihood function gives the likelihood, not the probability, that the parameter is θ given the data, X_i. The parameter is now random and X_i is fixed or non-random. The "given" is not actually correct, but will suffice. The likelihood function for an observation, i, is $L_i(\theta; X_i)$ which, mathematically, equals $f(X_i;\theta)$. So, the functions are the same since there is only one curve, but the interpretations are different. This is the crucial piece: same function, different interpretation. See Shao (2003) on the distinction between a probability density function and a likelihood function.

Formally, the likelihood function for a random sample on n independent observations is the product of n densities, not probabilities, by the product rule for independent random variables in elementary probability theory. The log-likelihood is the sum of the logs of the individual densities:

$$\ln(L(\theta; X)) = \sum_{i=1}^{n} \ln(L_i(\theta; X_i)). \tag{6.4.14}$$

It is better to use natural logs since maximizing the likelihood function requires taking the derivative of n products, while for the natural log it requires taking the derivative of n summation terms. This is easier to handle and gives the same result since the log is a monotonically increasing transformation of its argument. A function f is *monotonically increasing* if for all values of X and Y such that $X \leq Y$, then $f(X) \leq f(Y)$. Similarly, for monotonically decreasing. This monotonicity holds for the log function. This is what enables you to use the log transformation and get the same answer.

Since $AIC = 2 \times (p+1) - 2 \times ln(L)$, then $AIC < 0$ if $ln(L) > p + 1$. This requires $ln(L) > 1$. If $L = e$ $(= 2.71828)$, then $ln(L) = 1$. We can go further. Suppose the normal density is used with mean 0 and standard deviation 1. Then you get Panel (a) of Table 6.4. Suppose a normal density is used with mean 0 and standard deviation 1/100. Then you get Panel (b) in Table 6.4. Basically, the smaller standard deviation compresses the density function since the standard deviation is the distance from the mean to the inflection point of the curve (equidistant on both sides of the mean). The smaller the standard deviation, the smaller the distance and the more compressed the density curve. If the standard deviation is zero, then the density curve is degenerate at the mean. This is the case if all the values are the same. Since the density curve is compressed, the area under the curve, which must always equal 1.0, must go somewhere, and the only place for it to go is up. This implies that the curve can peak at a value greater than 1.0. If the standard deviation is very large, then the density curve is flatter around the mean and the height is less than 1.0. When (natural) logs are used, the log-density for the former is positive and it is negative for the latter. This is all evident in Table 6.4.

(a) Standard deviation 1.0			(b) Standard deviation 1/100		
X	Density	log-Density	X	Density	log-Density
−0.03	0.3987628	−0.9193885	−0.03	0.443184	−0.813768
−0.01	0.3989223	−0.9189886	−0.01	24.19707	3.186232
0.00	0.3989423	−0.9189385	0.00	39.89423	3.686232
+0.01	0.3989223	−0.9189886	+0.01	24.19707	3.186232
+0.03	0.3987628	−0.9193885	+0.03	0.443184	−0.813768

Table 6.4 Density vs log-density values for the normal density with mean 0 and standard deviation 1 vs standard deviation 1/100. Note that the values of the log-Density are negative around the mean 0 in the left panel but positive in the right panel

The implication of these positive and negative log-density possibilities is that the *AIC* can be either positive or negative. It is not guaranteed to be positive as is R^2. The sign depends on the standard deviation of the data; that is, it depends on the peakness or flatness of the likelihood function. The same holds for the *BIC*. The further implication is that a negative *AIC* says nothing about the badness-of-fit of the model because it has nothing to do with the fit of the model. The only aspect of one *AIC* value for one model that matters is its magnitude relative to other *AIC* values for other models. The smaller value indicates the better model. The same holds for *BIC*.

6.5 Case Study: Expanded Analysis

Let me expand the living room blinds Case Study to include several explanatory variables. Let the model now include the discounts and marketing region of each retailer. The marketing region variable is a discrete categorical concept with the four categories: Midwest, Northeast, South, and West. This variable has to be dummified which I do using the $C(\cdot)$ function. This takes the categorical variable as an argument, scans it, and creates dummy variables for the levels. These are called *Treatments* and are represented by a *T*. There are other ways to create indicator variables with Treatments being the default. Another one, called *effects coding*, is popular in market research and statistical design of experiments. This is called *Sum* in the $C(\cdot)$. See Paczkowski (2018) for a discussion and comparison of dummy coding and effects coding.

Remember that one level is selected as a base and a dummy is not created for the base to avoid the dummy variable trap.[2] When the $C(\cdot)$ function scans the levels, it puts the unique levels in alphanumeric order and drops the first as the base. The Midwest is first so it does not have a dummy variable. The dummy variable is represented by *C(Region)* followed by the level treatment designation and the level to which the dummy applies. So, the Northeast dummy variable is *C(Region)[T.Northeast]*.

I show the regression set-up in Fig. 6.6, Panel (a) and the *ANOVA* in Fig. 6.7. There may be some concern about a relationship between the pocket price and the discounts since the pocket price is a function of the discounts. The relationship is checked by examining the correlation matrix for the pocket price and the four discounts. I created a correlation matrix using the Pandas *corr* method and show it in Fig. 6.8. You can see that the correlations are all small.

Referring to Fig. 6.6, Panel (b), note the R^2 is virtually unchanged from the one in Fig. 6.3 Panel (b) and that the adjusted-R^2 is the same to three decimal places. So, as before, only 20% of the variation in (log) sales is accounted for in this expanded model. The F-statistic is also highly significant so this expanded model is better than the restricted model with only a constant term. Next, notice that the estimated coefficient for the (log) price is highly significant as before so conclusions about price (and elasticities) are unchanged. Notice, finally, that the discounts are all insignificant although the competitive discount (*Cdisc*) is marginally insignificant.

Now look at the marketing regions. How are the region coefficients interpreted? Each one shows the deviation from the base, which is Midwest in this case. The Midwest coefficient is the intercept. The Northeast coefficient shows the effect of moving from the Midwest to the Northeast so the total impact of the Northeast is the intercept plus the Northeast coefficient. Since the intercept is 6.4614 and the Northeast is 0.0026, then the Northeast is 6.4640. Similarly for the other two regions. The estimated dummy coefficients are interpreted as effects or shifts in the intercept due to the inclusion of that dummy variable. So, 6.4640 is the intercept if you looked at the Northeast region.

The dummy coefficients are all insignificant indicating that there is no regional effect on sales. This is difficult to understand. Nonetheless, I will do an F-test to test the significance of region even though the regression results indicate that Region as a concept is insignificant. I show the results in Fig. 6.9. A hypothesis statement is specified as a character string. Notice how this is written. The names of the variables in the regression output are used exactly as they appear there since this is

[2] In the case of a model with a constant term.

```
##
## OLS
##
## There are four steps for estimatng a model:
##
##   1. define a formula (i.e., the specific model to estimate)
##   2. instantiate the model (i.e., specify it)
##   3. fit the model
##   4. summarize the fitted model
##
## ===> Step 1: Define a formula <===
##
## The formula uses a "~" to separate the left-hand side from the right-hand side
## of a model and a "+" to add columns to the right-hand side.  A "-" sign (not
## used here) can be used to remove columns from the right-hand side (e.g.,
## remove or omit the constant term which is always included by default).
##
formula = 'log_Usales ~ log_Pprice + Ddisc + Odisc + Cdisc + Pdisc + C( Region )'
##
## Since Region is categorical, you must create dummies for the regions.  You
## do this using 'C( Region )' to indicate that Region is categorical.
##
## ===> Step 2: Instantiate the OLS model <===
##
mod = smf.ols( formula, data = df )
##
## ===> Step 3: Fit the instantiated model <===
##      Recommendation: number your fitted models
##
reg02 = mod.fit()
##
## ===> Step 4: Summarize the fitted model <===
##
reg02.summary()
```

(a)

OLS Regression Results

Dep. Variable:	log_Usales	R-squared:	0.197
Model:	OLS	Adj. R-squared:	0.197
Method:	Least Squares	F-statistic:	2151.
Date:	Mon, 25 May 2020	Prob (F-statistic):	0.00
Time:	15:53:41	Log-Likelihood:	-48263.
No. Observations:	70257	AIC:	9.654e+04
Df Residuals:	70248	BIC:	9.663e+04
Df Model:	8		
Covariance Type:	nonrobust		

	coef	std err	t	P>|t|	[0.025	0.975]
Intercept	6.4614	0.034	190.144	0.000	6.395	6.528
C(Region)[T.Northeast]	0.0026	0.006	0.425	0.671	-0.010	0.015
C(Region)[T.South]	-0.0064	0.006	-1.074	0.283	-0.018	0.005
C(Region)[T.West]	-0.0037	0.005	-0.810	0.418	-0.013	0.005
log_Pprice	-1.7302	0.015	-118.762	0.000	-1.759	-1.702
Ddisc	-0.0188	0.056	-0.335	0.737	-0.129	0.091
Odisc	0.0184	0.127	0.145	0.884	-0.230	0.267
Cdisc	-0.2431	0.091	-2.657	0.008	-0.422	-0.064
Pdisc	-0.1093	0.158	-0.691	0.489	-0.419	0.201

Omnibus:	47.728	Durbin-Watson:	1.992
Prob(Omnibus):	0.000	Jarque-Bera (JB):	47.713
Skew:	0.061	Prob(JB):	4.36e-11
Kurtosis:	2.964	Cond. No.	185.

(b)

Fig. 6.6 A multiple variable regression is shown here. (**a**) Regression setup. (**b**) Regression results

```
##
## ANOVA table
##
aov = anova_lm( reg02 )
aov.style.set_caption( 'ANOVA Table' ).\
    set_table_styles( tbl_styles )
```

ANOVA Table

	df	sum_sq	mean_sq	F	PR(>F)
C(Region)	3.000000	200.825208	66.941736	289.358104	0.000000
log_Pprice	1.000000	3777.996081	3777.996081	16330.526331	0.000000
Ddisc	1.000000	0.006224	0.006224	0.026905	0.869710
Odisc	1.000000	0.010535	0.010535	0.045537	0.831021
Cdisc	1.000000	1.616803	1.616803	6.988690	0.008204
Pdisc	1.000000	0.110589	0.110589	0.478024	0.489321
Residual	70248.000000	16251.568584	0.231346	nan	nan

Fig. 6.7 ANOVA table for the unit sales multiple regression model

what is stored in the regression output. The three expressions refer to the estimated coefficients and collectively each is assumed to be equal to zero. So, the expression is NE coefficient (i.e., effect) is zero AND the South coefficient (i.e., effect) is zero

```
##
## Check correlation between pocket price and discounts
##
cols = [ 'Pprice', 'Ddisc', 'Cdisc', 'Odisc', 'Pdisc' ]
df[ cols ].corr().style.set_caption( 'Correlation Matrix' ).\
    set_table_styles( tbl_styles )
```

Correlation Matrix

	Pprice	Ddisc	Cdisc	Odisc	Pdisc
Pprice	1.000000	-0.356307	-0.173863	-0.126673	-0.098220
Ddisc	-0.356307	1.000000	-0.000919	0.001962	-0.003748
Cdisc	-0.173863	-0.000919	1.000000	0.002217	0.000396
Odisc	-0.126673	0.001962	0.002217	1.000000	0.002888
Pdisc	-0.098220	-0.003748	0.000396	0.002888	1.000000

Fig. 6.8 Correlation matrix showing very little correlation

AND the West coefficient (i.e., effect) is zero. The Midwest coefficient (i.e., effect) is zero by implication. The *f_test* method associated with the regression model is then called with an argument that is the hypothesis character string. The results confirm that the region concept has no effect on (log) sales since the p-value > 0.05.

```
##
## Specify the joint Null hypothesis of no region effect.
##
hypothesis = ' ( C(Region)[T.Northeast] = 0, C(Region)[T.South] = 0, \
C(Region)[T.West] = 0 ) '
print( 'Null Hypothesis:\n\t{}'.format( hypothesis ) )
print( '\nAlternative Hypothesis:\n\tAt least one is not zero')
##
## Run an F-test
##
f_test = reg01.f_test( hypothesis )
##
## Retrieve the p-value
##
pval = round( float( f_test.pvalue ), 2 )
##
## Print results
##
print( '\np-value for F-Test: {}'.format( pval ) )
if pval < 0.05:
    print( '\nSignificant so reject H0' )
else:
    print( '\nInsignificant so do not reject H0' )
```

```
Null Hypothesis:
	( C(Region)[T.Northeast] = 0, C(Region)[T.South] = 0, C(Region)[T.West] = 0 )

Alternative Hypothesis:
	At least one is not zero

p-value for F-Test: 0.51

Insignificant so do not reject H0
```

Fig. 6.9 F-test showing no region effect

Why are these results so bad? One possibility is the data themselves. These are a combination of time series and cross-sectional data, or what is called a *panel data set*. Recall my discussion of a Data Cube in Chap. 1. This type of data is hard to analyze because there are two dynamics at work at once: time and space. I will discuss how to handle this data structure in Part III. Another possibility could be the use of non-logged discounts. I did not analyze their distributions, but if the sales and price are logged, then maybe the discounts should also be logged. Regarding the regions, they might have to be interacted with the price and discount terms to capture a richer interplay of the variables. Clearly, this now becomes complex. Incidentally, the regions may also have an effect via other factors I did not consider. The retail buyers are located or *nested* within the regions and each region has its own characteristics. Those characteristics may affect how the retailers' customers shop for and buy living room blinds which would impact the retailers' purchases. A *multilevel model* might be better. See Gelman and Hill (2007), Kreft and de Leeuw (1998), Luke (2004), Ray and Ray (2008), and Snijders and Bosker (2012).

6.6 Model Portfolio

A convenient way to summarize several models is to create a *model portfolio*. This is a somewhat standard table used in many academic publications. You have to import a *statsmodels* package as *from statsmodels.iolib.summary2 import summary_col* and define what summary measures you want displayed in the table. You can use the definitions I show in Fig. 6.10. Define both in your Best Practices section. Then you produce the portfolio as I show in Fig. 6.11 for the two regression models from this chapter.

```
##
## Format codes
##
## Regression portfolio
##
info_dict = { 'n':lambda x: "{0:d}".format( int( x.nobs ) ),
              'R2':lambda x: "{:0.3f}".format( x.rsquared ),
              'R2Adjust':lambda x: "{:0.3f}".format( x.rsquared_adj ),
              'AIC':lambda x: "{:0.3f}".format( x.aic )
             }
```

Fig. 6.10 You define the statistics to display in a portfolio using a setup like this

```
##
## Create a portfolio summary table
## This assumes you loaded:
##     from statsmodels.iolib.summary2 import summary_col
## in Best Practices.
## info_dict defines the table contents; see format Best Practices
##
model_names = [ 'Reg01', 'Reg02' ]
##
portfolio = summary_col( [ reg01, reg02 ],
            stars = True,
            model_names = model_names,
            info_dict = info_dict
)
print( portfolio )
```

```
============================================
                        Reg01      Reg02
--------------------------------------------
C(Region)[T.Northeast]              0.0026
                                   (0.0062)
C(Region)[T.South]                 -0.0064
                                   (0.0059)
C(Region)[T.West]                  -0.0037
                                   (0.0046)
Cdisc                              -0.2431***
                                   (0.0915)
Ddisc                              -0.0188
                                   (0.0561)
Intercept               6.4267***   6.4614***
                       (0.0238)    (0.0340)
Odisc                               0.0184
                                   (0.1268)
Pdisc                              -0.1093
                                   (0.1581)
R-squared               0.1966      0.1967
                        0.1966      0.1967
log_Pprice             -1.7249*** -1.7302***
                       (0.0132)    (0.0146)
n                       70270       70257
R2                      0.197       0.197
R2Adjust                0.197       0.197
AIC                     96581.385  96544.553
============================================
Standard errors in parentheses.
* p<.1, ** p<.05, ***p<.01
```

Fig. 6.11 This is the portfolio summary of the two regression models from this chapter

6.7 Predictive Analysis: Introduction

Recall from Chap. 1 that there is a cost to approximating the result of an action. This cost declines as you gather and process more Rich Information. That information is not about what did happen, but about what will happen under different circumstances, some of which you have control over and others that are beyond your control. Regardless of the nature of your control, you have to predict the outcome of your decision based on your Rich Information. In fact, part of that Rich Information for decision making is the predictions themselves since you never have just one prediction, but a series of them, each based on a different perspective of your other Rich Information. But this issue of predicting the likely outcome of an action begs three questions:

1. What is a prediction?
2. How do you develop one?
3. How do you assess the quality of a prediction?

I will lay the groundwork the first two questions in this chapter for answering. My answer to the first one will dispel a confusion most people have between predicting and forecasting. These two are often taken to be the same and so are treated as synonyms; but they are different in some regards. The second question is not straightforward to answer, and in fact requires more background and space than I can allocate in this chapter if it is to be done correctly. This is beyond the scope of this chapter which is to just lay the foundations for basic linear modeling in *BDA*. A detailed answer involves splitting a data set into two parts: a *training data set* and a *testing data set*. I will address it more fully in Chap. 10 after I develop more background on these two data sets in Chap. 9. Finally, the third question also requires more background that is also beyond the scope of this chapter. The answer will also be developed in Chap. 10.

6.7.1 Predicting vs. Forecasting

Let me compare and contrast prediction and forecasting. The confusion between the two centers on their similar use for producing a number for an unknown case or situation. I sometimes refer to this as filling in a hole in our knowledge. The issue is the hole. It could refer to an unknown case regardless of time or it could refer to an unknown case in a future time period. The distinction is critical. Forecasting is concerned with producing a number for an event or measure in a future time period. Predicting is concerned with producing a number for an unknown case regardless of time. You forecast sales for 2022 given historical data but you predict the impact on sales if you lower or raise your price, holding fixed other sales key drivers. Predicting encompasses forecasting, but not the other way around. It is comparable to saying that all thumbs are fingers but not all fingers are thumbs. All forecasts are predictions but not all predictions are forecasts.

BDA is usually concerned with predicting, although forecasting is certainly done. The skill sets for forecasting, however, are different because of the complexities of working with times series data. This sometimes means constructing the time series itself by collapsing the Data Cube on the spatial dimension. I will review issues with times series data, including collapsing the Data Cube, in Chap. 7.

6.7.2 Developing a Prediction

I will discuss prediction based on the *OLS* model developed in this chapter. There are two ways to develop a prediction once an *OLS* model has been estimated. One is to specify a *scenario* consisting of one set of specific values for the independent variables and then using the estimated model to calculate an outcome for it. This scenario approach is typically used in *BDA* and business decision making to develop a most likely view to reduce the Cost of Approximation.

As an example of a scenario, consider a predictive model for the sales of living room blinds as a function of price with an effect for marketing regions and values for discounts. A scenario could be:

- Price: $2.50
- Dealer Discount: 3%
- Order Size Discount: 3%
- Competitive Discount: 3%
- Pickup Discount: 3%
- Region: West

This would examine the effect in the Western region of setting the price at $2.50 and all discounts at a flat 3%. A number of different scenarios such as this could be tried and one selected as the "best" for pricing development.

It is not uncommon to develop several scenarios with a concomitant prediction for each. Quite often, one scenario is viewed as the most likely case (and is labeled *ML*) and the others are either optimistic or pessimistic cases (labeled *SO* and *SP*, respectively). The *ML* case could be specified by the client or executive management as what they believe is likely to occur. They could, in fact, have several most likely cases (labeled *ML01*, *ML02*, etc.) as their thoughts and opinions change. See Georgantzas and Acar (1995) for scenario analysis in policy decisions when uncertainty is an issue.

The second way to predict involves using the estimated model and a series of values for the independent variables so that the model will predict a separate value for each of the input values. This is actually different from a scenario analysis because it allows you to test the model's performance under a number of conditions, the goal being to assess the quality of the model. *Prediction Error Analysis* (*PEA*) is a vital part of this approach. This involves setting standards and determining whether or not they are met. The model can be used with some confidence in the scenario analysis I described above if the standards are met; it would be re-estimated otherwise.

The two approaches are not independent but work synergistically to "fill the hole." In fact, the second, using a series of values to test the estimated model, precedes the first, the scenario analysis. I depict this synergism in Fig. 6.12. If the *PEA* based on the testing of the estimated model shows that the model is inadequate, then the model has to be either re-estimated or discarded; otherwise, it is the used in the scenario analysis.

6.7.3 Simulation Tool for Prediction Application

It is one thing to produce a prediction with a single scenario, it is another to produce multiple predictions by changing numerous assumptions, either singularly or at one time. For the latter case, you need a program that allows a user, whether you as the analyst or your management/client, to specify different settings for key inputs and

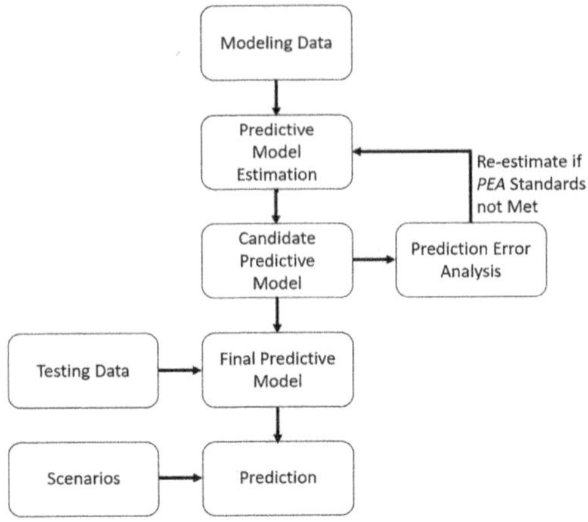

Fig. 6.12 This illustrates a framework for making predictions with a simulation tool

to generate output that describes the predictions. Such a tool is a *simulation tool*. A simulation tool allows you to produce a different prediction based on changeable inputs.

Python is an ideal framework for a simulation tool, especially when coupled with the Jupyter notebook paradigm. The notebook could contain simple instructions in Markdown cells and then code in code cells that is executed to ask for user-input and display predictions. The programming details for the simulation tool should be hidden from the user by writing Python scripts that are loaded into a Jupyter notebook by using Jupyter *magics*. Magics are short macro commands that perform various routine tasks. They streamline some coding by making repetitive routines more transparent to the user. Magics to import a script file are *%load script.py* and *%run script.py* (magic commands begin with %). The *load* magic simply loads a script file while the *run* magic loads and executes the script file. The problem with the *load* magic for a simulation tool is that the Python code in the script file becomes visible in the code cell. If the objective is to hide the code from the user, then this magic may not be the right one to use. The *run* magic executes the script file code which may make it a better selection.

Chapter 7
Time Series Analysis

The time dimension of the Data Cube is a major complication you will eventually face in analyzing your business data because time is a part of most business data sets. The data, a *time series*, could be for each second because of sensor readings, each minute for a production process, daily for accounting recording, monthly for sales and revenue processing and reporting, quarterly for financial reporting to legal and regulatory agencies, or annually for shareholder meetings. To complicate matters, the time series could be commingled with cross-sectional elements. Cross-sectional data are data collected at one point in time for multiple units or objects. For example, you could collect data by warehouses your company owns or leases, each one in a different state or section of a state. The warehouses are the cross-sectional units.

Cross-sectional data have their own problems, usually a different variance for each unit. This is called *heteroskedasticity*. The implication of heteroskedasticity is that the *OLS* estimators I outlined in Chap. 6 are not efficient. If whatever you are measuring varies by these cross-sectional units and over time, then you have a *panel data set* that reflects complications due to time and cross-sections. Transactions data, as for the living room blinds data set, are a panel data set: orders are collected by customers who are in different locations (e.g., cities, states, marketing regions) and at different times (e.g., daily). In fact, all transactions data are panel data. This is the Data Cube from Chap. 1. I will discuss heteroskedasticity and panel data in Chap. 10. My focus here is strictly on time series.

7.1 Time Series Basics

Defining a times series is not as simple as it might seem. In fact, the whole notion of time is quite complex from a practical and philosophical perspective. From a practical perspective, you need to be aware of the finer and finer divisions of a point

W. R. Paczkowski, *Business Analytics*,
https://doi.org/10.1007/978-3-030-87023-2_7

in time. Python divides a point in time into nine components whose specific values are collected into an immutable *time tuple*:

1. year (four-digit representation such as 2020;
2. month (1–12);
3. day (1–31);
4. hour (military hour: 0–23);
5. minute (0–59);
6. second (0–60);
7. day of the week (0–6; 0 is Monday)
8. day of the year (1 to 366); and
9. daylight savings indicator.

The daylight savings indicator is 1 if daylight savings time is in effect; 0 if standard time is in effect; and −1 if is unknown whether daylight savings time is in effect.

Time keeping is further complicated when you introduce time zones, leap year, business day, business week, fiscal year, and type of calendar. The calendar type refers to the Julian and Gregorian calendars; the Gregorian calendar is in common use today. There is also an issue of how dates are represented. In the U.S., a date is in the MM/DD/YYYY format; it is DD/MM/YYYY in Europe; and YYYY/MM/DD in *ISO* conventional format. Python, and especially Pandas, can handle any date and time format.

7.1.1 Time Series Definition

A time series is based on a set, S, of time points at which measurements are made. In many applications, S is a set of discrete equidistant time points, $S = \{0, 1, 2, \ldots, T\}$, where T is the number of time observations and interpreted as the last time point. If S is an interval on the real time axis, then $S = \{0 \leq t \leq T\}$ where t is a specific time point. The time point $t = 0$ is the beginning of time called the *epoch*. An observation or measurement, called a *realization*, of an event made at time t is denoted Y_t. The set of observations $\{Y_t, t \in S\}$ is called a *time series*.[1]

Some points about times series are:

- The realizations, the data you observe and record, are the result of sampling a true time series which is a continuous process. The process does not operate in fits and starts. The series is generated in real, continuous time, but you cannot observe this. Hence, you sample or measure. The result is that your realizations, your sample, will deviate from the true continuous series. This is a subtle point often

[1] See Parzen (1962, p. 5).

overlooked when describing time series data. Later, I will describe resampling to convert from one time series frequency to another.

- A time series, $\{Y_t, t \in S\}$, is a set of random variables, not a single random variable, each with a separate distribution. There is a joint distribution across the random variables in the set. This means there are variances, covariances, and correlations between and among the random variables (i.e., between and among the Y at different *lags* and *leads*) from some point in time t. A lag is a past period; a lead is the next period. Most of the focus is on using lagged data to guide forecasting. In simple terms: any observed value at one point in time (e.g., *sales*) is related to previous values; no data point is independent of other data points. The influence of prior values diminishes the further apart are the observations.
- A univariate time series is modeled as a realization of a sequence of random variables called a *time series process*. In the modern time series literature, the term "time series" refers to both the data and the process that generated the data realization.
- The sequence $\{Y_t, t \in S\}$ may be regarded as a sub-sequence of what is called a doubly infinite collection: $\{Y_t : t = \ldots, -2, -1, 0, 1, 2, \ldots\}$ with $t = 0$ as *today*. The negative indexes are history used to build a model. The positive indexes are the future to forecast, sometimes written as $Y_T(h), h = 1, 2, \ldots$. Context tells you how to interpret $\{Y_t, t \in S\}$. See Parzen (1962).
- A time series consists of contiguous time labels by an accepted calendar convention. This means if you have, say, monthly data, then each month's label logically follows the previous label. So, February follows January, March follows February, and so forth. Similarly, for annual data 2020 follows 2019, 2021 follows 2020. There is *time continuity*.

7.1.2 Time Series Concepts

There are two terms you need to know to effectively analyze time series data in Python. Pandas has the same concepts but with more functionality. Most of my discussion is about the Pandas concepts since your data will be in a Pandas DataFrame.

The first concept is *datetime*. This is a date and time stamp when an event occurs with two parts consistent with its name:

1. date: the date of the occurrence of the event; and
2. time: the time of the occurrence of the event.

It is a point in time; the time tuple or, better yet, a *timestamp*. Depending on the event and its importance, as well as the potential use of the measurement, the stamp could be to the minute, second, or even finer detail. The best way to view a datetime is as a frozen moment in time: a point signifying an event. Coinciding with that event is a measure which could simply be, and at a minimum is, a flag indicating that an event took place. Usually, there are more complicated measures that are either categorical or quantitative. As an example, a purchase placed online is recorded (i.e., time stamped) as of a specific date and time, with the time perhaps to the minute. The measure recorded at that moment could be the product ordered (categorical), the price point (quantitative), and any discount or special promotion (categorical or quantitative).

The second concept is *period*. This is an interval of dates and times, a logical grouping, although what is an interval as distinguished from a date or time is not always clear. First, a week is an interval: it consists of 7 days, so a week is a period; a month is a period; so is a quarter and a year. A day could be a period but a date is not since a date is a single entity. The period involves aggregating datetime measures in a logical manner. If a datetime measure is simply an indicator that a product was purchased, then a logical aggregation is the count of the orders in a day; the day is the period. If the measure on the datetime order event is the price net of special promotions, then a logical aggregation is the average price of the orders in a day.

There are four concepts involved in this discussion:

1. an event;
2. a datetime value as a point in time for that event;
3. a measure made or observed at that point in time for that event; and
4. a period as a span of datetime values.

The measures can be aggregated in each period as I just mentioned and will further discuss below. I illustrate the relationships among these four concepts in Fig. 7.1. The first three are recorded by the data collection system, which, in most instances, is maintained by your *IT* department. The period can certainly be provided by your *IT* department but typically you will have to create the periods you need for your analysis. You will import a data file, in *CSV* or other format, containing the date, the measures, and other data, and then parse the *IT*-provided date into a Pandas datetime value. From that point, you can group the datetimes into logical periods (e.g., months) and also aggregate the measures for your needs. I will explain and illustrate below how to parse datetimes, group the datetime values, and aggregate the measures.

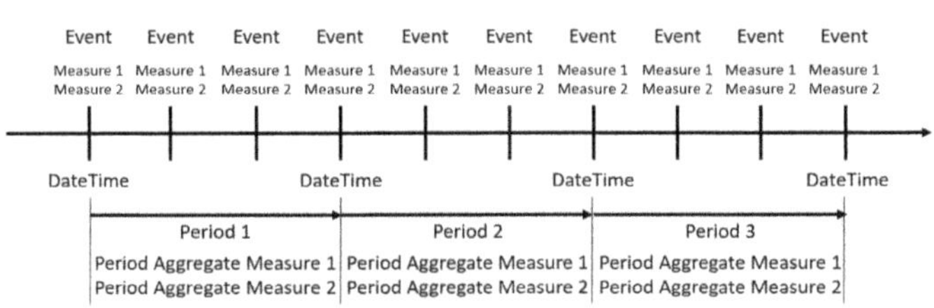

Fig. 7.1 The relationships among the four concepts are shown here

7.2 Importing a Date/Time Variable

If you import a *CSV* file that has a string variable that is supposed to be interpreted as a date and time, you could use the *parse_dates* parameter with that variable. Pandas will automatically interpret the string as dates and times and create the appropriate datetime variable.

7.3 The Data Cube and Time Series Data

I described the Data Cube in Chap. 1 and illustrated it in Fig. 1.5. It has three dimensions: a measure, space, and time. As I noted in Chap. 1, if you collapse the spatial dimension, you produce a time series for the measure. I illustrate one way to collapse the Data Cube in Fig. 7.2. There are three steps:

1. Create a datetime variable if one does not exist.
2. Access a period inside the datetime variable using the accessor *dt* and an appropriate period option (e.g., month).
3. Aggregate the data.

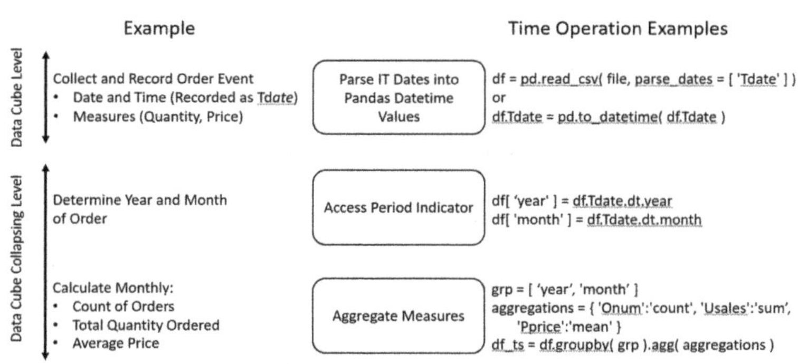

Fig. 7.2 The Data Cube can be collapsed by aggregating the measures for periods that were extracted from a datetime value using the accessor *dt*. Aggregation is the done using the *groupby* and *aggregate* functions

In order to aggregate datetime measures, you may have to access the logical date or time to do the aggregation. You can access or extract important parts of a datetime value, such as the month, the year, the day-of-week, and so forth, using the Pandas *accessor dt* and the appropriate part. You simply chain the datetime value, the accessor, and the part you want. For example, to get the month from a datetime value, x, use *x.dt.month*. I list some accessor possibilities in Table 7.1. Basically, the accessor gives you periods such as week, month, quarter, and year. I will describe the accessor *dt* later.

Time to extract	Command for extraction
Day	x.dt.day
Day name	x.dt.day_name( )
Week number	x.dt.week
Week-of-year	x.dt.weekofyear
Month	x.dt.month
Month name	x.dt.month_name( )
Quarter	x.dt.quarter
Year	x.dt.year

Table 7.1 These are examples of the datetime accessor command, *dt*. The symbol x is a datetime such as x = *pd.to_datetime(pd.Series(['06/15/2020']))*. The accessor is applied to a datetime variable created from a series. NOTE: Month as January = 1, December = 12; Day as 1, 2, . . ., 31

7.4 Handling Dates and Times in Python and Pandas

Python and Pandas, especially Pandas, have a rich array of date and time functionalities. The basic unit is a *datetime* which is a time stamp at a particular moment. That moment is identified by a date (e.g., June 17, 2020), an hour, a minute, and a second. The format is a string: *YYYY-MM-DD 00:00:00*. The hours are 24-h or "military time" format. For example, '2020-06-17 01:23:05' is June 17, 2020 at 1 h, 23 min and 5 s past midnight. This string is actually a representation of the number of seconds since a base time, the *epoch*. The Pandas epoch, January 1, 1970, can be found using a function to convert the datetime value.[2] I provide an example in Fig. 7.3.

Using the datetime value allows you to do a wide array of calendrical calculations. For example, you could calculate the day of week, day of month, month, year, and so forth given any datetime value. See Dershowitz and Reingold (2008) on calendrical calculations.

You can write dates in a wide assortment of formats. Pandas is smart enough to interpret them all as dates. For example, you could use any of the following:

- 06/15/2020
- 6/16/2020
- 2020-06-17
- June 18, 2020

[2] Function source: StackOverFlow: https://stackoverflow.com/questions/15203623/convert-pandas-datetimeindex-to-unix-time. Accessed June 18, 2020.

```
def datetime_to_epoch( ser ):
    """Don't convert NaT to large negative values.
    Source: StackOverFlow: Convert pandas DateTimeIndex to Unix Time?
    """
    if ser.hasnans:
        res = ser.dropna().astype('int64').astype('Int64').reindex( index = ser.index )
    else:
        res = ser.astype( 'int64' )
    return res // 10**9
x = pd.to_datetime( pd.Series( [ "1970-01-01", "1970-01-02" ] ) )
z = datetime_to_epoch( x )
##
## Pandas epoch: 01/01/1970
##
print( f'Datetime value at Pandas Epoch: {z[ 0 ]}\nDatetime value 1 day later {z[ 1 ]}' )
print( f'One day is {60 * 60 *24} seconds' )
```

```
Datetime value at Pandas Epoch: 0
Datetime value 1 day later 86400
One day is 86400 seconds
```

Fig. 7.3 This function in this example, returns date as a datetime integer. This integer is the number of seconds since the Pandas epoch which is January 1, 1970. The Unix epoch is January 1, 1960

```
##
## Create a list of consecutive dates
##
lst = [ '06/15/2020', '6/16/2020', '2020-06-17', 'June 18, 2020', '19Jun2020' ]
##
## Convert to datetime
##
x = pd.to_datetime( pd.Series( lst ) )
##
## Display the corresponding datetime values
##
value = datetime_to_epoch( pd.to_datetime( pd.Series( lst ) ) )
df = pd.DataFrame( {'Time Value':value } )
df[ 'Time Delta' ] = df[ 'Time Value' ] - df[ 'Time Value' ].shift()
df.style.set_caption( 'Datetime Values' ).set_table_styles( tbl_styles )
```

Datetime Values

	Time Value	Time Delta
0	1592179200	nan
1	1592265600	86400.000000
2	1592352000	86400.000000
3	1592438400	86400.000000
4	1592524800	86400.000000

Fig. 7.4 These are consecutive dates, each written in a different format. Each format is a typical way to express a date. Pandas interprets each format the same way and produces the datetime value, which is the number of seconds since the epoch. The column labeled "Time Delta" is the day-to-day change. Notice that it is always 86,400 which is the number of seconds in a day

- 19 Jun 2020

and get the correct consecutive datetime values. I show this in Fig. 7.4. A date written as June, 2020 is interpreted a June 1, 2020, the first day of the month.

7.4.1 Datetimes vs. Periods

A datetime is a specific point in time as I stated above. It is when an event, such as an order or delivery or manufacturing robotic failure, occurs. A succession of datetimes forming a coherent, logical whole or entity is a *period*. A period spans a series of datetimes. The events can then be logically grouped by a period. For example, you could have a series of product orders made each day for 7 days. They can be logically grouped and analyzed as occurring within a week. The week is the period. You could have the orders for a month so a month is a period. You could also have sensor readings every second for a production robot but have the sensor readings aggregated into 5-min intervals which are periods. You can view periods as containers for a group of datetimes.

Pandas has the ability to handle periods as well as datetimes. In fact, datetimes and periods are two fundamental time concepts in Pandas.[3]

7.4.2 Aggregating Datetime Measures

You aggregate datetime measures once a logical grouping has been accessed from the datetime values. There is a very useful function named *groupby* that does exactly what its name says: it *groups* data *by* something. This function actually groups the rows of a DataFrame by one or more variables in the DataFrame. It returns a grouping variable that has all the information about the grouping, but it does not return a grouped DataFrame *per se*.

7.4.3 Converting Time Periods in Pandas

In addition to accessing periods, you could also convert from one period designation to another. You may have to do this before merging two DataFrame if one is at a monthly frequency and the other is at a quarterly frequency. You could have a large number of transaction records for the same date, just the transaction times vary. This may be impractical to use because of the sheer volume of data, not to forget the statistical issues involved with the fine time granularity of the data. You may want to aggregate the transactions to a lower frequency level, say monthly. You can accomplish this by using the *resample* method or the *groupby* method. These are different methods that, yet, can be used together. The *resample* method groups rows of a DataFrame based on a datetime variable. I summarize some of the available options in Table 7.2. For example, if your orders data are daily, it will group together all the daily records in the same month. The *groupby* method also groups records, but it groups them based on variables in the DataFrame that are not necessarily datetime variables. You can use the two together with *resample* following *groupby*. In this order, you group your spatial data and then collapse the temporal measures to more convenient levels. I illustrate this approach in Fig. 7.5.

[3] Two other concepts are time deltas and date offsets. Time deltas are absolute time duration while date offsets relative time duration.

Frequency alias	Description
B	Business day
D	Calendar day
W	Weekly
M	Month end
Q	Quarter end
A or Y	Year end

Table 7.2 This is a short list of available frequencies and aliases for use with the "freq" parameter of the date_range function. A complete list is available in McKinney (2018, p. 331)

```
##
## Groupby CID and resample the daily price and discount data to month end
##
lst = [ 'CID', 'Tdate', 'Pprice', 'Ddisc', 'Odisc', 'Cdisc', 'Pdisc' ]
df_grouped = df_orders[ lst ].groupby( 'CID' ).resample( 'M', on = 'Tdate' )
df_grouped[ lst ].mean().head().style.set_caption( 'Groupby and Resample' ).\
    set_table_styles( tbl_styles )
```

Groupby and Resample

CID	Tdate	CID	Pprice	Ddisc	Odisc	Cdisc	Pdisc
14	2003-08-31 00:00:00	14.000000	5.400400	0.131820	0.051113	0.068248	0.038241
17	2003-10-31 00:00:00	17.000000	5.704396	0.137429	0.052000	0.077357	0.032429
	2003-11-30 00:00:00	17.000000	5.842836	0.124828	0.049172	0.069138	0.039069
26	2003-11-30 00:00:00	26.000000	5.744514	0.136357	0.046714	0.071179	0.040036
28	2003-10-31 00:00:00	28.000000	5.629682	0.138714	0.051464	0.076964	0.041250

Fig. 7.5 The *groupby* method and the *resampling* method can be combined in this order: the rows of the DataFrame are first grouped by the *groupby* method and then each group's time frequency is converted by the *resample* method

A more efficient way to group your data, both spatially and temporally, is to use the Pandas *Grouper* function to group the time dimension. It takes as an argument a key which is a datetime variable and a frequency indicator. I illustrate this approach in Fig. 7.6. This *Grouper* method is not restricted to grouping datetime variables. It can be used as a convenience for grouping any type of data. I illustrate this in Fig. 7.7.

The arguments for the *resample* method are the rules to convert your data and the object the conversion should be based on. The period you want to convert to is the rule represented by a symbol for the new period: "M" for month, "Q" for quarter, "A" for annual, and so forth. If you are converting daily data to monthly data, it is ambiguous which point in the month the data should be converted to. You have options. The "M" is for the calendar month end but you could use the calendar month begin ("MS"). The same holds for quarter and year. See McKinney (2018, p. 331) for a partial, yet comprehensive, list. I list some in Table 7.2. The object for the conversion could be the index or a variable in the DataFrame, but in either case it must be a datetime object otherwise there is no date to key on for the conversion.

The resample method produces a new object of type *DatetimeIndexResampler* that has the information about the resampling, but it does not display the resampled

```
##
## Groupby CID and resample the daily price and discount data to month end
## Use Grouper function
##
lst = [ 'CID', 'Tdate', 'Pprice', 'Ddisc', 'Odisc', 'Cdisc', 'Pdisc' ]
grpr = pd.Grouper( key = 'Tdate', freq = 'M' )
df_grouped = df_orders[ lst ].groupby( [ 'CID', grpr ] )
df_grouped[ lst ].mean().head().style.set_caption( 'Grouper Only' ).\
    set_table_styles( tbl_styles )
```

Grouper Only

CID	Tdate	CID	Pprice	Ddisc	Odisc	Cdisc	Pdisc
14	2003-08-31 00:00:00	14	5.400400	0.131820	0.051113	0.068248	0.038241
17	2003-10-31 00:00:00	17	5.704396	0.137429	0.052000	0.077357	0.032429
	2003-11-30 00:00:00	17	5.842836	0.124828	0.049172	0.069138	0.039069
26	2003-11-30 00:00:00	26	5.744514	0.136357	0.046714	0.071179	0.040036
28	2003-10-31 00:00:00	28	5.629682	0.138714	0.051464	0.076964	0.041250

Fig. 7.6 The *groupby* method is called with an additional argument to the variable to group on. The additional argument is *Grouper* which groups by a datetime variable. This method takes two arguments: a key identifying the datetime variable and a frequency to convert to. The *Grouper* can be placed in a separate variable for convenience as I show here

```
##
## Grouping using Grouper without a datetime variable
##
lst = [ 'CID', 'Tdate', 'Pprice', 'Ddisc', 'Odisc', 'Cdisc', 'Pdisc' ]
grpr = pd.Grouper( key = 'CID' )
df_grouped = df_orders[ lst ].groupby( grpr )
df_grouped[ lst ].mean().head().style.set_caption( 'Grouper and Groupby' ).\
    set_table_styles( tbl_styles )
```

Grouper and Groupby

CID	CID	Pprice	Ddisc	Odisc	Cdisc	Pdisc
1015	1015	3.986153	0.140930	0.051894	0.069262	0.039694
1766	1766	3.849132	0.130703	0.049154	0.069857	0.037484
818	818	3.797609	0.133464	0.052928	0.069529	0.040819
1235	1235	3.964647	0.132605	0.049395	0.072163	0.041395
1547	1547	3.819484	0.133647	0.050931	0.069009	0.039101

Fig. 7.7 The *groupby* method is called with the *Grouper* specification only

data. You have to operate on the object, perhaps by applying the sum or mean functions, and saving the aggregated data in a new DataFrame. I illustrate this in Fig. 7.8.

7.4.4 Date-Time Mini-Language

Pandas, and also Python, has a *mini-language* that allows you to read (i.e., *parse*) or write any specialized data and time formats. For example, you may have a date format as 2021*M*01 or 2021*Q*01 which stand for the third month (March) of the year 2021 and the first quarter of the same year, respectively. To read either one,

```
##
## Resample the daily price and discount data to month end
##
lst = [ 'Pprice', 'Ddisc', 'Odisc', 'Cdisc', 'Pdisc' ]
df_orders_resmpl = df_orders.resample( 'M', on = 'Tdate' )
df_orders_resmpl[ lst ].mean().head().style.set_caption( 'Resampled Data' ).\
    set_table_styles( tbl_styles )
```

Resampled Data

Tdate	Pprice	Ddisc	Odisc	Cdisc	Pdisc
2003-06-30 00:00:00	4.513912	0.108419	0.050751	0.070453	0.039970
2003-07-31 00:00:00	4.548639	0.107782	0.050133	0.070193	0.039937
2003-08-31 00:00:00	5.528507	0.112491	0.050237	0.069798	0.040039
2003-09-30 00:00:00	6.018170	0.115927	0.050152	0.069765	0.039786
2003-10-31 00:00:00	6.114422	0.122851	0.050216	0.069959	0.039900

Fig. 7.8 The furniture daily transactions data are resampled to monthly data and then averaged for the month. The rule is "M" for end-of-month, the object is *Tdate* and the aggregation is *mean*

use the Python function *strptime* or to write either one to a file use *strftime*. Each has two parameters: the date as a string and the format as a mini-language string. For example, *strptime('2021M01', '%YM%m')* will parse the monthly string and *strptime('2021Q01', '%YQ%q')* will parse the quarter string. Notice the use of the percent sign; this indicates the mini-language element. See McKinney (2018) for a list of the mini-language elements. Also see Table 7.3 for a summary.

You could create a custom parser (or writer) as *custom_date_parser = lambda x: datetime.strptime(x, "%YM%m")* and then use it in a Pandas read statement: *pd.read_csv(path + file, parse_dates = ['date'], date_parser = custom_date_parser)*.

Code	Meaning	Example
%a	Weekday as locale's abbreviated name	Mon, Tue
%A	Weekday as locale's full name	Monday
%d	Day of the month as a zero-padded decimal number	30
%b	Month as locale's abbreviated name	Sep
%B	Month as locale's full name	September
%m	Month as a zero-padded decimal number	09
%-m	Month as a decimal number (Platform specific)	9
%q	Quarter as a decimal number	01 ... 04
%y	Year without century as a zero-padded decimal number	21
%Y	Year with century as a decimal number	2021
%H	Hour (24-h clock) as a zero-padded decimal number	00, 01, ..., 23
%I	Hour (12-h clock) as a zero-padded decimal number	01, 02, ..., 12
%M	Minute as a zero-padded decimal number	00, 01, ..., 59
%S	Second as a zero-padded decimal number	00, 01, ..., 59

Table 7.3 This is an abbreviated listing of the Python/Pandas date-time mini-language. See McKinney (2018) for a larger list

7.5 Some Calendrical Calculations

There are some routine calendrical calculations you can and will do using the *datetime* variable. These are:

- shifting or lagging a time series;
- differencing a time series; and
- calculating a period-period percentage change.

Lagging means the whole series is shifted by the number of periods you specify, thus creating a new series; the default lag is one period. When you lag a series, the first observation becomes the second observation for the new series; the second observation becomes the third observation in the new series; and so forth. The first observation in the new series is filled with an NaN value. You lag a series using the *shift* method with a parameter for the number of lags. For example, *df['lag_X_1'] = df.X.shift(periods = 1)* lags the variable X in the DataFrame df one period and places that new lagged series in the DataFrame with the name *'lag_X_1'*. The *shift* method is *vectorized* meaning that it operates on the original series all at once without the need for a *for* loop. A number of Pandas functions are vectorized which is a very convenient feature.

Differencing involves finding the difference between an observation and the previous observation of the same variable: $X_t - X_{t-1}$. The difference becomes the second observation in the new variable, and so on. An NaN fills in the first observation of the new series. The default differencing is one period. A one period difference is called a *first difference*; a two period difference is a *second difference*, and so on.

Percent changes are simply the first difference divided by the first value. You could use the *pct_change* method. The default is a one period percent change.

Lags and differences are a key part of time series models, especially the stochastic models I will briefly discuss later in this chapter.

7.6 Time Series Generation Process: AR(1) Model

This is certainly not the place to discuss, or even begin to discuss, the concept of time. There are many deep philosophical issues associated with what may seem simple, everyday phenomenon. But what I can discuss is how events in time are related. Consider a physical system such as a musical instrument, say a guitar.[4] Initially, at some base time, the guitar is in a state of rest; a steady state. No sound emanates from the guitar while it is in this state. Place a decibel (*db*) meter next to the guitar and then pluck a single string on the guitar. A sound, created by the

[4] This example was inspired by Kmenta (1971, p. 270).

string's vibration, is registered on the meter as db_0. The string was disturbed from its initial state where it was at rest. Let a_0 be this disturbance which is a random shock. Then $db_0 = a_0$. Assume $a_0 \sim \mathcal{N}(0, \sigma_a^2)$.

The time when you pluck the string is t_0, the *epoch* or the beginning of time. Wait 1 second to time t_1 and pluck the string again. A sound registered on the meter as db_1 at t_1 is due to the new sound plus what is left over from the previous sound at t_0. The left-over sound is the result of a decay in the string's vibration due to air friction which slows the vibration. Let the proportion of the previous sound remaining be ρ and assume it is a constant regardless of the time period. Clearly, $0 \leq \rho < 1$. The total sound is $db_1 = \rho \times db_0 + a_1$ where a_1 is the meter reading for the new sound created at t_1. This is a random shock to the overall sound brought about by the plucking of the string. If the second string was not touched, then the total sound at t_1 is simply $db_1 = \rho \times db_0 = \rho \times a_0$.

If $\rho = 0$, then there is no dampened or left-over sound from the previous period; any sound is totally new. If $\rho = 1$, there is no dampening so there is a complete left-over from the previous period. Note that $\rho > 1$ is not possible due to friction; otherwise, the sound will get louder and louder and become infinitely deafening and life threatening.

Assume that you pluck the string once again at t_2. The total sound is a remainder from t_0, plus a remainder from t_1, plus a new sound or disturbance at t_2. That is, the total at t_2 is

$$db_2 = \rho \times db_1 + a_2 \tag{7.6.1}$$

$$= \rho \times (\rho \times db_0 + a_1) + a_2 \tag{7.6.2}$$

$$= \rho^2 \times db_0 + \rho \times a_1 + a_2 \tag{7.6.3}$$

$$= \rho^2 \times a_0 + \rho \times a_1 + a_2. \tag{7.6.4}$$

This is generalized to any time, t, as

$$db_t = \rho \times db_{t-1} + a_t \tag{7.6.5}$$

$$= \sum_{i=0}^{t} \rho^i \times a_{t-i}. \tag{7.6.6}$$

The total sound at time t is the weighted disturbance all the way back to the epoch. The weights, raised to higher and higher powers, guarantee that the earlier sounds have little if any effect. The process followed here is called *backward substitution*. Notice that I never mentioned the case for $\rho < 0$. If $\rho < 0$, then clearly the sound will oscillate, based on (7.6.6), between getting louder and then softer which we do not see in reality.

Now redefine the notation to any time series measure such as monthly sales or weekly raw material deliveries. Let Y_t be this measure. Then

$$Y_t = \rho \times Y_{t-1} + a_t \tag{7.6.7}$$

$$= \sum_{i=0}^{t} \rho^i \times a_{t-i}. \tag{7.6.8}$$

The base model is $Y_t = \rho \times Y_{t-1} + a_t$. This is a regression model without an intercept that involves regressing the random variable Y_t on itself (albeit with a lag). Consequently, this is called an *autoregressive model*. The lag of one period on the right-hand side is the order of the model. Therefore, the model is an *autoregressive of order 1 model* or *AR(1)*. As I showed above, this can be written as the sum of disturbances stretching back to the epoch, $t = 0$.

The *AR(1)* model can be applied to the disturbance term in a regression model so that the complete specification of a regression model (with a single variable for simplicity) is

$$Y_t = \beta_0 + \beta_1 \times X_t + \epsilon_t \tag{7.6.9}$$

$$\epsilon_t = \rho \times \epsilon_{t-1} + a_t \tag{7.6.10}$$

with $\rho <\mid 1 \mid$ and $a_t \sim \mathcal{N}(0, \sigma_a^2)$. If $\rho = 0$, then $\epsilon_t = a_t$ and the *OLS* model from Chap. 6 results. The ρ is called the *autocorrelation coefficient*. For most, if not all, business data, $0 \leq \rho < 1$ which indicates positive autocorrelation.

Since (7.6.10) is a regression model, an estimator for ρ is written in terms of residuals and using the results from Chap. 6. You can immediately write

$$\hat{\rho} = \frac{\sum_{t=2}^{T} e_t \times e_{t-1}}{\sum_{t=1}^{T} e_t^2}. \tag{7.6.11}$$

This *AR(1)* model for the disturbance term has implications for *OLS* estimation. In particular, it can be shown that the *OLS* estimators, $\hat{\beta}_0$ and $\hat{\beta}_1$, are linear and unbiased but they no longer have minimum variance in the class of linear, unbiased estimators. The Gauss-Markov Theorem is violated. The implication is that hypothesis testing will lead you to the wrong conclusions which means you will have Poor, not Rich, Information. See Greene (2003), Gujarati (2003), Hill et al. (2008), and Goldberger (1964) for discussions.

Since the *OLS* estimators are no longer efficient under the *AR(1)* disturbance terms, you need to check if, in fact, your time series model has $\rho > 0$. There are two checks: graphical and a formal statistical test. I review both in the following sections.

7.7 Visualization for *AR(1)* Detection

The issue for (7.6.10) is the *AR(1)* specification for the disturbance term. A proxy for the disturbance is the residual, $e_t = Y_t - \hat{Y}_t$. Recall from Chap. 6 that $e_i \approx \epsilon_i$. The simplest visual, therefore, is a plot of the residuals against time. There are two signature patterns to look for: a sine wave and a jagged, saw-tooth appearance.

The sine wave signature is indicative of positive autocorrelation (i.e., $\rho > 0$) because positive residuals tend to follow positive residuals and negative ones tend to follow negative residuals, thus giving the sine wave appearance. If the residuals have been increasing, their tendency is to continue to increase. However, recall that part of the *AR(1)* model for the disturbance term, (7.6.10), is a separate disturbance term, a_t. If this term is large at any point in time, it could overpower the positive (or negative) pattern and turn the series around. As long as these added disturbances are small, then the series as a whole will trend upward or downward.

If, however, there is negative autocorrelation (i.e., $\rho < 0$), then the time pattern for the residuals is erratic with a jagged sawtooth pattern without an underlying sine wave. The negative ρ implies that a positive residual is followed by a negative one, which, in turn, is followed by a positive residual, and so on. This is unlike the case where $\rho > 0$. The added disturbance term in (7.6.10) will still have an influence that could add more jaggedness to the pattern.

I illustrate the extraction of residuals in Fig. 7.9 for a regression model of log unit sales on log pocket price but for time series data. I estimated this same model with for cross-sectional data in Chap. 6. I next plotted the time series residuals in Fig. 7.10. Notice the sine wave pattern suggesting a positive autocorrelation for the disturbance term. This reflects the Gestalt *Common Fate* and *Connection Principles*.

```
##
## Create DataFrame of the residuals
## The residuals are a Pandas series so use to_frame
##
resid = reg02.resid.to_frame( name = 'Residual' )
resid.reset_index( inplace = True )
##
## Add a datetime variable and set as index
##
resid[ 'day' ] = 1
resid[ 'Date' ] = pd.to_datetime( resid[ ['Year', 'month', 'day' ] ] )
resid.drop( [ 'day' ], axis = 1, inplace = True )
resid.set_index( 'Date', inplace = True )
resid.head().style.set_caption( 'Residuals for Time Series Regression' ).\
    set_table_styles( tbl_styles )
```

Residuals for Time Series Regression

Date	Year	month	Residual
2003-06-01 00:00:00	2003	6	-0.661957
2003-07-01 00:00:00	2003	7	0.520765
2003-08-01 00:00:00	2003	8	0.934218
2003-09-01 00:00:00	2003	9	0.872695
2003-10-01 00:00:00	2003	10	1.002666

Fig. 7.9 The residuals for a times series model of log unit sales on log pocket price are retrieved

A second type of graph uses the residuals (Y-axis) against their one-period lagged values (X-axis). Since the average of the residuals is zero (because the sum of

```
##
## Residual Plot: Against Time
## Sine wave pattern
##
ax = resid.plot( y = 'Residual' )
ax.set_title( 'Residual Plot', fontsize = font_title )
ax.set( xlabel = '', ylabel = 'Residuals' );
```

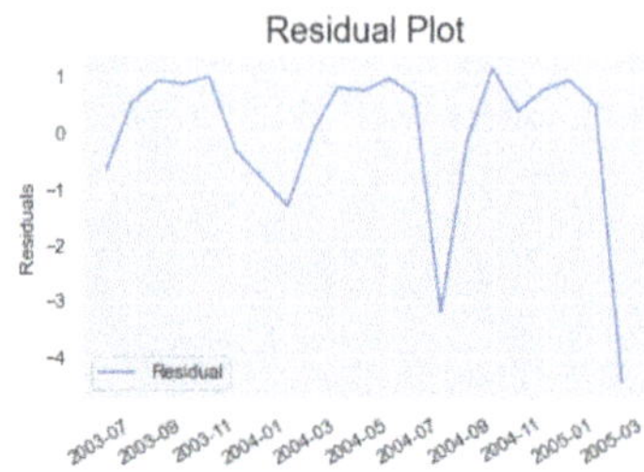

Fig. 7.10 The residuals from Fig. 7.9 are plotted against time. A sine wave appearance is evident

residuals is zero from Chap. 6), you could draw vertical and horizontal lines at zero which divides the plot area into four quadrants. Plot-points in the upper left quadrant are indicative of negative autocorrelation (i.e., $\rho < 0$) since a negative lagged residual is associated with a positive unlagged residual. The other three quadrants have obvious interpretations. I list the four possibilities in Table 7.4 and then provide a plot in Fig. 7.11 of the residuals in Fig. 7.9. This reflects the Gestalt *Proximity* and *Similarity Principles*.

Quadrant	Location	Autocorrelation
I	Upper left	Negative ($\rho < 0$)
II	Upper right	Positive ($\rho > 0$)
III	Lower left	Positive ($\rho > 0$)
IV	Lower right	Negative ($\rho < 0$)

Table 7.4 A graph of the residuals (Y-axis) vs one-period lagged residuals (X-axis) can be divided into four quadrants. The autocorrelation is identified by a signature: the quadrant most of the points fall into. There will, of course, be random variation among the four quadrants, but it is where the majority of points lie that helps to identify the autocorrelation

7.8 Durbin-Watson Test Statistic

Graphs are easy to produce and they can be suggestive of patterns based on signatures such as those I just discussed. They are not, however, infallible. Two people can look at the exact same graph and see something different. This is not new; it has been observed and discussed many times before. Samuelson (1973, p. 11, emphasis in original), for example, argues for an "irreducible subjective element in *any* science" by presenting two images that people interpret differently. Samuelson (1973, p. 11) comments that a shape that is actually "objectively reproducible"

```
##
## Lag the residuals one period
##
resid[ 'lag_Residual' ] = resid.Residual.shift()
##
## Residual Plot: Against Lagged Value
##
ax = resid.plot( y = 'Residual', x = 'lag_Residual', kind = 'scatter' )
ax.set_title( 'Residual Plot', fontsize = font_title)
ax.set( xlabel = 'Lagged Residuals', ylabel = 'Residuals' )
plt.axvline( x = 0 )
plt.axhline( y = 0 )
##
plt.annotate( 'Quadrant I', xy = ( -3, 1 ), ha = 'center' )
plt.annotate( 'Quadrant II', xy = ( 0.90, 1 ), ha = 'center' )
plt.annotate( 'Quadrant III', xy = ( -3, -4 ), ha = 'center' )
plt.annotate( 'Quadrant IV', xy = ( 0.90, -4 ), ha = 'center' );
```

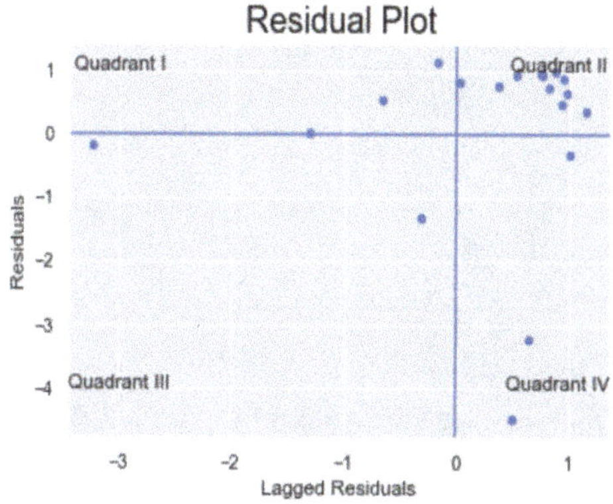

Fig. 7.11 The residuals from Fig. 7.9 are plotted against their lagged values. Most of the points fall into the upper right quadrant suggesting positive autocorrelation based on Table 7.4. This graph can also be produced using the Pandas function *pd.plotting.lag_plot(series)* where "series" is the residual series

would yet look "subjectively different depending on the context in which it appears." So even one person could have different interpretations depending on their context. See, also, Peebles and Ali (2015) and Stewart (2019, p. 196) for some other discussions of this phenomenon. This implies for our purposes that, although graphs are simple to create, useful, and can be informative, they are just the first step in identifying a pattern which is exactly what I emphasized in Chap. 4. A formal test may still be required.

One of the oldest tests for disturbance term autocorrelation is the *Durbin-Watson Test*. See any econometrics text such as Gujarati (2003), Hill et al. (2008), and Greene (2003). There are many such tests available, but they all have the Durbin-Watson Test as their base. More importantly, almost all statistical software automatically calculate and present this test statistic. You can see it, for example, in Figs. 6.3 and 6.6.

The Durbin-Watson test statistic is defined as

$$d = \frac{\sum_{t=2}^{T} (e_t - e_{t-1})^2}{\sum_{t=1}^{T} e_t^2} \tag{7.8.1}$$

where T is the total number of time series observations. The Null Hypothesis is H_O: $\rho = 0$ and the Alternative Hypothesis is H_A: $\rho \neq 0$. There are three assumptions:

1. the disturbances are generated by an $AR(1)$ process;
2. the model does not contain a lagged dependent variable; and
3. the model has a constant term.

To evaluate the Durbin-Watson d-statistic, observe that you can write it as

$$d = \frac{\sum_{t=2}^{T} e_t^2 + \sum_{t=2}^{T} e_{t-1}^2 - 2 \times \sum_{t=2}^{T} e_t \times e_{t-1}}{\sum_{t=1}^{T} e_t^2}. \tag{7.8.2}$$

In large samples, $\sum_{t=2}^{T} e_t^2 \approx \sum_{t=2}^{T} e_{t-1}^2$ so

$$d = \frac{2 \times \sum_{t=2}^{T} e_t^2 - 2 \times \sum_{t=2}^{T} e_t \times e_{t-1}}{\sum_{t=1}^{T} e_t^2} \tag{7.8.3}$$

$$= 2 \times \left(1 - \frac{\sum_{t=2}^{T} e_t \times e_{t-1}}{\sum_{t=1}^{T} e_t^2}\right) \tag{7.8.4}$$

$$= 2 \times (1 - \hat{\rho}). \tag{7.8.5}$$

where $\hat{\rho}$ comes from (7.6.11). As a rule-of-thumb, a value of the d-statistic "close" to 2 is an indication of no autocorrelation. I provide ranges for the Durbin-Watson d-statistics and their implied autocorrelation in Table 7.5. There are formal tables available to help you decide about the autocorrelation, but the rule-of-thumb of "close" to 2 usually suffices in empirical work.

Durbin-Watson value	ρ	Autocorrelation
$d = 2$	$\hat{\rho} = 0$	None (Desirable)
$d = 4$	$\hat{\rho} = -1$	Negative
$2 < d < 4$	$-1 < \hat{\rho} < 0$	Negative
$d = 0$	$\hat{\rho} = +1$	Positive
$0 < d < 2$	$0 < \hat{\rho} < 1$	Positive

Table 7.5 These are some guides or rules-of-thumb for the Durbin-Watson test statistic. The desirable value for d is clearly 2

There are four problems, however, with the Durbin-Watson d-statistic:

1. It is applicable for $AR(1)$ only. Some disturbances may be generated by $AR(2)$ or other more complicated processes.
2. There is an indeterminate region around 2—this is the basis for the "close".
3. The model cannot contain a lagged dependent variable. Many models have this variable to capture dynamics. In this case, the d-statistic is biased toward 2. A variant of the Durbin-Watson, known as *Durbin's h-statistic*, can be used. Durbin's h-statistic is defined as

$$h = \left(1 - \frac{d}{2}\right) \times \sqrt{\frac{T}{1 - T \times s^2_{\hat{\beta}_1}}} \tag{7.8.6}$$

   where d is the Durbin-Watson d-statistic and $\hat{\beta}_1$ is the coefficient for the lagged dependent variable with variance $s^2_{\hat{\beta}_1}$. Note that $h \sim \mathcal{N}(0, 1)$.
4. It assumes no missing observations.

The last problem is actually a restriction, and an insidious one. If you have a long time series, then it is possible you may have missing values for any time periods. They must be "filled" as I discussed in Chap. 6. I illustrate this in Fig. 7.12 by first resampling the unit sales and pocket price data to a monthly series and then aggregating sales by summing and aggregating price by averaging. If there are any missing value in the original orders data for a particular month, then the sum of sales for that month should be zero and the mean price should be NaN. Then I use the *info( )* method in Fig. 7.13 to check for missing values, of which there is one in December, 2003. I use the Pandas *interpolate( )* method to fill-in the NaN values as I show in Fig. 7.14.

Once the times series DataFrame is prepared, a regression model can be estimated and the Durbin-Watson d-statistic rechecked. I show a regression estimation in Fig. 7.15. Notice that the Durbin-Watson statistic is 1.387 which indicates positive autocorrelation. This autocorrelation issue can be corrected using the *Cochrane-Orcutt* procedure, which is one of several correction methods available. This particular one involves estimating the autocorrelation coefficient, ρ, using (7.6.11). This is then used in the following steps. First, lag the model one period to get

$$Y_{t-1} = \beta_0 + \beta_1 \times X_{t-1} + \epsilon_{t-1}. \tag{7.8.7}$$

Now multiply through by ρ so that

$$\rho \times Y_{t-1} = \rho \times \beta_0 + \beta_1 \times \rho \times X_{t-1} + \rho \times \epsilon_{t-1}. \tag{7.8.8}$$

```
##
## Resample the daily price and discount data to month end
##
lst = [ 'Usales', 'Pprice', 'Tdate' ]
df_ts = df_orders[ lst ].resample( 'M', on = 'Tdate' ).agg( {'Usales':np.sum, 'Pprice':np.mean } )
##
## Replace 0 sales with NaN
##
df_ts[ 'Usales' ] = df_ts[ 'Usales' ].replace( 0, np.nan )
##
## Display first 7 records
df_ts.head( 7 ).style.set_caption( 'Aggregated Resampled Data' ).set_table_styles( tbl_styles )
```

Aggregated Resampled Data

Tdate	Usales	Pprice
2003-06-30 00:00:00	57181.000000	4.513912
2003-07-31 00:00:00	180917.000000	4.548639
2003-08-31 00:00:00	124559.000000	5.528507
2003-09-30 00:00:00	83182.000000	6.018170
2003-10-31 00:00:00	88856.000000	6.114422
2003-11-30 00:00:00	28275.000000	5.846772
2003-12-31 00:00:00	nan	nan

Fig. 7.12 The unit sales and pocket price data were resampled to a monthly frequency and then aggregated. The sum of sales would be zero for a particular month if there were no sales in that month. That zero value was replaced by NaN

```
##
## Check for missing values
##
df_ts.info()
```

```
<class 'pandas.core.frame.DataFrame'>
DatetimeIndex: 21 entries, 2003-06-30 to 2005-02-28
Freq: M
Data columns (total 2 columns):
 #   Column  Non-Null Count  Dtype
---  ------  --------------  -----
 0   Usales  20 non-null     float64
 1   Pprice  20 non-null     float64
dtypes: float64(2)
memory usage: 1.1 KB
```

Fig. 7.13 The resampled and aggregated orders data are checked for missing values. Notice that there are 21 records but 20 have non-null data

Subtract the lagged model from the original one to get

$$Y_t - \rho \times Y_{t-1} = \quad \beta_0 \times (1 - \rho) + \beta_1 (X_t - \rho \times X_{t-1}) + (\epsilon_t - \rho \times \epsilon_{t-1}). \tag{7.8.9}$$

Or

$$Y_t^* = \beta_0^* + \beta_1 \times X_t^* + u_t. \tag{7.8.10}$$

You can now apply *OLS* to this transformed model to estimate the unknown parameters. This is the *Cochrane-Orcutt* procedure. Most software that implements this procedure has it iterate several times for an optimal estimation solution. This estimator is in a broader family of estimators called the *Generalized Least Squares*

```
##
## Interpolate missing values
##
df_ts.interpolate( inplace = True )
df_ts.head( 7 ).style.set_caption( 'Interpolated Missing Values' ).\
    set_table_styles( tbl_styles )
```

Interpolated Missing Values

Tdate	Usales	Pprice
2003-06-30 00:00:00	57181.000000	4.513912
2003-07-31 00:00:00	180917.000000	4.548639
2003-08-31 00:00:00	124559.000000	5.528507
2003-09-30 00:00:00	83182.000000	6.018170
2003-10-31 00:00:00	88856.000000	6.114422
2003-11-30 00:00:00	28275.000000	5.846772
2003-12-31 00:00:00	43226.500000	4.838774

Fig. 7.14 The missing values are filled-in using the Pandas *Interpolate( )* method

```
##
## OLS
##
## ===> Step 1: Define a formula <===
##
formula = 'np.log( Usales ) ~ np.log ( Pprice )'
##
## ===> Step 2: Instantiate the OLS model <===
##
mod = smf.ols( formula, data = df_ts )
##
## ===> Step 3: Fit the instantiated model <===
##      Recommendation: number your fitted models
##
reg01 = mod.fit()
##
## ===> Step 4: Summarize the fitted model <===
##
display( reg01.summary() )
```

OLS Regression Results

Dep. Variable:	np.log(Usales)	**R-squared:**	0.141
Model:	OLS	**Adj. R-squared:**	0.095
Method:	Least Squares	**F-statistic:**	3.111
Date:	Mon, 18 Jan 2021	**Prob (F-statistic):**	0.0938
Time:	13:10:23	**Log-Likelihood:**	-37.025
No. Observations:	21	**AIC:**	78.05
Df Residuals:	19	**BIC:**	80.14
Df Model:	1		
Covariance Type:	nonrobust		

	coef	std err	t	P>|t|	[0.025	0.975]
Intercept	17.5004	3.676	4.760	0.000	9.806	25.195
np.log(Pprice)	-3.9341	2.230	-1.764	0.094	-8.602	0.734

Omnibus:	19.786	**Durbin-Watson:**	1.387
Prob(Omnibus):	0.000	**Jarque-Bera (JB):**	22.269
Skew:	-1.986	**Prob(JB):**	1.46e-05
Kurtosis:	6.160	**Cond. No.**	25.6

Warnings:
[1] Standard Errors assume that the covariance matrix of the errors is correctly specified.

Fig. 7.15 The Durbin-Watson statistic is low, 1.387

(*GLS*) family of estimators. I show an example in Fig. 7.16 which is a continuation of the results in Fig. 7.15.

```
##
## GLS( Cochrane-Orcutt) Procedure
## Notice the rho is set to 1 for a one period lag: AR(1)
## Also notice the new command: GLSAR.from_formula
##
formula = 'np.log( Usales ) ~ np.log ( Pprice )'
reg02 = sm.GLSAR.from_formula( formula, rho = 1, data = df_ts )
##
## Iterate and print the results
##
result = reg02.iterative_fit(maxiter = 10)
display( result.summary() )
```

```
                           GLSAR Regression Results
==============================================================================
Dep. Variable:         np.log(Usales)   R-squared:                       0.137
Model:                          GLSAR   Adj. R-squared:                  0.089
Method:                 Least Squares   F-statistic:                     2.846
Date:                Mon, 18 Jan 2021   Prob (F-statistic):              0.109
Time:                        13:12:38   Log-Likelihood:                -35.593
No. Observations:                  20   AIC:                             75.19
Df Residuals:                      18   BIC:                             77.18
Df Model:                           1
Covariance Type:            nonrobust
==================================================================================
                     coef    std err          t      P>|t|      [0.025      0.975]
----------------------------------------------------------------------------------
Intercept         17.7179      3.982      4.450      0.000       9.353      26.083
np.log(Pprice)    -4.0554      2.404     -1.687      0.109      -9.106       0.995
==============================================================================
Omnibus:                       20.424   Durbin-Watson:                   1.399
Prob(Omnibus):                  0.000   Jarque-Bera (JB):               22.941
Skew:                          -2.047   Prob(JB):                     1.04e-05
Kurtosis:                       6.281   Cond. No.                         24.8
==============================================================================

Warnings:
[1] Standard Errors assume that the covariance matrix of the errors is correctly specified.
```

Fig. 7.16 After the *GLS* correction, the Durbin-Watson statistic is improved only slightly to 1.399

7.9 Lagged Dependent and Independent Variables

The model I consider had one independent variable that was contemporaneous with the dependent variable. This is a *static model*: the effect of the independent variable is immediate without any carry-over or long-term effects. You could certainly include many more contemporaneous independent variables, but the implication will be the same. You still have a static model. You could introduce a *dynamic model* but including lagged dependent and independent variables. You could lag the dependent variable a maximum of p periods and lag the independent variable a maximum of q periods. The overall model is referred to as a *autoregressive distributed lag model* of order p and q (*ARDL(p, q)*). The static model is *ARDL(0, 0)*. I will briefly mention several variations.

7.9.1 *Lagged Independent Variable:* ARDL(0, 1)

The simplest extension to the static model is the addition of the independent variable lagged one period. You use the Pandas *shift* method with its default to create the one-period lag. The model would then be:

$$Y_t = \beta_0 + \beta_1 \times X_t + \beta_2 \times X_{t-1} + \epsilon_t. \tag{7.9.1}$$

This is analyzed by examining its behavior when the equilibrium values are obtained. These values are the long-run values found by taking the expectations of both sides of the model. Let $\tilde{Y}$ and $\tilde{X}$ be the long-run, equilibrium values. The model, after collecting like terms, is then,:

$$\tilde{Y} = \beta_0 + \beta_1 \times \tilde{X} + \beta_2 \times \tilde{X} \tag{7.9.2}$$

$$= \beta_0 + (\beta_1 + \beta_2) \times \tilde{X}. \tag{7.9.3}$$

7.9.2 *Lagged Dependent Variable:* ARDL(1, 0)

You can include lag the dependent variable only so that the model is:

$$Y_t = \beta_0 + \beta_1 \times X_t + \beta_2 \times Y_{t-1} + \epsilon_t. \tag{7.9.4}$$

7.9.3 *Lagged Dependent and Independent Variables:* ARDL(1, 1)

You can include lag the dependent variable and lagged independent variable so that the model is:

$$Y_t = \beta_0 + \beta_1 \times X_t + \beta_2 \times X_{t-1} + \beta_3 \times Y_{t-1} + \epsilon_t. \tag{7.9.5}$$

It should be clear that the extensions quickly become more complicated. See Dhrymes (1971) for a detailed technical discussion and (Dougherty, 2016, Chapter 11) for more basic discussions of these models including estimation issues.

7.10 Further Exploration of Time Series Analysis

Time series modeling can become very complex because of the lag structures of the dependent variable, the independent variables, and the disturbance term. A discussion of the issues and methods for time series modeling is beyond the scope of this book. For a detailed introduction, see Box et al. (1994). For a high level introduction with a management focus see Nelson (1973). For the use of time series analysis for new product development with some background on methods, see Paczkowski (2020).

Although time series modeling is an extremely complex subject, I can still highlight some features. I will consider a very broad, general class of models sometimes called *stochastic time series models*, *time series models*, or *Box-Jenkins models* which is named after Box and Jenkins who popularized the methodology. See Box et al. (1994) for a complete discussion and development. In this class of models, the random element, ϵ, plays a dominant role rather than being just an add-on error (i.e., disturbance) to a strictly deterministic model as in econometrics. Also, this class of models does not contain an independent variable. The complete explanation of a time series is based on the lag structure of the series itself and the disturbance term.

I will follow a procedure to

1. develop the model: *AR(p)*, *MA(q)*, *ARMA(p, q)*, or *ARIMA(p, d, q)*;
2. calculate the mean of the time series;
3. calculate the variance and covariances of the times series; and
4. calculate the time series correlations: the *autocorrelation function* (*ACF*) and *partial autocorrelation function* (*PACF*).

The last two are instrumental in identifying the type of time series model, one of the four listed in the first point above: the autoregressive model of order p ($AP(p)$), moving average of order q ($MA(q)$), autoregressive moving average of order p, q ($ARMA(p, q)$), and autoregressive integrated moving average of order p, d, q ($ARIMA(p, d, q)$). The p and q orders are for lags of the variable and disturbance, respectively. The d is the number of times the series is differenced. I will also discuss a procedure for building a model that involves several steps.

The autocorrelations are the keys to identifying a model or set of candidate models. They are based on the autocovariances, γ_k, of a times series. If a time series Y_t comes from a distribution with mean μ, then

$$\gamma_k = COV(Y_t, Y_{t-k}) \tag{7.10.1}$$

$$= E[(Y_t - \mu) \times (Y_{t-k} - \mu)] \tag{7.10.2}$$

where γ_0 is the variance, σ^2, of Y_t. The variance is a constant so $V(Y_t) = V(Y_{t-k}) = \gamma_0$, regardless of k. The autocorrelations are

$$\rho_k = \frac{COV(Y_t, Y_{t-k})}{[V(Y_t) \times V(Y_{t-k})]^{\frac{1}{2}}} \tag{7.10.3}$$

$$= \frac{\gamma_k}{\gamma_0}. \tag{7.10.4}$$

Clearly, $\rho_0 = 1$. Autocorrelations depend only on the lag (or time difference) k. Therefore, autocorrelations are referred to as the *autocorrelation function* (*ACF*), or, sometimes, as the *correlogram*, because they are a function of k. The *ACF* is plotted with different values of k along with 95% confidence bounds. Note that, since

$$\gamma_k = COV(Y_t, Y_{t-k}) = COV(Y_{t-k}, Y_t) \quad (7.10.5)$$

$$= \gamma_{-k}, \quad (7.10.6)$$

it follows that $\rho_k = \rho_{-k}$ and so only the positive half of the ACF is usually used.

The parameters μ, γ_0, and the ρ_k are unknown. You can estimate them as expected:

$$\mu: \quad \bar{Y} = n^{-1} \times \sum_{t=1}^{n} Y_t \quad (7.10.7)$$

$$\gamma_0: \quad s^2 = n^{-1} \times \sum_{t=1}^{n} (Y_t - \bar{Y})^2 \quad (7.10.8)$$

The lag k autocorrelation, ρ_k, is estimated using

$$r_k = \frac{\sum_{t=k+1}^{n} (Y_t - \bar{Y}) \times (Y_{t-k} - \bar{Y})}{\sum_{t=1}^{n} (Y_t - \bar{Y})^2} \quad (7.10.9)$$

$$k = 1, 2, \ldots. \quad (7.10.10)$$

The *partial autocorrelation function* (*PACF*) adjusts or controls for the effects of other periods in the lag structure of the time series. This is what partial correlations do in Stat 101 descriptive statistics: they measure the degree of association between two random variables with the effect of a set of other variables removed or "partialed out." The partial autocorrelations are. Some statistical software packages estimate the partial correlations, but most data analysts ignore them. In time series analysis, the partial autocorrelation function (*PACF*), represented by ϕ_{kk}, is a graph of these partials against the lags and so it cannot be ignored. This plays an important role in identifying the lag in an autoregressive process—the p in *AR(p)*. The $PACF$ is also usually plotted with 95% confidence bounds. I illustrate both graphs in Fig. 7.17.

These two functions are instrumental in identifying a times series model. I will outline a modeling process due to Box and Jenkins that uses them. See Box et al. (1994) and Wei (2006) for details. This process has four steps:

Step 1: Identification of a model.
Step 2: Estimation of the model.
Step 3: Validation of the model.
Step 4: Forecasting with the model.

```
##
## Create an autocorrelation plot
## Use 10 period lag since series is short
##
sm.graphics.tsa.plot_acf( df_ts.log_Pprice, lags = 10 )
plt.show();
```

```
##
## Create a partial autocorrelation plot
## Use 10 period lag since series is short
##
sm.graphics.tsa.plot_pacf(df_ts.log_Pprice, lags = 10)
plt.show();
```

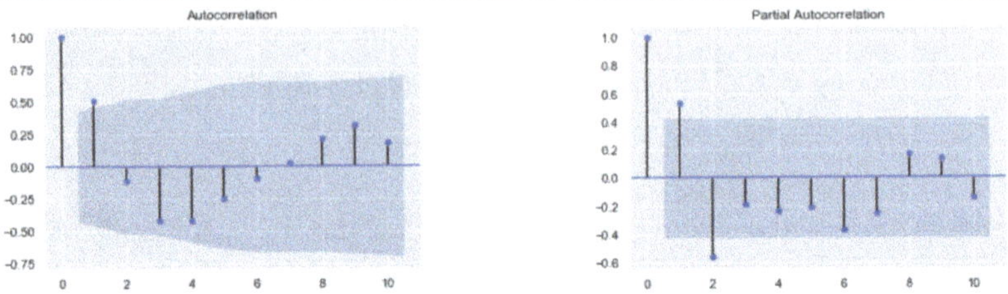

(a) (b)

Fig. 7.17 This illustrates the two time series plots instrumental in identifying a times series model. Panel (**a**) is an autocorrelation plot for 10 lags; (**b**) is a partial autocorrelation plot for the same lags. The shaded areas are the 95% confidence interval

Model identification deals with identifying an initial model or a set of initial models that will be used for model estimation in the second step. It is important to be aware that these are tentative, candidate models that, once estimated, have to be validated against data in the third step. This is an iterative process of identification $\Rightarrow$ estimation $\Rightarrow$ validation. Once you settle on a final model, then you can use that model for forecasting.

7.10.1 *Step 1: Identification of a Model*

There are different graph signatures that help identify a candidate model, or perhaps several models. I summarize these in the next few subsections.

7.10.1.1 AR(p) Model

You already know about the autoregressive model for the disturbance term in an *OLS* model. This is defined as

$$\epsilon_t = \rho\epsilon_{t-1} + u_t. \tag{7.10.11}$$

This is an $AR(1)$ model. Now, for a times series model, the disturbance term, u_t, is referred to as *white noise* and is represented by a_t to distinguish it from the regression model's disturbance. A white noise process is a random process with $\{a_t : t = 1, 2, \ldots\}$, $E(a_t) = 0, \forall t$, and

$$\gamma_k = COV(a_t, a_{t+k}) = \begin{cases} \sigma^2 & \text{if } k = 0 \\ 0 & \text{otherwise.} \end{cases} \tag{7.10.12}$$

An important example of a white noise process is a sequence of independent normal random variables with common mean 0 and variance σ^2.

Consider the $AR(1)$ model

$$Y_t = \phi Y_{t-1} + a_t \tag{7.10.13}$$

where a_t is white noise. Through backward substitution, you get

$$Y_t = a_t + \phi a_{t-1} + \ldots + \phi^t Y_0 \tag{7.10.14}$$

so Y_t is a weighted sum of past white noise terms and an initial value of Y at the epoch. A general form for this model of practical value with only p lags is

$$Y_t = \phi_1 Y_{t-1} + \ldots + \phi_p Y_{t-p} + a_t. \tag{7.10.15}$$

This is an $AR(p)$ model. The question is: "*What is the order or lag p?*" This question is answered by looking at the *ACF* and the *PACF* graphs which provide signatures for the lag structure but also for whether or not the process is autoregressive. The derivation of the signatures is complex and beyond the scope of this book, but basically the *ACF* decays exponentially while the *PACF* spikes at lags 1 to p. If the *PACF* spikes at lag 1, then an AR(1) model is suggested. See Box et al. (1994) and Wei (2006) for details. I summarize the signatures in Table 7.6. Using these results and the graphs in Fig. 7.17, an $AR(2)$ is suggested as a first iteration of model identification.

Process	*ACF*	*PACF*
$AR(p)$	Decays exponentially	Spikes at lags 1 to p, then cuts off

Table 7.6 These are the signatures for an $AR(p)$ model based on the *ACF* and *PACF*

7.10.1.2 MA(q) Model

Suppose you now write

$$Y_t = a_t + \theta_1 a_{t-1} + \ldots + \theta_q a_{t-q} \tag{7.10.16}$$

This is an $MA(q)$ model. The name "MA" is misleading since the weights do not necessarily sum to 1.0. So, do not confuse this with a moving average smoothing process often applied to time series data to smooth out the irregularities to reveal the general trend or pattern in the series.

There is a duality between the $AR(p)$ and $MA(q)$ models. An $MA(q)$ is equivalent to $AR(\infty)$, an AR process with an infinite order. Conversely, an $AR(p)$ process is equivalent to an $MA(\infty)$. Therefore, if a low-order model of one type explains a time series, then a high-order model of the other type should also explain the series. This leads to an application of the *Principle of Parsimony* (i.e., *Occam's Razor*): use the simplest model which is the one with the lowest order. This includes the model with the fewest parameters and fewest, simplest assumptions. Also, the MA model is good when shocks (i.e., information or innovations measured as a) are temporary (i.e., one period) while the AR is good when there are carry-over or lingering effects (e.g., consumption spending).

It can be shown that the ACF cuts-off after lag 1 for an $MA(q)$ model while the $PACF$ does not; instead, it decays exponentially to zero. This leads to signatures for the $MA(q)$ that I summarize in Table 7.7. Based on the graphs in Fig. 7.17, an $MA(2)$ might be suggested in addition to the $AR(2)$ as a first iteration of model identification.

Process	*ACF*	*PACF*
$AR(p)$	Decays exponentially	Spikes at lags 1 to p, then cuts off
$MA(q)$	Spikes at lags 1 to q, then cuts off	Decays exponentially

Table 7.7 These are the signatures for the $AR(p)$ and $MA(q)$ models. This table is an extension of Table 7.6

7.10.1.3 ARMA(p, q) Model

You can extend your model to include both AR and MA components to capture lingering effects and temporary shocks, respectively. This enhanced model is written as

$$Y_t = \phi_1 Y_{t-1} + \ldots + \phi_p Y_{t-p} + a_t + \theta_1 a_{t-1} + \ldots + \theta_q a_{t-q} \quad (7.10.17)$$

This is called an $ARMA(p, q)$ model. You can view it as a combination of an $AR(p)$ amd $MA(q)$ process suggesting that the ACF will decay geometrically from lag 1 onward like the $AR(1)$ process but the $PACF$ will decay in magnitude from an initial value $\phi_{11} = \rho_1$. Like for the $MA(q)$ process. I extend the signature summary tables in Table 7.8 to include these observations.

In practice, many times series possibilities are special cases of $ARMA(p, q)$ models with $p + q \leq 2$. For examples, $ARMA(1, 0) \rightarrow AR(1)$, $ARMA(2, 0) \rightarrow AR(2)$, $ARMA(0, 1) \rightarrow MA(1)$, $ARMA(0, 2) \rightarrow MA(2)$, and $ARMA(1, 1)$ are possibilities. Clearly, $AR(p)$ and $MA(q)$ models are special cases of the more general $ARMA(p, q)$.

Process	*ACF*	*PACF*
$AR(p)$	Decays exponentially	Spikes at lags 1 to p, then cuts off
$MA(q)$	Spikes at lags 1 to q, then cuts off	Decays exponentially
$ARMA(p, q)$	Decays exponentially; may oscillate	Decays exponentially; may oscillate

Table 7.8 These are the signatures for the three models: $ARMA(p, q)$, $AR(p)$ and $MA(q)$ models. This table is an extension of Table 7.7

7.10.1.4 ARIMA(p, d, q) Model

The final extension of our base model is the $ARIMA(p, d, q)$ model, which stands for *Autoregressive, Integrated, Moving Average*. The "Integrated" component arises from the observation that many time series have a rising (falling) trend which implies that the mean increases (decreases) over time. In addition, the variance normally increases as time passes implying more volatility from a more complex underlying generating process. The models I considered so far assume a constant mean and constant variance; the models are said to be *stationary*. When these conditions do not hold, the series is said to be *non-stationary*. Transforming a time series using a first difference is usually sufficient to make it mean stationary. Using a natural log transformation makes its variance stationary. See Wei (2006) for a proof of the variance transformation and discussions about stationarity. Also, see Box et al. (1994).

Differencing by itself is sometimes insufficient to achieve stationarity, especially for a series that spans long periods. Changes in logs may be appropriate since a log difference is a percentage change. Frequently, economic times series exhibit nonstationarity and the first or second difference will be still be stationary. Sample autocorrelations can be large and remain large even at long lags. If nonstationarity is suspected, then the sample autocorrelations of the first differences should be examined. Only occasionally in economic time series data will sample autocorrelations of first differences fail to damp out, in which case second differences are examined.

When a time series is first differenced, that differencing is denoted as $d = 1$. Although the series may be first differenced to induce stationarity for statistical purposes and forecasting, it is the original, undifferenced series that you are concerned with so the differenced series has to be converted back to the original once the forecast is complete. This is called "Integration", represented as "I". The amount of differencing, and, therefore, integration is represented by "d". The $ARIMA(p, d, q)$ model has characteristics of an $ARMA(p, q)$ model but with d differencing of the data before statistical work is done.

7.10.1.5 Digression: Time Series Stationarity—An Overview

A major feature of any time series is its stationarity. A time series is stationary if its mean is invariant with respect to time. That is, no matter what time period you look at, the mean is the same. The mean, in this case, is the expected value of the random variable, Y_t. This definition is extended to the variance and covariance which are also time invariant.

As an example, consider the simple $AR(1)$ model: $Y_t = \beta_0 + \rho \times Y_{t-1} + a_t$ where a_t is white noise with mean zero and variance σ^2. It can be shown that the mean is $E(Y_t) = \beta_0/1-\rho$ and the variance is $V(Y_t) = \sigma^2/1-\rho^2$. In both instances, these are constants with respect to time. I derive these results in the Appendix to this chapter.

Notice that the variance is also finite as long as $\mid \rho \mid < 1$. If $\mid \rho \mid = 1$, then the variance is undefined. In this case, the series is said to have a *unit root*. A condition for stationarity is $\mid \rho \mid < 1$. See Hill et al. (2008) for details.

There are two tests for stationarity:

1. Dickey-Fuller Test and
2. Kwiatkowski-Phillips-Schmidt-Shin Test

I will give an overview of each and their application in Python in the following two subsections.

Dickey-Fuller Test The $AR(1)$ process $Y_t = \rho \times Y_{t-1} + a_t$ is stationary when $\mid \rho \mid < 1$. When $\rho = 1$ it becomes a nonstationary *random walk process*: $Y_t = Y_{t-1} + a_t$. You should, therefore, test whether ρ equals one or is significantly less than one. These tests are known as unit root tests for stationarity. The most popular is the *Dickey–Fuller Test*. For the simple $AR(1)$ model, the Null Hypothesis is $\rho = 1$ and the Alternative Hypothesis is $\mid \rho \mid < 1$. The Null Hypothesis states that you have nonstationarity.

You can show that

$$Y_t - Y_{t-1} = \rho \times Y_{t-1} - Y_{t-1} + a_t \tag{7.10.18}$$

or

$$\Delta Y_t = (\rho - 1) \times Y_{t-1} + a_t \tag{7.10.19}$$

$$= \gamma \times Y_{t-1} + a_t \tag{7.10.20}$$

where $\gamma = \rho - 1$. The Null Hypothesis, which is $\rho = 1$, is then equivalent to $\gamma = 0$ and the Alternative Hypothesis is equivalent to $\gamma < 1$. This is the simplest version of the Dickey-Fuller Test, or *Case I*. A second case, *Case II*, includes a constant so $\Delta Y_t = \beta_0 + \gamma \times Y_{t-1} + a_t$, but the Null and Alternative Hypotheses remain the same. A third case, *Case III*, includes a constant and a deterministic trend: $\Delta Y_t = \beta_0 + \gamma \times Y_{t-1} + \lambda \times t + a_t$ with the Null and Alternative Hypotheses the same. These three cases are estimated using *OLS*, but with one modification. The t-distribution cannot be used to test the hypotheses about the γ coefficient since it is not applicable for this problem. Instead, a statistic called the τ-statistic, which has tabled values, is used. If τ_c is the critical value, then you reject the Null Hypothesis of nonstationarity if $\tau \leq \tau_c$ and not reject otherwise.

The Dickey-Fuller Test is extended even further by allowing for the possibility of an autocorrelated disturbance term. In this case, the test is called the *Augmented Dickey-Fuller Test*. As noted by Hill et al. (2008), this is almost always used in practice.

Hill et al. (2008) recommend the following steps:

1. Plot the variable of interest against time series.
   - If the series fluctuates around a zero sample average, use the Case I model.
   - If the series fluctuates around a nonzero sample average, use the Case II model.
   - If the series fluctuates around a linear trend, use the Case III model.
2. There is a fourth possibility: a constant plus linear and quadratic trend.

The Dickey-Fuller Tests are implemented in the statsmodels' *tsa* package. There is a submodule, *adfuller*, for doing them, although it has an argument that allows you to do the three I cited above. The argument is "regression" with four possible settings which I list in Table 7.9. I illustrate the use of this function in Fig. 7.18.

Case	Argument setting
I: No constant	'nc'
II: Constant only	'c' (default)
III: Constant and linear trend	'ct'
IV: Constant and linear and quadratic trend	'ctt'

Table 7.9 These are the possible argument settings for the Augmented Dickey-Fuller Test. The argument name is 'regression'. So, regression = 'nc' does the Dickey-Fuller Test without a constant

KPSS Test Another test for stationarity is the *Kwiatkowski-Phillips-Schmidt-Shin Test* (*KPSS*). See Kwiatkowski et al. (1992) for details. The Null Hypothesis for this test is that the series is either constant or trend stationary, so there are two cases. The statsmodels submodule *kpss* has an argument "regression" that has two options that I show in Table 7.10. I show an application for the constant stationarity case in Fig. 7.19.

7.10.2 Step 2: Estimation of the Model

Times series model estimation usually involves a maximum likelihood approach, not an *OLS* approach in which a residual sum of squares is minimized with respect to the unknown parameters of a linear model. In the *OLS* framework, the estimators are the same as the maximum likelihood estimators under what I call the Classical Assumptions. These assumptions are that the disturbance terms in the linear model

```
##
## Log pocket price
##
ax = df_ts.log_Pprice.plot()
ax.set_title( 'Time Series Plot\nLog Pocket Price', fontsize = font_title )
ax.set( ylabel = 'Log Pocket Price', xlabel = '' );
```

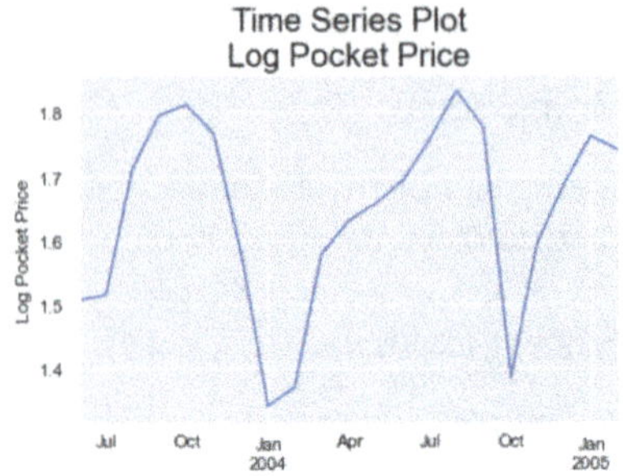


```
##
## Augmented Dickey-Fuller Test
## Case II: Constant
##
from statsmodels.tsa.stattools import adfuller
adf, pval, lags, obs, crit, icbest = adfuller( df_ts.log_Pprice, regression = 'c' )
print( f'Dickey-Fuller Statistic: {adf:.3f}\np-value: {pval:.3f}' )
if pval > 0.05:
    print( 'Conclusion: Reject Null Hypothesis' )
else:
    print( 'Conclusion: Do Not Reject Null Hypothesis' )
```

```
Dickey-Fuller Statistic: 0.299
p-value: 0.977
Conclusion: Reject Null Hypothesis
```

Fig. 7.18 This illustrates the application of the Augmented Dickey-Fuller Test to the pocket price time series. Notice that the time series plot shows that the series varies around 1.6 on the log scale. This suggests Case II which includes a constant but no trend. The test suggests there is stationarity since the Null Hypothesis is that the series is nonstationary

are uncorrelated, have constant variance (i.e., *homoskedasticity*, and have zero mean (i.e., zero expected value). This is the basis for the *Gauss-Markov Theorem*. See Hill et al. (2008) and Greene (2003) for discussions. Under the Classical Assumptions, the maximum likelihood procedure will give the same results. For time series model estimation, the maximum likelihood procedure is used almost exclusively.

Case	Argument setting
I: Constant	'c'
II: Linear Trend	'ct'

Table 7.10 These are the possible argument settings for the *KPSS* Test. The argument name is 'regression'

The *AR(1)* model is estimated using the *statsmodels*' *AutoReg* function in the *tsa* submodule package. I illustrate how to do this in Fig. 7.20 for the pocket price times series. The model is first instantiated and then fit using the *fit( )* function. The results are stored and they can be retrieved as before.

```
##
## KPSS Test
## Case I: Constant or Level Stationarity
##
from statsmodels.tsa.stattools import kpss
kpss, pval, lags, crit = kpss( df_ts.log_Pprice, regression = 'c' )
print( f'KPSS Statistic: {kpss:.3f}\np-value: {pval:.3f}' )
if pval < 0.05:
    print( 'Conclusion: Reject Null Hypothesis' )
else:
    print( 'Conclusion: Do Not Reject Null Hypothesis' )
```

```
KPSS Statistic: 0.325
p-value: 0.100
Conclusion: Do Not Reject Null Hypothesis
```

Fig. 7.19 This illustrates the application of the *KPSS* Test to the pocket price time series. The time series plot in Fig. 7.18 suggests constant or level stationarity. The test suggests there is level stationarity

```
##
## Import the AutReg module
##
from statsmodels.tsa.ar_model import AutoReg
##
## Instantiate the model
## Specify a lag of one period (for the AR(1) and
## a constant term.
##
mod = AutoReg( df_ts.log_Pprice, lags = 1, trend = 'c' )
##
## Fit the instantiated model
##
ar01 = mod.fit( )
print( ar01.summary() )
```

```
                            AutoReg Model Results
==============================================================================
Dep. Variable:                log_Pprice   No. Observations:                   21
Model:                        AutoReg(1)   Log Likelihood                  13.397
Method:                  Conditional MLE   S.D. of innovations              0.124
Date:                   Mon, 10 Aug 2020   AIC                             -3.878
Time:                           12:46:53   BIC                             -3.728
Sample:                       07-31-2003   HQIC                            -3.848
                            - 02-28-2005
================================================================================
                  coef    std err          z      P>|z|      [0.025      0.975]
--------------------------------------------------------------------------------
intercept       0.7988      0.310      2.580      0.010       0.192       1.406
log_Pprice.L1   0.5191      0.188      2.756      0.006       0.150       0.888
                                    Roots
=============================================================================
                  Real          Imaginary           Modulus         Frequency
-----------------------------------------------------------------------------
AR.1            1.9263           +0.0000j            1.9263            0.0000
-----------------------------------------------------------------------------
```

Fig. 7.20 The *AR(1)* model for the pocket price times series

7.10.3 Step 3: Validation of the Model

Validating a model means checking its predictive ability. Just because a model has good estimation results does not mean it will predict well. In fact, it could be a very poor predictor. The reason is that the model was trained on a specific data set so it "knows" that data. A forecast, however, requires that the model venture into unknown territory. There may be trends or patterns that were not in the past data set which is what the model "knows." Validation is a complex problem which I discuss in Chap. 10. So, it is best to postpone discussion of this important topic until then.

7.10.4 Step 4: Forecasting with the Model

Once a time series is converted to a stationary series (if necessary) and a model identified and estimated, there is one final task that have to be completed: the model must then be used to produce a forecast. A major use of the estimated model is forecasting. The periods you forecast are sometimes called *steps*. If you forecast one period, then it is a 1-step head forecast; two periods are 2-steps ahead; *h* periods are h-steps ahead. You forecast *h-steps* ahead using the estimated model. The procedure to develop a forecast depends on the model you estimated: one member in the general class of *ARIMA(p, d, q)* models. I only explored the simple *AR(1)* model in the previous section. That model can be used to forecast using the *predict* method associated with the estimated model. I show in Fig. 7.21 how this would be done for 4-steps ahead of the last recorded date in the time series. The results are shown in Fig. 7.22. For more details on forecasting see Box et al. (1994), Wei (2006), and Nelson (1973).

```
from dateutil.relativedelta import *
from datetime import datetime
##
## Set steps ahead
##
fct_start = df_ts.Pprice.index[ - 1 ]+ relativedelta(months =+ 1 )
fct_end = fct_start + relativedelta(months =+ 3 )
##
## Print some data
##
print( f'Historical Period End: {df_ts.Pprice.index[ - 1 ]:%m/%d/%Y}' )
print( f'\nForecast Start Period: {fct_start:%m/%d/%Y}' )
print( f'\nForecast End Period: {fct_end:%m/%d/%Y}' )
##
## Forecast using the predict method
##
fct = ar01.predict( start = fct_start, end = fct_end )
##
## Print results
##
print( f'\nBase Series: {df_ts.Pprice}' )
print( f'\nForecast Series:\n{fct}' )
x = df_ts.Pprice.append( fct )
print( f'\nCombined Series:\n{x}' )
ax = x.plot()
ax.set_title( 'Pocket Price\nActuals and Forecast', fontsize = font_title )
ax.set( ylabel = 'Pocket Price ($)' )
ax.axvline( df_ts.Pprice.index[ - 1 ], linestyle = 'dashed', linewidth = 1, color = 'red' );
```

Fig. 7.21 The *AR(1)* model is used to forecast the pocket price times series. In this case, I forecast 4-steps ahead, or four periods into the future

```
Historical Period End: 02/28/2005

Forecast Start Period: 03/28/2005

Forecast End Period: 06/28/2005

Base Series: Tdate
2003-06-30    4.513912
2003-07-31    4.548639
2003-08-31    5.528507
2003-09-30    6.018170
2003-10-31    6.114422
2003-11-30    5.846772
2003-12-31    4.836774
2004-01-31    3.826775
2004-02-29    3.938385
2004-03-31    4.843687
2004-04-30    5.101348
2004-05-31    5.243045
2004-06-30    5.431462
2004-07-31    5.792135
2004-08-31    6.250363
2004-09-30    5.901477
2004-10-31    3.998509
2004-11-30    4.918601
2004-12-31    5.401024
2005-01-31    5.825170
2005-02-28    5.697888
Freq: M, Name: Pprice, dtype: float64

Forecast Series:
2005-03-31    1.702183
2005-04-30    1.682501
2005-05-31    1.672283
2005-06-30    1.666978
Freq: M, dtype: float64
```

(a)

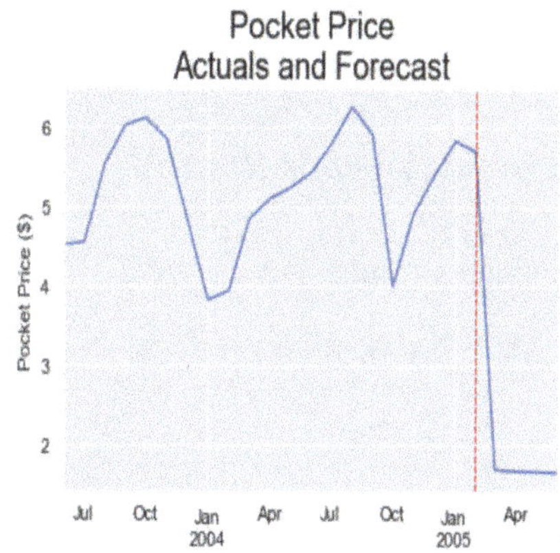

(b)

Fig. 7.22 These are the 4-steps ahead forecasts for the pocket prices. (**a**) Forecast values. (**b**) Forecast plot

7.11 Appendix

I will show the key time series stationarity results in this Appendix. But first, I will introduce an operator, called the *backshift operator* that will make the derivations easier.

7.11.1 Backshift Operator

Consider a time series Y_t. The one-period lag of this series is just Y_{t-1}. You can define an operator B, called the *backshift operator*, that, when applied to the original time series, produces the lagged series. That is, $BY_t = Y_{t-1}$. This has a useful property: $B^2Y_t = B(BY_t) = BY_{t-1} = Y_{t-2}$. The "power" 2 does not mean "square"; it means apply the operator twice. Consequently, $B^nY_t = Y_{t-n}$. Also, if c is a constant, then $Bc = c$.

Now consider the $AR(1)$ model: $Y_t = \beta_0 + \rho \times Y_{t-1} + a_t$. Using the backshift operator, this can be written as

$$Y_t = \beta_0 + \rho \times Y_{t-1} + a_t \tag{7.11.1}$$

$$= \beta_0 + \rho \times BY_t + a_t \tag{7.11.2}$$

$$= \frac{\beta_0}{1-\rho} + \frac{a_t}{1-\rho} \tag{7.11.3}$$

where the Y_t terms were merely collected. For more information on the backshift operator from a mathematical perspective, see Dhrymes (1971, Chapter 2).

7.11.2 Useful Algebra Results

Two useful algebra results are:

1. $\frac{1}{1-\rho} = \sum_{i=0}^{\infty} \rho^i$; and
2. $\frac{1}{1-\rho^2} = \sum_{i=0}^{\infty} \rho^{2\times i}$.

   Each can be shown by doing long division.

7.11.3 Mean and Variance of Y_t

Using the result in (7.11.3), the mean and variance of the $AR(1)$ model are easy to find. The mean, that is, the expected value, is

$$E(Y_t) = \frac{\beta_0}{1-\rho} + \frac{E(a_t)}{1-\rho} \tag{7.11.4}$$

$$= \frac{\beta_0}{1-\rho} \tag{7.11.5}$$

since a_t is white noise. Therefore, $\mu = \frac{\beta_0}{1-\rho}$ or $\beta_0 = \mu \times (1-\rho)$ which is invariant with respect to time.

The variance of Y_t is

$$V(Y_t) = V\left(\frac{\beta_0}{1-\rho}\right) + V\left(\frac{a_t}{1-\rho}\right) \tag{7.11.6}$$

$$= \frac{\sigma^2}{1-\rho^2} \tag{7.11.7}$$

where the first term is zero since it is a constant. The second term simplifies to $V(a_t) \times \sum_{i=0}^{\infty} \rho^{2 \times i}$. The power of 2 results from each coefficient having to be squared when finding the variance. I used a result I stated above which gives the result I stated in the text. You can also get (7.11.7) by noting that $V(\beta_0) = 0$, $V(Y_t) = V(Y_{t-1}) = \sigma^2$, and the backshift operator applied to a constant returns the constant. Simplifying give (7.11.7).

7.11.4 Demeaned Data

Consider the $AR(1)$ model. Suppose you subtracted the mean, μ, from both sides where you know that $\mu = \beta_0/1-\rho$. This gives you

$$Y_t - \mu = \beta_0 + \rho \times Y_{t-1} + a_t - \mu \tag{7.11.8}$$

$$= \mu \times (1 - \rho) + \rho \times Y_{t-1} + a_t - \mu \tag{7.11.9}$$

$$= \rho \times (Y_{t-1} - \mu) + a_t. \tag{7.11.10}$$

This is a "demeaned" version of the time series: the mean is subtracted from each value of the series. It is easy to show that $E(Y_t - \mu) = 0$. So, the demeaned version is stationary. This implies that you can work with either the original series or the demeaned series. Most time series analysts work with the demeaned series.

7.11.5 Time Trend Addition

You can add a linear time trend to the $AR(1)$ model as $Y_t = \beta_0 + \rho \times Y_{t-1} + a_t + \delta \times t$ where δ is a constant. Just as you demeaned the time series, you could also *detrend* the series by subtracting the trend term in addition to the mean. The expected value of the detrended series is zero so the series with the linear trend is said to be *deterministic trend stationary* because the trend is deterministic. See Hill et al. (2008).

Chapter 8
Statistical Tables

Statistical tables supplement scientific data visualization for the analysis of categorical data. This type of data is exemplified by, but not restricted to, survey data. Any categorical data can be analyzed via tables. I will continue to use the bread baking company Case Study data. I will use the customers at the class level where I had defined the classes in Table 3.1, which I repeat here for convenience as Table 8.1.

Group	Class	Example
Grocery	Mass merchandise	Target, Walmart
	Supermarkets	Krogers, Shoprite
	Club stores	BJs, Costco, Sams Club
Convenience	Convenience stores	7-Eleven
	Dollar stores	Dollar store
	Drug stores	CVS, Walgreen, Rite Aid
Restaurants	Restaurants	Dinners, etc.

Table 8.1 This is a listing of the bakery's customers by groups and classes within a group I previously defined in Chap. 3

8.1 Data Preprocessing

As the data scientist for the baking company, you import data for the Case Study from two *CSV* files into two separate DataFrames. One has the accounts data and the other the customer specific data. You merged them into one using the methods I discussed in Chap. 3. Unfortunately, after the data are merged, you learn of a major problem. The *IT* department discovered that many customers were incorrectly classified regarding their classes and payment status. This is not unusual and reflects the fact that real world data are often very messy. The *IT* staff informed you that

W. R. Paczkowski, *Business Analytics*,
https://doi.org/10.1007/978-3-030-87023-2_8

they need several weeks to correct the errors but your analysis is needed now. So, you have to recode what you have based on a mapping they provided. I show the mapping in Table 8.2 and how this remapping could be done in Fig. 8.1. I chose to do this by first creating a copy of my DataFrame, calling it *tmp*, and then using a series of list comprehensions to do the recoding. The remapped data will be used in this chapter.

Old classification	New classification
SuperCenter	Restaurant
Mass merchandise	Convenience store
Convenience store	Mass merchandise
Past due	Current

Table 8.2 These are the new mappings to correct incorrect labeling. You can see the code to implement these mappings in Fig. 8.1

```
##
## Create temporary copy of DataFrame
##
tmp = df.copy()
##
## Create new variables
##
tmp[ 'storeTypeNew' ] = tmp.storeType
tmp[ 'paymentStatusNew' ] = tmp.paymentStatus
##
## Remap store types
##
tmp.storeTypeNew = [ 'Restaurant' if x == 'SuperCenter' else x for x in tmp.storeTypeNew ]
tmp.storeTypeNew = [ 'Temp' if x == 'Mass Merchandise' else x for x in tmp.storeTypeNew ]
tmp.storeTypeNew = [ 'Mass Merchandise' if x == 'Convenience' else x for x in tmp.storeTypeNew ]
tmp.storeTypeNew = [ 'Convenience' if x == 'Temp' else x for x in tmp.storeTypeNew ]
##
## Remap payment status
##
tmp.paymentStatusNew = [ 'Temp' if x == 'Past Due' else x for x in tmp.paymentStatusNew ]
tmp.paymentStatusNew = [ 'Past Due' if x == 'Current' else x for x in tmp.paymentStatusNew ]
tmp.paymentStatusNew = [ 'Current' if x == 'Temp' else x for x in tmp.paymentStatusNew ]
##
## Overwrite old DataFrame
##
df = tmp.copy()
```

Fig. 8.1 This illustrates the code to remap values in a DataFrame

8.2 Categorical Data

Not all the data analysis in *BDA* involves ratio data such as prices, discounts, units sold, time for order delays, credit scores, and so forth. There are instances when an analysis requires *categorical data*, that is, nominal or ordinal data. For instance, the bread baking company has an accounting database that classifies accounts receivable for its 400 customers as "Current" (i.e., paid on time), "1–30 Days Past Due", "31–60 Days Past Due", "Need Collection." This is clearly ordered.

As an example, customers are in marketing and sales regions each of which is managed by a vice president. The analysis problem may be to identify regional vice presidents with the worst accounts receivable performance with respect to past-due classification. A variant focuses on the distribution of past due status to determine if it is independent of any particular regional vice president so that an accounts receivable problem is systemic to the business and not a particular region. The regions are a nominal classification of customers and the past due payment status is an ordinal classification.

A first step you should follow for analyzing ordinal data is to create a categorical data type. You do this by first importing the Pandas *CategoricalDtype* module using *from pandas.api.types import CategoricalDtype*. Using this module allows you to declare a variable as data type *categorical* as well as specify an order for the levels. It is not necessary to specify an order since order will not affect any computations or results, but you should do this for an ordinal variable since, by definition of this type of variable, order counts for interpretation. Also, the printed output will be clear and logical.

Technically, a variable with an object data type can be analyzed without regard to its categorical nature. Specifying it as categorical economizes on internal storage and data processing since a categorical data type is stored and handled differently than one that is just an object data type. An object variable has each level repeated "as-is", which is inefficient. As an example, consider an object variable *Region* that has four levels: Midwest, Northeast, South, and West. These are in your DataFrame as strings, each one repeated a potentially large number of times depending on the depth of your DataFrame. If the levels are assigned integer values, however, say 1 for Midwest, 2 for Northeast, etc., then only those integers have to be stored with a simple look-up table containing the translation. There is considerable storage and processing saved using this scheme. This is the essence of what the Pandas *CategoricalDtype* module allows you to do with an object variable. You could achieve the same result, by the way, by label encoding the Region categories using the *LabelEncoder* I discussed in Chap. 5. Using the *CategoricalDtype* is more efficient.

You create the categorical data type by creating an instantiation of the *CategoricalDtype* class as a variable and then using it as an argument to the *astype( )* method applied to the object variable to be categorized. I illustrate this in Fig. 8.2.

8.3 Creating a Frequency Table

A simple first summary table is a *frequency table* comparable to the first table you were taught in a basic statistics class. I show such a table in Fig. 8.3 and another in Fig. 8.4 which is subsetted on the Midwest region. In both instances, I added a bar chart style to the DataFrame display to highlight patterns in keeping with the pattern identification I discussed in Chap. 4. The *stb* command in Fig. 8.3 is an accessor method that accesses the data in the DataFrame. This accessor has a function named *freq* that produces the tables I show. You install *stb* using *pip install sidetable* or *conda install -c conda-forge sidetable*.

```
##
## Set paymentStatus order
##
lst_odr = [ 'Current', 'Past Due', 'Warning', 'Collection'  ]
##
## Instantiate the CategoricalDtype class
##
cat = CategoricalDtype( lst_odr, ordered = True )
##
## Apply the instantiated variable through astype( )
##
df.paymentStatusNew = df.paymentStatusNew.astype( cat )
##
## Display the DataFrame info
##
df.info()
```

```
<class 'pandas.core.frame.DataFrame'>
Int64Index: 730000 entries, 0 to 729999
Data columns (total 19 columns):
 #   Column            Non-Null Count   Dtype
---  ------            --------------   -----
 0   cid               730000 non-null  object
 1   fids              730000 non-null  int64
 2   invoiceNumber     730000 non-null  float64
 3   deliveryDate      730000 non-null  datetime64[ns]
 4   invoiceAmount     730000 non-null  float64
 5   invoiceDate       730000 non-null  datetime64[ns]
 6   dueDate           730000 non-null  datetime64[ns]
 7   paidDate          730000 non-null  datetime64[ns]
 8   daysToPay         730000 non-null  int64
 9   daysLate          730000 non-null  int64
 10  paymentStatus     730000 non-null  object
 11  fid               730000 non-null  int64
 12  region            730000 non-null  object
 13  state             730000 non-null  object
 14  stateCode         730000 non-null  object
 15  location          730000 non-null  object
 16  storeType         730000 non-null  object
 17  storeTypeNew      730000 non-null  object
 18  paymentStatusNew  730000 non-null  category
dtypes: category(1), datetime64[ns](4), float64(2), int64(4), object(8)
memory usage: 106.5+ MB
```

Fig. 8.2 A Categorical data type is created using the *CategoricalDtype* method. In this example, a list of ordered levels for the *paymentStatus* variable is provided. The categorical specification is applied using the *astype( )* method

You should immediately notice in both tables that the payment categories are in the order I specified in Fig. 8.2. This order makes logical sense. But more importantly, this order allows you to make sense of the cumulative count and cumulative percent columns; without the order, these columns would not make sense. These two columns reveal that 88% of the customers both nationally and in the Midwest region are current or past due, but that there is a big gap between both levels. In particular, the overlayed bar chart highlights that 81% of the customers are past due. This should be cause for alarm.

```
##
## Frequency Distribution Table
##
df.stb.freq( [ 'paymentStatusNew' ], sort_cols = True ).style.set_caption( 'Payment Status' ).\
    set_table_styles( tbl_styles ).hide_index().bar( align = 'zero', color = 'lightblue' ).\
    format( {'count':'{:,d}', 'percent': '{:0.1f}%', 'cumulative_count':'{:,d}', 'cumulative_percent': '{:0.1f}%'} )
```

Payment Status

paymentStatusNew	count	percent	cumulative_count	cumulative_percent
Current	51,150	7.0%	51,150	7.0%
Past Due	591,151	81.0%	642,301	88.0%
Warning	65,968	9.0%	708,269	97.0%
Collection	21,731	3.0%	730,000	100.0%

Fig. 8.3 The variable with a declared categorical data type is used to create a simple frequency distribution of the recoded payment status. Notice how the levels are in a correct order so that the cumulative data make logical sense

```
##
## Frequency Distribution Table, MW only
##
tmp = df.query( 'region == "MW"' )
tmp.stb.freq( [ 'region', 'paymentStatusNew' ], sort_cols = True ).style.set_caption( 'Payment Status: MW Only' ).\
    set_table_styles( tbl_styles ).hide_index().bar( align = 'zero', color = 'lightblue' ).\
    format( {'count':'{:,d}', 'percent': '{:0.1f}%', 'cumulative_count':'{:,d}', 'cumulative_percent': '{:0.1f}%'} )
```

Payment Status: MW Only

region	paymentStatusNew	count	percent	cumulative_count	cumulative_percent
MW	Current	5,093	7.0%	5,093	7.0%
MW	Past Due	59,170	81.1%	64,263	88.0%
MW	Warning	6,530	8.9%	70,793	97.0%
MW	Collection	2,207	3.0%	73,000	100.0%

Fig. 8.4 The variable with a declared categorical data type is used to create a simple frequency distribution, but this time subsetted on another variable, *region*

8.4 Hypothesis Testing: A First Step

Knowing the frequency distribution is insightful, but there is more you can do. In particular, you could test the hypothesis that the frequency distribution equals a known distribution. For example, suppose you know the industry payment status for drug stores in California. Are your customers' payment status statistically different from the industry in this segment and area? You can use a *chi-square test* to compare your distribution against the known industry distribution. The hypotheses are

$$H_O : \text{Equality with the Industry Distribution} \tag{8.4.1}$$

$$H_A : \text{Inequality with the Industry Distribution.} \tag{8.4.2}$$

The *chi-square test statistic* is

$$\chi^2 = \sum_{i=1}^{k} \frac{(O_i - E_i)^2}{E_i} \tag{8.4.3}$$

where O_i is the observed frequency for level i, k is the number of levels, and E_i is the expected frequency for level i. The statistic in (8.4.3) is called the *Pearson Ch-Square Statistic*. I discuss the distribution of this statistic in this chapter's Appendix.

Payment status	Industry proportion
Current	0.07
1–30 days late (pass due)	0.80
31–60 days late (warning)	0.10
Collection	0.03
Total	1.00

Table 8.3 This is the (hypothetical) distribution for the industry for drug stores in California. This corresponds to the distribution in the dictionary named *industry* in Fig. 8.6

Assume that the industry distribution is the one I show in Table 8.3. The observed frequencies are from the frequency table, such as the one in Fig. 8.5. The expected frequency is calculated assuming the Null Hypothesis is true. The calculation is simple: multiply the total frequency by the respective Null Hypothesis distribution's relative frequency (e.g., the proportion column in Table 8.3).

I first show the frequency distribution for the relevant data subset (i.e., drug stores in California) in Fig. 8.5. The chi-square calculation is in Fig. 8.6. The chi-square function has two parts: observed and expected frequencies. The observed frequency is calculated using the Pandas *value_counts* method. The expected frequency is the number of observations times the expected proportion in each cell. These are themselves the product of the respective marginal proportions because of independence. The chi-square function returns a 2-tuple: the chi-square statistic and its p-value. This 2-tuple is "unpacked" (i.e., split into two separate variables) as I show in Fig. 8.6. The p-value is compared to the conventional significance level, $\alpha = 0.05$. There is nothing scared about this level, however. In an industrial setting, $\alpha = 0.01$ (or less) might be used because of the precision required in most manufacturing contexts. In marketing, $\alpha = 0.05$ is typically used.

```
##
## Frequency Distribution Table, drug stores in California
##
tmp = df.query( "( storeTypeNew == 'Drug' ) & ( stateCode == 'CA')" )
tmp.stb.freq( [ 'paymentStatusNew' ], sort_cols = False ).\
    style.set_caption( 'Payment Status' ).\
    set_table_styles( tbl_styles ).hide_index().bar( align = 'zero', color = 'lightblue' ).\
    format( {'count':'{:,d}', 'percent': '{:0.1f}%', 'cumulative_count':'{:,d}', 'cumulative_percent': '{:0.1f}%'} )
```

Payment Status

paymentStatusNew	count	percent	cumulative_count	cumulative_percent
Past Due	69,625	81.2%	69,625	81.2%
Warning	7,697	9.0%	77,322	90.1%
Current	5,819	6.8%	83,141	96.9%
Collection	2,634	3.1%	85,775	100.0%

Fig. 8.5 This is the frequency table for drug stores in California. Notice that 81.2% of the drug stores in California are past due

```
##
## Chi-Square Test comparing two distributions
##
## Subset the data and use value-counts method for observed frequency distribution
## Put result in dictionary and create a DataFrame
##
tmp = df.query( "(storeTypeNew == 'Drug') & (stateCode == 'CA')" )
obs_freq = tmp.paymentStatusNew.value_counts( sort = False ).to_dict( )
df_chi = pd.DataFrame.from_dict( obs_freq, orient = 'index', columns = [ 'Observed' ] )
##
## Create DataFrame of expected frequencies
##
industry = { 'Current':0.07, 'Past Due':0.80, 'Warning':0.10, 'Collection': 0.03 }
industry = pd.DataFrame.from_dict( industry, orient = 'index', columns = [ 'Expected' ] )
##
## Merge
## Put in logical order
##
df_chi = df_chi.merge( values, left_index = True, right_index = True )
ord = [ 'Current', 'Past Due', 'Warning', 'Collection' ]
df_chi = df_chi.reindex( ord )
nobs = tmp.shape[ 0 ]
df_chi[ 'Expected_Freq' ] = df_chi.Expected * nobs
display( df_chi.style.set_caption( 'Observed and Expected Data' ).set_table_styles( tbl_styles ).\
        format( { 'Expected':'{:0.1%}', 'Expected_Freq':'{:0.2f}' } ) )
##
## Use chisquare function with observed frequency and expected frequency
##
stat, pval = chisquare( df_chi.Observed, df_chi.Expected_Freq )
print( 'Chi-Square Test:' )
print( f'\tStatistic: {stat:0.3f}\n\tpval: {pval:0.3f}' )
if pval < 0.05:
    print( '\tDecision: Reject HO' )
else:
    print( '\tDecision: Do Not Reject HO' )
```

Observed and Expected Data

	Observed	Expected	Expected_Freq
Current	5819	7.0%	6004.25
Past Due	69625	80.0%	68620.00
Warning	7697	10.0%	8577.50
Collection	2634	3.0%	2573.25

```
Chi-Square Test:
        Statistic: 112.254
        pval: 0.000
        Decision: Reject HO
```

Fig. 8.6 This illustrates a chi-square test comparing an observed frequency distribution and an industry standard distribution. The industry distribution is in Table 8.3. The Null Hypothesis is no difference in the two distributions. The Null is rejected at the $\alpha = 0.05$ level of significance

8.5 Cross-tabs and Hypothesis Tests

In the previous section, I presented an analysis using one categorical variable. What if you have two? Or two that are categorical and a third that is quantitative? I address the first question in this section and the second question in the following section.

You can create a simple *cross tabulation* (*a.k.a.*, *cross-tab* or *tab*) using the Pandas *crosstab* method. You have to specify the row and column variables separately as Pandas series or Numpy arrays. I recommend the Pandas series since you will most likely have your data stored in a Pandas DataFrame. The method returns a new DataFrame. The series specified for the rows of the cross-tab define the index of the returned DataFrame. The cell values of the tab are the frequencies at the intersection of the index and the columns. This is the default aggregation, but you can change this as I will show shortly. I show an example of a simple tab in

Fig. 8.7. This shows the cross tabulation of customer class and payment status for all customers located in California.

```
##
## Crosstab of payment status by stores in California
##
## Query DataFrame
##
tmp = df.query( "stateCode == 'CA'" )
##
## Create crosstab: payment status is row index
##
pd.crosstab( tmp.paymentStatusNew, tmp.storeTypeNew, margins = True ).\
    style.set_caption( 'Crosstab of Payment Status by Store Type in CA' ).\
    set_table_styles( tbl_styles )
```

Crosstab of Payment Status by Store Type in CA

storeTypeNew	Club	Convenience	Dollar	Drug	Mass Merchandise	Restaurant	Supermarket	All
paymentStatusNew								
Current	915	357	256	5819	382	901	481	9111
Past Due	10261	4411	2962	69625	4414	10390	5932	107995
Warning	1206	539	332	7697	521	1115	672	12082
Collection	393	168	100	2634	158	369	215	4037
All	12775	5475	3650	85775	5475	12775	7300	133225

Fig. 8.7 This illustrates a basic cross-tab of two categorical variables. The payment status is the row index of the resulting tab. The argument, *margins* = *True* instructs the method to include the row and column margins. The sum of the row margins equals the sum of the column margins equals the sum of the cells. These sums are all equal to the sample size

Proportions, or *normalized frequencies*, are more informative and easier to interpret than frequencies. Normalized frequencies are frequencies divided by the sample size so they are counts per unit sampled; the bases are all the same. These are more useful when comparing tabs for two (or more) groups because proportions are automatically normalized by the sample sizes. Unnormalized frequencies have the drawback that their magnitudes depend on the sample you are using; different sample sizes give different frequencies. Comparisons across samples are difficult if not impossible because the bases are all different.

You can create a cross-tab of proportions by including the *normalize* argument that can have one of three values:

- 'all' or True to normalize all values (divide all cell values by the table total);
- 'index' to normalize each row (divide all cells in a row by the row total); or
- 'columns' to normalize each column (divide all cells in a column by the column total).

Including *normalize* = *'all'* will result in a table with each frequency divided by the sample size. Another argument is *margins* which is either True or False. If *margins* = *True*, then the margins are included and appropriately normalized based on the *normalize* parameter. The normalized row marginal distribution is $p_{i.} = \sum_{j=1}^{c} p_{ij}, \sum_{i=1}^{r} p_{i.} = 1$ with $p_{ij} = {}^{n_{ij}}/_{n_{..}}$. The normalized column marginal distribution is $p_{.j} = \sum_{i=1}^{r} p_{ij}, \sum_{j=1}^{r} p_{.j} = 1$. The table is assumed to have r rows and c columns so its size is $r \times c$.

You can include a variable to be aggregated in some fashion at the intersections of the rows and columns rather than have frequencies. For example, you might want to know the average days late by payment status and store type for the customers located in California. You use the same cross-tab set-up as in Fig. 8.7 but you include an argument for the values to average (i.e., aggregate) and the function to use for the aggregation (i.e., the mean). The resulting table has the aggregated values for the third variable in the cells. I show an example in Fig. 8.8.

```
##
## Crosstab of payment status by stores in California
## Average days Late
##
## Query DataFrame
##
tmp = df.query( "stateCode == 'CA'" )
##
## Create crosstab: payment status is row index
## Use aggfunc to calculate average;
##
pd.crosstab( tmp.paymentStatusNew, tmp.storeTypeNew, values = tmp.daysLate, aggfunc = 'mean', margins = True ).\
    style.set_caption( 'Crosstab of Payment Status by Store Type in CA Average Days Late' ).\
    set_table_styles( tbl_styles ).format( '{:0.1f}' )
```

Crosstab of Payment Status by Store Type in CA Average Days Late

storeTypeNew	Club	Convenience	Dollar	Drug	Mass Merchandise	Restaurant	Supermarket	All
paymentStatusNew								
Current	7.9	8.2	7.6	7.9	7.9	7.8	8.0	7.9
Past Due	0.0	0.0	0.0	0.0	0.0	0.0	0.0	0.0
Warning	17.7	17.7	17.7	17.9	17.8	17.9	17.9	17.9
Collection	41.4	42.0	42.0	41.5	41.4	41.2	41.9	41.5
All	3.5	3.6	3.3	3.4	3.4	3.3	3.4	3.4

Fig. 8.8 This illustrates a basic tab but with a third variable, "daysLate", averaged for each combination of the levels of the index and column variables

You can concatenate the frequency and proportion tables into one table that basically has the two tables interweaved: a row of the frequency table followed by the corresponding row of the proportions table. This interweaving requires coding that I show in Fig. 8.9. You have to create the two tables separately with indexes used for sorting, vertically concatenate the two tables, and then sort this table on the index so that the combined table looks interweaved. I show the final result in Fig. 8.10. You could create a table like Fig. 8.10 but with the two original tables horizontally concatenated rather than vertically concatenated. Use *axis* = *1* for this.

Creating a cross-tab may not be the end of analyzing two categorical variables. You may want to complete two more tasks to aid your analysis. The first is to test the independence, or check association, between the two variables that form the rows and columns of the table, and the second is to plot the table if the two variables are not independent so that some association exists. The second is important because

```
##
## Combine frequency and normalized tabs
##
## Create freq tab (use researchpy)
## Set index for sorting
##
x = pd.crosstab( tmp.paymentStatusNew, tmp.storeTypeNew )
x.rename_axis( None, axis = 'index', inplace = True )
x.rename_axis( None, axis = 'columns', inplace = True )
idx_1 = list(x.index) ## save index values as list
x.index = [ str( x ) + '.1' for x in range( 1, x.shape[ 0 ] + 1 ) ]
##
## Create normalized tab (use researchpy)
## Set index for sorting
## Normalize to table total
##
xprop = pd.crosstab( tmp.paymentStatusNew, tmp.storeTypeNew, normalize = 'all' )
xprop.rename_axis( None, axis = 'index', inplace = True )
xprop.rename_axis( None, axis = 'columns', inplace = True )
xprop = xprop.applymap( lambda x: str( round( x*100, 1 ) ) + '%' )
xprop.index = [ str( x ) + '.2' for x in range( 1, xprop.shape[ 0 ] + 1 ) ]
##
## Concatenate and sort tabs
##
df_inter = pd.concat( [ x, xprop ], axis = 0 )
df_inter.sort_index( axis = 0,  inplace = True )
##
## Create MultiIndex
##
idx_2 = [ 'Freq', 'Table Percent' ]
##
idx = pd.MultiIndex.from_product( [ idx_1, idx_2 ], names = [ 'Status', 'Measure' ] )
df_inter.index = idx
df_inter.style.set_caption( 'Frequency Summary of Payment Status by Store Type in CA' ).\
    set_table_styles( tbl_styles )
```

Fig. 8.9 This is the Python code for interweaving a frequency table and a proportions table. There are two important steps: (1) index each table to be concatenated to identify the respective rows and (2) concatenate based on axis 0

Frequency Summary of Payment Status by Store Type in CA

Status	Measure	Club	Convenience	Dollar	Drug	Mass Merchandise	Restaurant	Supermarket
Current	Freq	915	357	258	5819	382	901	481
	Table Percent	0.7%	0.3%	0.2%	4.4%	0.3%	0.7%	0.4%
Past Due	Freq	10261	4411	2962	69625	4414	10390	5932
	Table Percent	7.7%	3.3%	2.2%	52.3%	3.3%	7.8%	4.5%
Warning	Freq	1206	539	332	7697	521	1115	672
	Table Percent	0.9%	0.4%	0.2%	5.8%	0.4%	0.8%	0.5%
Collection	Freq	393	168	100	2634	158	369	215
	Table Percent	0.3%	0.1%	0.1%	2.0%	0.1%	0.3%	0.2%

Fig. 8.10 This is the result of interweaving a frequency table and a proportions table using the code in Fig. 8.9. This is sometimes more compact than having two separate tables

if there is an association between the two, then you should know what it is and the implications of that association. Examining the raw table is insufficient for revealing any association, especially as the size of the table increases. I will consider the hypothesis testing first, followed by the graphing of a table.

8.5.1 Hypothesis Testing

I reviewed a chi-square test above for comparing an observed distribution against another one which is usually a target or population distribution. The Null Hypothesis is that the observed distribution is not different from the target distribution. You can expand this to test the independence of the rows and columns of a frequency table; i.e., to test the independence of the two categorical variables. There is an adjustment to the test statistic: you have to account for the rows and columns whereas before there was only one "row" (or "column"). For a frequency table with r rows and c columns, so the size of the table is $r \times c$, the chi-square test statistic is

$$\chi^2 = \sum_{i=1}^{r} \sum_{j=1}^{c} \frac{(O_{ij} - E_{ij})^2}{E_{ij}} \tag{8.5.1}$$

where O_{ij} is the observed frequency in cell $ij, i = 1, 2, \ldots, r; j = 1, 2, \ldots, c$, and E_{ij} is the corresponding expected value. Under the Null Hypothesis of independence, E_{ij} is the product of the respective marginal proportions times the table total. If $n = \sum_{i=1}^{r} \sum_{j=1}^{c} O_{ij}$ is the table total, then $E_{ij} = n \times p_{.j} \times p_{i.}$ where $p_{.j} = \sum_{i=1}^{r} O_{ij}/n_{.j}$ is the column j marginal. Similarly for $p_{i.}$. This is based on the elementary probability theory result that under independence, $Pr(A \cup B) = Pr(A) \times Pr(B)$ for two events, A and B. See, for example, Weiss (2005). Equation (8.5.1) is the *Pearson Chi-Square Test Statistic* which is distributed following a chi-square distribution for $(r - 1) \times (c - 1)$ degrees-of-freedom. See Agresti (2002) for a discussion. Also see Paczkowski (2016) for an explanation of the degrees-of -freedom as well as Bilder and Loughin (2015).

A variation of this test statistic is the *Likelihood-Ratio Chi-Square Test Statistic*, defined as

$$G^2 = 2 \times \sum_{i=1}^{r} \sum_{j=1}^{c} \left[O_{ij} \times ln \left(\frac{O_{ij}}{E_{ij}} \right) \right] \tag{8.5.2}$$

This is asymptotically distributed as a chi-square also with $(r - 1) \times (c - 1)$ degrees-of-freedom. See Agresti (2002) for a discussion. You can get both statistics for an $r \times c$ frequency table by specifying the test parameter for the cross-tab function in the *researchpy* package.[1] I provide an example of the Pearson test in Fig. 8.11. You install *researchpy* using *pip install researchpy* or *conda install -c researchpy researchpy*.

[1] Two other tests are available: Fisher's Exact Test and the McNemar chi-square test for paired nominal data. See Agresti (2002) for a discussion for these tests.

Value	Association
>0.25	Very strong
>0.15	Strong
>0.10	Moderate
>0.05	Weak
>0	Non or very weak

Table 8.4 Guidelines for interpreting Cramer's V statistic. Source: Akoglu (2018)

The *researchpy* output also displays the *Cramer's V statistic* based on Pearson's chi-square statistic.[2] The statistic is defined as

$$V = \sqrt{\frac{\chi^2/n}{\min(r-1, c-1)}} \tag{8.5.3}$$

This is a measure of association between two categorical variables and as such ranges between 0 and +1 (inclusive). It is interpreted as a correlation measure with values close to zero indicating no or little association. You can see this statistic in Fig. 8.11. The value 0.0069 indicates no association between the two variables. See Cramer (1946) for the original development of this statistic. Akoglu (2018) provides guidelines for interpreting the statistic which I summarize in Table 8.4.

8.5.2 Plotting a Frequency Table

If you conclude from the chi-square test that there is a relationship between the two categorical variables, then you need to investigate this relationship beyond the table. Looking at a table *per se* is insufficient because it becomes more challenging, if not impossible, to spot relationships as it becomes larger. Your analysis is helped a lot by graphing the table. The simplest graph is a *heatmap*. It is called a "map" because it guides you in your understanding of the table. The heatmap resembles the frequency table in that it is composed of cells such that the size of the map is the same as the table: $r \times c$ so the number and arrangement of cells is the same as the table. Rather than have numbers in the cells, the cells are color coded, the color in a cell showing the intensity or density of the data in that cell. The frequency numbers can be included in the cells, but this may just overwhelm the image which would defeat the purpose of mapping the table. A color gauge or thermometer is usually included to show how the intensity of the data varies by the colors. I illustrate a heatmap in Fig. 8.12 for the tab in Fig. 8.7.

[2] This is also called *Cramer's phi-statistic*, denoted as ϕ_c.

```
##
## Crosstab with Pearson Chi-Square Test
##
## Query DataFrame
##
tmp = df.query( "stateCode == 'CA'" )
##
## Create crosstab with chi-square test
##
xtab = rp.crosstab( tmp.paymentStatusNew, tmp.storeTypeNew, test = "chi-square" )
display( xtab[ 0 ], xtab[ 1 ] )
```

storeTypeNew	storeTypeNew							
storeTypeNew	Club	Convenience	Dollar	Drug	Mass Merchandise	Restaurant	Supermarket	All
paymentStatusNew								
Current	915	357	256	5819	382	901	481	9111
Past Due	10261	4411	2962	69625	4414	10390	5932	107995
Warning	1206	539	332	7697	521	1115	672	12082
Collection	393	168	100	2634	158	369	215	4037
All	12775	5475	3650	85775	5475	12775	7300	133225

	Chi-square test	results
0	Pearson Chi-square (18.0) =	18.8573
1	p-value =	0.4007
2	Cramer's V =	0.0069

Fig. 8.11 This illustrates the Pearson Chi-Square Test using the tab in Fig. 8.7. The p-value indicates that the Null Hypothesis of independence should not be rejected. The Cramer's V statistic is 0.0069 and supports this conclusion

Another graphical tool, often called an exploratory tool, is a *correspondence map*. I disagree with the "exploratory" characterization. First, all *BDA* is exploratory, actually iterative, in that you are seeking to extract the Richest Information from your data. Second, correspondence maps are extremely informative and revealing in their own right. And finally, they are a different view of a table.

A map allows you to see relationships among the rows of a table, among the columns of a table, and associations among rows and columns. This is a more complicated tool to develop and the resulting map is more challenging to interpret. Nonetheless, it is powerful and insightful enough to warrant your time and effort for analyzing tables such as a cross-tab. Its usefulness becomes very evident when the table underlying the map becomes large.

The basis for a correspondence map is a decomposition or "splitting up" of a table (i.e., a matrix) into three submatrices based on the *Singular Value Decomposition* (*SVD*) that I summarized in Chap. 5. The three submatrices are simply referred to as the left matrix, the center matrix, and the right matrix. The left submatrix is associated with the rows of the table; the right submatrix is associated with the columns of the table, and the center connects the other two but also provides insight about the variance of the table. The columns of the left and right submatrices give

```
##
## Crosstab with Pearson Chi-Square Test
##
## Query DataFrame
##
tmp = df.query( "stateCode == 'CA'" )
##
## Create crosstab with Pandas
## Remove index names for clarity
##
xtab = pd.crosstab( tmp.paymentStatusNew, tmp.storeTypeNew, normalize = 'all' )
xtab.rename_axis( None, axis = 'index', inplace = True )
xtab.rename_axis( None, axis = 'columns', inplace = True )
##
## Use Seaborn
##
ax = sns.heatmap( xtab, cmap = "Greens" )
ax.set_title( 'Payment Status by Store Type\nHeatmap\nCalifornia', fontsize = font_title );
```

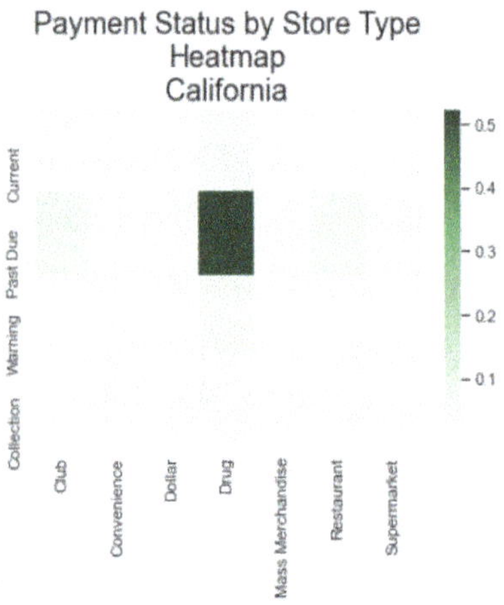

Fig. 8.12 This illustrates a heatmap using the tab in Fig. 8.7. It is clear that the majority of Grocery stores is current in their payments

plotting coordinates. In particular, a function of the columns of the left submatrix give columns of plotting coordinates for the rows of the main matrix while a function of the columns of the right submatrix give columns of plotting coordinates for the columns of the main matrix. In both cases, the columns of coordinates are referred to as *dimensions*. The maximum number of dimensions that can be created or extracted from the main matrix is $Min(r - 1, c - 1)$ where r is the number of rows of the main matrix and c is the number of columns of the main matrix. This is a restriction on the usability of correspondence analysis: the main matrix must have more than two rows and more than two columns. For example, if $r = 3$ and $c = 2$, then $Min(2, 1) = 1$ and so only one set of plotting coordinates can be extracted—but you need two since a graph has two axes: X and Y.

The plotting coordinates are used to create the map. Typically, two columns of coordinates for the rows are plotted to produce a map (i.e., the X and Y axes). Then, two columns of coordinates for the columns are plotted on top of this map so

```
##
## Instantiate CA
##
n_components = min( xtab.shape[ 0 ], xtab.shape[ 1 ] )
ca = prince.CA( n_components = n_components, random_state=42 )
##
## Fit the CA
##
xtab_ca = ca.fit( xtab )
```

Fig. 8.13 This is the main function for the correspondence analysis of the cross-tab developed in Fig. 8.7. The function is instantiated with the number of dimensions and a random seed or state (i.e., 42) so that results can always be reproduced. The instantiated function is then used to fit the cross-tab

that one plot is overlaid on another. This overlaying results in a *biplot* which is the correspondence map.

The center submatrix is a rectangular *diagonal matrix*. It has the same number of rows as columns with nonzero (mostly) elements on the main diagonal cells and zeros on the off-diagonal cells. The diagonal values are called *singular values*. The square of each singular value is an *eigenvalue* of the original matrix. See Lay (2012) and Strang (2006) for discussions about eigenvalues. These eigenvalues are also called the *inertia* of the matrix. If SV_i is the ith singular value, then SV_i^2 is the inertia for dimension i. The chi-square value for that dimension is $SV_i^2 \times n$ where n is the sample size. The sum of the chi-square values is the total chi-square used in the test of independence. This means that *SVD* applied to a cross-tab provides not only the chi-square value for the cross-tab but also a decomposition of that chi-square to dimensions of the cross-tab. The decomposed components of the total chi-square are usually expressed as percentages of the total as well as cumulative percentages. The cumulative percentages help you determine the dimensions to plot, which are usually the first two. These are the two that account for the most variation in the cross-tab.

Correspondence analysis is done using the set-up I show in Fig. 8.13. I used the *prince* package that has a function *CA* for correspondence analysis. You install *prince* using *pip install prince* or *conda install -c bioconda prince*. I used the cross-tab I created earlier with the Pandas *crosstab* method. The resulting tab, called *xtab*, was used as an argument to the *fit* property of the correspondence analysis function *CA* along with the number of dimensions (i.e., components to extract). This number is the minimum of the number of rows of *xtab* less 1 and the number of columns of *xtab* less 1. The fitted information was saved an object called *xtab_ca*. The cross-tab and the correspondence information from the *CA* function were used as inputs to three functions that:

1. summarize the chi-square statistics;
2. summarize the dimension information from the Singular Value Decomposition; and
3. display the biplot.

I show these in Fig. 8.14. A separate function uses these three to produce the final display of the analysis results which I show in Fig. 8.15.

```
def ca_dimensions( tab, ca ):
    """CA Plotting Dimensions with Singular Values and Chi-square"""
    smpl_size = tab.sum().sum()
    eigenvals = ca.eigenvalues_ ## = np.sqrt( ca.s_ )
    chisq = [ x * smpl_size for x in eigenvals ]
    sv = ca.s_ ## = np.sqrt( eigenvals )
    pct = [ x * 100 for x in xtab_ca.explained_inertia_ ]
    ##
    data = {
      'Singular Values': sv,   'Inertia': eigenvals, 'Chi-Square': chisq,
      'Percent': pct,   'Cum. Percent': np.cumsum( pct ) ##pct.cumsum()
    }
    ##
    df = pd.DataFrame( data )
    df.index.set_names( [ 'Dimension' ], inplace = True )
    df.index += 1
    ##
    fmt = {'Percent':'{:,.2f}%', 'Cum. Percent':'{:,.2f}%', 'Singular Values':'{:,.5f}', 'Inertia':'{:,.5f}',
           'Chi-Square':'{:,.5f}'}
    return df.style.set_caption( 'CA Plotting Dimension Summary' ).format( fmt ).set_table_styles( tbl_styles )
```

```
def ca_plot( tab, ca, ttl ):
    """CA biPlot"""
    ax = ca.plot_coordinates(
        X = tab, ax = None, figsize = ( 8, 6 ), x_component = 0,
        y_component = 1, show_row_labels = True, show_col_labels = True
    )
    ax.set_title( ttl, fontsize = font_title )
```

```
def ca_display( xtab, ca, ttl = 'Correspondence Analysis Plot' ):
    """Display CA Results
    Parameters:
        xtab: crosstab; ca: correspondence results from CA function in prince module; ttl: title for biplot
    """
    from IPython.display import display
    display( ca_summary( xtab, ca ) )
    display( ca_dimensions( xtab, ca ) )
    display( ca_plot( xtab, ca, ttl ) )
ca_display( xtab, ca )
```

Fig. 8.14 The functions to assemble the pieces for the final correspondence analysis display are shown here. Having separate function makes programming more manageable. This is *modular programming*

First, look at the *CA Chi-Square Summary* table in Fig. 8.15, Panel (a). The total inertia is 0.0001 and the sample size is 133,225 (which you can see in Fig. 8.11).[3] The product of these two is the total chi-square 18.86 which equals the Pearson chi-square value I reported in Fig. 8.11. The components of the total chi-square are reported in the *CA Plotting Dimension Summary*. You can see that the first dimension or component extracted from the cross-tab accounts for 56.01% of the total chi-square and the second accounts for 30.71%. These two percentages are reflected in the biplot in Panel (b). The cumulative percentage is 86.72%.

The biplot in Panel (b) is the main diagnostic tool for analyzing the cross-tab. But how is it interpreted? There are two tasks you must complete in order to extract the most insight from this plot and, therefore, the cross-tab:

[3] The total inertia is actually 0.00014154489494587526 so what is shown is obviously rounded.

```
def ca_display( xtab, ca, ttl = 'Correspondence Analysis Plot' ):
    """Display CA Results
    Parameters:
        xtab: crosstab; ca: correspondence results from CA function in prince module; ttl: title for biplot
    """
    from IPython.display import display
    display( ca_summary( xtab, ca ) )
    display( ca_dimensions( xtab, ca ) )
    display( ca_plot( xtab, ca, ttl ) )
ca_display( xtab, ca )
```

CA Chi-Square Summary

	Sample Size	Total Inertia	Total Chi-Square
Summary	133225	0.000142	18.86

CA Plotting Dimension Summary

Dimension	Singular Values	Inertia	Chi-Square	Percent	Cum. Percent
1	0.00890	0.00008	10.56221	56.01%	56.01%
2	0.00659	0.00004	5.79148	30.71%	86.72%
3	0.00434	0.00002	2.50363	13.28%	100.00%
4	0.00000	0.00000	0.00000	0.00%	100.00%

(a)

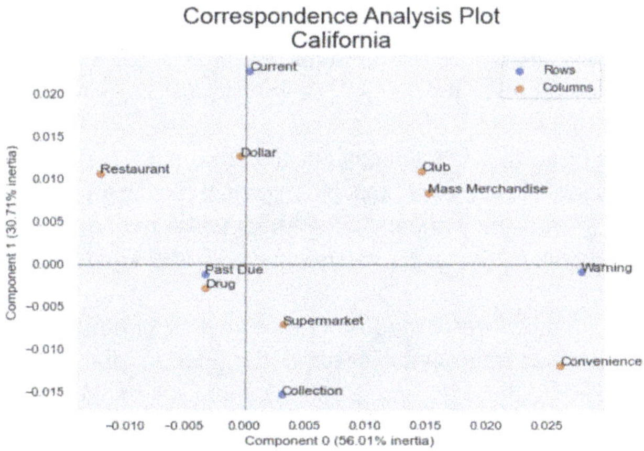

(b)

Fig. 8.15 The complete final results of the correspondence analysis are shown here. Panel (a) shows the set-up function for the results and the two summary tables. Panel (b) shows the biplot

1. interpret the two axes; and
2. interpret the proximities of the row and column points.

To interpret the map, first look at the X-axis dimension. Two customer classes stand out: Restaurants on the left and Convenience stores on the right. The fact that they are on the left and right is not important, but the fact that they are at extremes is important. Their positions can be reversed by multiplying the plotting coordinates by -1: nothing is impacted except position. Restaurants can be viewed as a relaxing type of business for their customers: they come into a restaurant to enjoy a meal with family or friends or perhaps to conduct business but with less pressure, more leisure, than in an office environment. Time is probably less important to them. Convenience store customers, however, are most likely rushed, in a hurry to pick up something simple and without fuss; time is important to them. This explains why these stores are called "convenience stores." So, the horizontal axis reflects a contrast between leisure and speed. It is a value of time dimension. Supermarkets and drug stores fit in the middle. Generally, they cater to people who have to spend time shopping more as a chore than as a relaxation. Notice also that Club stores and Mass Merchandising stores are very close in proximity which may make sense since they are basically the same type of stores: very large, almost warehouse-type. Their proximity, incidentally, reflects the Gestalt *Proximity Principle*.

Now look at the Y-axis dimension. Notice that the Current payment status on the top and the Collection status on the bottom stand out. These are two extremes of payment. Customers who are good to do business with are those who make payments on-time. They are customers with low-risk of default. Those out for collection are a challenge and run the risk of defaulting. This suggests that the second dimension is a contrast not only of the payment status but also, and more importantly, a contrast of the chance for risk default: high vs. low. One extreme is relatively risky customers and the other relatively riskless customers. This concept of risky customers is important and begs a question: "*How can you classify customers by their risk of default*?" I will discuss classification methods in Chap. 11.

You can also see that the first dimension, the store class contrast, accounts for 56.01% of the variation in the cross-tab. The second dimension, the risk contrast, accounts for 30.71% of the variance. Together, these two dimensions account for 86.72% of the cross-tab's variance agreeing with the *CA Plotting Dimension Summary*.

There are several more observations. Notice the cross-hairs at the (0.0, 0.0) intersection. These reflect the row and column marginal distribution of the original cross-tab. These marginals are interpreted as the centroids or averages of the table. The fact that drug stores and past due notices are at the center of the cross-hairs suggests that they are the average characteristic for their respective parts of the cross-tab. Further, notice that they are right next to each other so drug stores are, on average, past due on their accounts to the bakery company. Supermarkets are also close to the past due point while convenience stores are closer to the warning point. These reflect the Gestalt *Principle of Proximity*.

This analysis was for California. What about the entire nation? I reran the map construction without subsetting on California. You can see the map in Fig. 8.16. The placement of the points is different but the messages, the Rich Information, are the same.

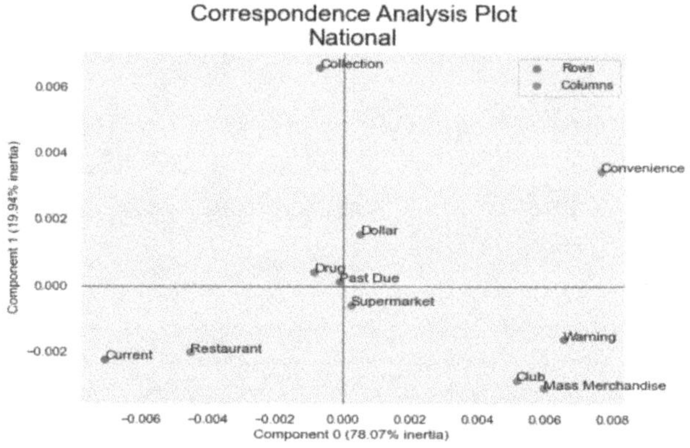

Fig. 8.16 This is the map for the entire nation for the bakery company

8.6 Extending the Cross-tab

The basic cross-tab has cells that are just the frequencies of occurrence of two intersecting categorical levels. I illustrated this basic functionality, the default, above. You could, if needed, have measures other than counts in the cells but the values have to be based on a third variable which, of course, you must specify. For example, you could have the mean for a third variable contingent on the levels of two categorical variables.

To illustrate this enhanced functionality, suppose you need the mean days-late for invoice payments for the accounts receivable data for stores in California by store type and payment status. In addition to specifying the two categorical variables, *storeType* and *paymentStatus*, you also need the method of aggregation (mean in this case) for the number of days. I show you how this is done in Fig. 8.17.

There is another way to get the same data without using the Pandas cross-tab function. You could group your data in the DataFrame by the two categorical variables and apply the *mean* function to the third variable. I show this in Fig. 8.18.

For this particular application, the cross-tab function, in my opinion, is a better way to aggregate the data. However, the use of *groupby* coupled with another function, *agg*, adds more functionality if you want to work with multiple variables and calculate multiple summary measures for each. The *agg* function allows you to apply one or more functions, such as a mean or standard deviation, to one or more variables resulting from the *groupby*. I illustrate how to find the mean and standard deviation of two variables based on groupings of two categorical variables in Fig. 8.19. The *agg* function in this case takes a dictionary as an argument, with the variable to aggregate as the key and the aggregation method (i.e., mean and standard deviation) as the value for the key. The value could be just a single function or a list of functions as I show in Fig. 8.19.

```
##
## Crosstab with mean for Days-to-Pay
##
## Query DataFrame
##
tmp = df.query( "stateCode == 'CA'" )
##
## Create crosstab with Pandas
## Remove index names for clarity
##
xtab = pd.crosstab( tmp.paymentStatusNew, tmp.storeTypeNew, values = tmp.daysToPay, aggfunc = 'mean' )
xtab.rename_axis( None, axis = 'index', inplace = True )
xtab.rename_axis( None, axis = 'columns', inplace = True )
xtab.style.set_caption( 'Aggregation Using Crosstab Function').format( '{:0.4}' ).set_table_styles( tbl_styles )
```

Aggregation Using Crosstab Function

	Club	Convenience	Dollar	Drug	Mass Merchandise	Restaurant	Supermarket
Current	37.91	37.91	38.27	37.98	37.82	37.68	38.0
Past Due	14.99	14.68	14.74	14.96	14.74	14.88	14.69
Warning	47.96	47.65	48.34	47.86	47.84	47.95	47.93
Collection	66.42	66.28	67.22	66.69	67.01	67.08	66.51

Fig. 8.17 The cross-tab in Fig. 8.7 is enhanced with the mean of a third variable, *days-late*

```
##
## Use groupby function
##
## Query DataFrame
##
tmp = df.query( "stateCode == 'CA'" )
##
x = tmp.groupby( by = [ 'paymentStatusNew', 'storeTypeNew' ] ).daysToPay.mean()
x.to_frame().style.set_caption( 'Aggregation Using Groupby' ).format( '{:0.4}' ).\
    set_table_styles( tbl_styles )
```

Aggregation Using Groupby

		daysToPay
paymentStatusNew	storeTypeNew	
Current	Club	37.91
	Convenience	37.91
	Dollar	38.27
	Drug	37.98
	Mass Merchandise	37.82
	Restaurant	37.68
	Supermarket	38.0
Past Due	Club	14.99
	Convenience	14.68
	Dollar	14.74
	Drug	14.96
	Mass Merchandise	14.74

Fig. 8.18 The cross-tab in Fig. 8.17 can be replicated using the Pandas *groupby* function and the mean function. The values in the two approaches are the same; just the arrangement differs. This is a partial display since the final table is long

```
##
## Groupby with multiple variables and aggregations
##
aggregation = { "daysToPay": [ np.mean, np.std ], "daysLate": [ np.mean, np.std ] }
x = tmp.groupby( by = [ 'paymentStatusNew', 'storeTypeNew' ] ).agg( aggregation )
x.style.set_caption( 'Summary Statistics Based on Groupby and a Dictionary' ).\
    format( '{:0.4}' ).set_table_styles( tbl_styles )
```

Summary Statistics Based on Groupby and a Dictionary

		daysToPay		daysLate	
		mean	std	mean	std
paymentStatusNew	**storeTypeNew**				
Current	Club	37.91	4.099	7.93	4.115
	Convenience	37.91	4.152	8.224	4.11
	Dollar	38.27	4.145	7.586	3.98
	Drug	37.98	4.161	7.932	4.14
	Mass Merchandise	37.82	4.006	7.866	4.175
	Restaurant	37.68	4.051	7.831	4.076
	Supermarket	38.0	4.004	7.979	4.194
Past Due	Club	14.99	8.211	0.0001949	0.01974
	Convenience	14.68	8.351	0.0	0.0
	Dollar	14.74	8.16	0.0006752	0.02598
	Drug	14.96	8.183	0.0002729	0.02177
	Mass Merchandise	14.74	8.31	0.0002266	0.01505

Fig. 8.19 The cross-tab in Fig. 8.17 is aggregated using multiple variables and aggregation methods. The *agg* method is used in this case. An aggregation dictionary has the aggregation rules and this dictionary is passed to the *agg* method

8.7 Pivot Tables

A cross-tab is the fundamental tabulation of categorical data. You can create a DataFrame as in Fig. 8.18, but then *pivot* the table for a different perspective of your data. This adds a dynamic element to your toolkit; the simple cross-tab is a static tool. The arrangement in Fig. 8.18 is sometimes described as a *long-form* arrangement because the number of rows is greater than the number of columns. An alternative is *wide-form* which is the long-form pivoted along the major axes. The Pandas *pivot* function is used to switch from long- to wide-form. I illustrate this in Fig. 8.20.

The pivot function is a complicated way to create a summary table. A more convenient way is with the *pivot_table* function which uses the *groupby* function behind the scenes. I illustrate the use of this function in Fig. 8.21. This new function is quite flexible, allowing you to create summary tables in a multitude of ways. As an example, see the pivot table in Fig. 8.22.

```
##
## Use pivot function to rearrange the DataFrame
##
## Query DataFrame
##
tmp = df.query( "stateCode == 'CA'" )
##
x = tmp.groupby( by = [ 'paymentStatusNew', 'storeTypeNew' ] ).daysToPay.mean()
x = x.reset_index( )
x.pivot( index = 'paymentStatusNew', columns = 'storeTypeNew', values = 'daysToPay' )
```

storeTypeNew	Club	Convenience	Dollar	Drug	Mass Merchandise	Restaurant	Supermarket
paymentStatusNew							
Current	37.914754	37.910364	38.265625	37.976972	37.819372	37.678135	38.004158
Past Due	14.989865	14.675810	14.735652	14.962111	14.740598	14.876420	14.694707
Warning	47.956053	47.653061	48.343373	47.856957	47.838772	47.951570	47.928571
Collection	66.424936	66.279762	67.220000	66.686029	67.006329	67.081301	66.511628

Fig. 8.20 The DataFrame created by a *groupby* in Fig. 8.18, which is a *long-form* arrangement, is pivoted to a *wide-form* arrangement using the Pandas *pivot* function. The DataFrame is first reindexed

```
##
## Use pivot_table to create aggregated data
##
## Query DataFrame
##
tmp = df.query( "stateCode == 'CA'" )
##
x = tmp.pivot_table( index = 'paymentStatusNew', columns = 'storeTypeNew', values = 'daysToPay', aggfunc = [ 'mean' ] )
x.style.set_caption( 'Pivoted DataFrame' ).format( '{:0.4}' ).set_table_styles( tbl_styles )
```

Pivoted DataFrame

	mean						
storeTypeNew	Club	Convenience	Dollar	Drug	Mass Merchandise	Restaurant	Supermarket
paymentStatusNew							
Current	37.91	37.91	38.27	37.98	37.82	37.68	38.0
Past Due	14.99	14.68	14.74	14.96	14.74	14.88	14.69
Warning	47.96	47.65	48.34	47.86	47.84	47.95	47.93
Collection	66.42	66.28	67.22	66.69	67.01	67.08	66.51

Fig. 8.21 The *pivot_table* function is a more convenient way to pivot a DataFrame

The Pandas *pivot_table* function differs somewhat from the *crosstab* function. Some of these differences are:[4]

1. *pivot_table* uses a DataFrame as the base for the data while *crosstab* uses series. Since most of your data will be in a DataFrame, this could be important, but only as a convenience.
2. You can name the margins in *pivot_table*.
3. You can use a function called *Grouper* for the index in *pivot_table*.
4. *crosstab* allows you to normalize the frequencies: row conditional distribution; column conditional distribution; each cell divided by the total.

[4] This list builds on comments on StackOverflow. See https://stackoverflow.com/questions/36267745/how-is-a-pandas-crosstab-different-from-a-pandas-pivot-table.

```
##
## Use pivot_table to create aggregated data - alternative arrangement
##
## Query DataFrame
##
tmp = df.query( "stateCode == 'CA'" )
##
x = tmp.pivot_table( index = [ 'paymentStatusNew', 'storeTypeNew' ], values = 'daysToPay', aggfunc = [ 'mean', 'std' ] )
x.style.set_caption( 'Pivoted DataFrame: Alternative Arrangement' ).format( '{:0.4}' ).set_table_styles( tbl_styles )
```

Pivoted DataFrame: Alternative Arrangement

		mean	std
		daysToPay	daysToPay
paymentStatusNew	storeTypeNew		
Current	Club	37.91	4.099
	Convenience	37.91	4.152
	Dollar	38.27	4.145
	Drug	37.98	4.161
	Mass Merchandise	37.82	4.006
	Restaurant	37.68	4.051
	Supermarket	38.0	4.004

Fig. 8.22 The *pivot_table* function is quite flexible for pivoting a table. This is a partial listing of an alternative pivoting of our data

My recommendation: use which ever one meets your needs for a particular task. It is as simple as that.

8.8 Appendix

In this Appendix, I will describe the Pearson chi-square statistic and some of its properties.

8.8.1 Pearson Chi-Square Statistic

The Pearson chi-square statistic, χ^2, in (8.4.3) is related to the standard normal distribution. Let Z be a random variable such that $Z \sim \mathcal{N}(0, 1)$. The formula for this, known as the *probability density function* (*pdf*) is

$$f(x) = \frac{1}{\sqrt{2\pi}} e^{-(x)^2/2} \tag{8.8.1}$$

It can be shown, which is outside the scope of this book, that Z^2 follows a new distribution which is called the chi-square distribution, represented as χ_1^2. That is, $Z^2 \sim \chi_1^2$. The "1" is called the *degrees-of-freedom* (*dof*) which defines the shape of the distribution. See Fig. 8.23. Consequently, the dof is also referred to as a *shape parameter*. The standard normal does not have any degrees-of-freedom since there is only one shape, but this is not the case for the chi-square.

Suppose there are two independent standard normal random variates, Z_1 and Z_2, and let $Z = Z_1 + Z_2$. Then $Z \sim \chi_2^2$ with two degrees-of-freedom. In general, if $Z = \sum_{i=1}^{k} Z_i^2$, then $Z \sim \chi_k^2$. The $dof \geq 1$ is a positive integer. The mean of a chi-square random variable is k and the variance is $2 \times k$. See Dobson (2002, p. 7).

If $k = 1$ or $k = 2$, then the chi-square distribution is a *negative exponential*. In fact, for $k = 2$, the chi-square is an *exponential distribution* which is a member of the *gamma distribution* family. If $k \rightarrow \infty$, then the chi-square random variable approaches the standard normal distribution (i.e., $\mathcal{N}(0, 1)$).

Other distributions are related to the chi-square distribution. If $Z \sim \mathcal{N}(0, 1)$ and $X \sim \chi_k^2$, then $Z/\sqrt{X/k} \sim t_k$ is a Student's t-random variable. Furthermore, the ratio of a χ^2 random variable with k_1 degrees-of-freedom to an independent χ^2 random variable with k_2 degrees-of-freedom, each divided by its respective degrees-of-freedom, follows an F-distribution with k_1 and k_2 degrees-of-freedom. Note that the F_{1,k_2} is t^2. You can see this from the definition of a t with k degrees-of-freedom: $Z/\sqrt{\chi_k^2/k} \sim t_k$. This is the basis for the *ANOVA* table I discussed above. See Dobson (2002, pp. 8-9).

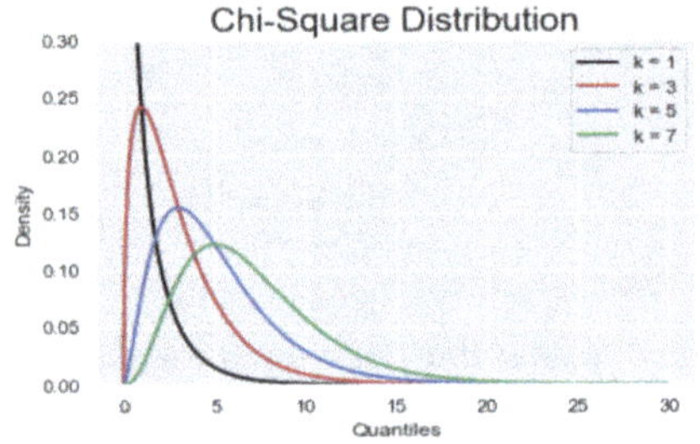

Fig. 8.23 This illustrates the chi-square distribution for several value of k. Notice how the shape changes as k increases and begins to approach the standard normal curve

It can be shown that the Pearson chi-square is a function of the squared distance between a row profile and the row centroid in a cross-tab. A *row profile* is the row conditional distribution. If n_{ij} is the frequency in the cell at the intersection of row i and column j and $n_{i.}$ is the marginal frequency for row i summed over the J columns of the cross-tab, then the ith profile is $n_{ij}/n_{i.}$. The marginal frequency is called the *row mass*. If each term in this last expression is divided by the sum of frequencies across all cells, $n_{..}$, then you can see that the row profile is just a conditional probability distribution. The *row centroid* is the weighted average of the row profiles where the weights are the respective row masses. It can be shown that the row centroid is the column mass. All of this applies to columns as well.

Part III
Advanced Analytics

This third part of the book extends the intermediate analytical methodologies to include advanced regression analysis which includes a discussion of regression as a family of methods. *OLS* is one member. Logistic regression is introduced as another member. Machine learning methods are also introduced. All of this hinges on the type of data which is built from an expanded view of data depicted as a cube. The methods discussed will allow you to become a more advanced user capable of more detailed analysis of business data. After reading this part of the book, you will be able to do very sophisticated analyses of business data.

Chapter 9
Advanced Data Handling for Business Data Analytics

In this chapter, I will set the stage for analysis beyond what I discussed in the previous chapters. I covered that material at a high level. Specialized books cover them in greater detail; in fact, whole volumes are written on each of those topics. The ones in this chapter are different. They cover advanced data handling topics.

9.1 Supervised and Unsupervised Learning

In machine learning, we do not say that the parameters of a model are estimated using data because not all procedures involving data are concerned with estimating parameters. But there are models where there may not even be parameters. Models with parameters are *parametric models*; those without are *nonparametric models*. Parametric models have a finite number of parameters that are said to exist in the population being investigated and are fixed and unknown. The task is to use data to estimate them. Since the number of parameters is finite, parametric models are constrained. No amount of data will change the number of parameters. This is strongly noted by Russell and Norvig (2020, p. 737): "*A ... model that summarizes data with a set of parameters of fixed size ... is called a parametric model. No matter how much data you throw at a parametric model, it won't change its mind about how many parameters it needs.*"[1] Nonparametric models are not dependent on parameters and so, therefore, they are not constrained like parametric models. They only rely on data for the clues as to how to proceed. Also noted by Russell and Norvig (2020, p. 757): "*Nonparametric methods are good when you have a lot of data and no prior knowledge, and when you don't want to worry too much about*

[1] Cited by Jason Brownlee at https://machinelearningmastery.com/parametric-and-nonparametric-machine-learning-algorithms/. Last accessed on October 7, 2020.

W. R. Paczkowski, *Business Analytics*,
https://doi.org/10.1007/978-3-030-87023-2_9

choosing just the right features."[2] To avoid confusion regarding to-estimate or not-to-estimate, both cases involving the use of data, machine learning practitioners say that a method *learns* from the data.

When parameters are involved, a method learns about them from data. But because they are finite and fixed, they constrain what can be revealed by the data. Parametric models, in addition to the parameters, have a *target* or *dependent variable* or *label* defined by the parameters. The definition could be linear, for example. Nonparametric models do not have a target. The target directs the method in the learning of the parameters. So, in a sense, the target, through the parameters, controls or *supervises* what the method is capable of learning. Formal models involving a finite set of fixed but unknown parameters are *supervised learning methods*. Methods that do not involve parameters are *unsupervised learning methods*

In machine learning, the independent variables are called *features*. In regression analysis, an example of supervised learning, the goal is to estimate the unknown parameters based on the features to give the best estimate of the target. Since unsupervised learning has neither a target nor a set of parameters, the goal is to find *clusters* or *groups* of features using algorithms. Unsupervised learning methods are clustering, pattern recognition, and classification identifiers unlike supervised learning methods which are parameter-identifying methods like *OLS*.

A college class analogy helps to clarify the distinction and terminology. In a college setting, a professor guides a student (i.e., the "learner") in learning course material (i.e., data) and then tests the learner's performance. The professor establishes a set of parameters for getting a target course grade such as an "A." There is a teacher, a learner, data, a set of parameters, a performance measure, and a target. This is supervised educational learning. For the statistical analogy, there is a model with a target dependent variable or label that guides a learner (i.e., estimation technique) in processing data with the processing performance measured by a goodness-of-fit measure such as R^2 to give the best prediction of the target. The parameters constrain how the learning technique hits the target. This is *supervised learning* in statistics and machine learning.

Without a professor, students are on their own; they are unsupervised, self-paced, their only data are some books they find and, perhaps, the Internet. This is the case for people who do self-study, say, for a professional license. They are autodidactic. Without a professor setting constraints (i.e., parameters), they are free to define their own best learning and accomplishment. There may still be a test, but it is not formal as for a college course. The test may be the application of what is learned in an informal setting on the job or in the public domain (e.g., voting). In statistics and machine learning, an algorithm, not a model, operates on data with (maybe) a performance criterion; the algorithm is unsupervised and unconstrained by parameters and a target.

[2] Also cited by Jason Brownlee at https://machinelearningmastery.com/parametric-and-nonparametric-machine-learning-algorithms/. Last accessed on October 7, 2020.

There are many supervised and unsupervised methods designed to address different problems and data, both target (if there is one) and features. I provide a small list in Table 9.1. I will discuss the methods in the remaining chapters.

Supervised Learning Methods
• Generalized Linear Model (*GLM*)
• *OLS*
• Logistic Regression
• Classification
• Naive Bayes
• Decision Trees
• Support Vector Machines
Unsupervised Learning
• Clustering
• Hierarchical
• K-Means
• Mixture

Table 9.1 This is a list of supervised and unsupervised methods by functionality

9.2 Working with the Data Cube

I introduced the Data Cube paradigm in Chap. 1 and referred to it several times in succeeding chapters. A major point I made about the Cube is that you can collapse either the spatial or temporal dimension to get one of the traditional data sets you may be familiar with from a statistics or econometrics course: cross-sectional data set or a time series data set. I provided some examples illustrating this collapsing. In fact, all of Chap. 7 is based on collapsing the Cube on the spatial dimension to get a time series. There is more to working with the Data Cube, however, than what I showed you until now. You could work with the entire Cube in which case you would work with a combination of time series and cross-sectional data. I noted in Chap. 2 and again in Chap. 6 that this is called a *panel data set*, sometimes also referred to as a *longitudinal data set*. Time series and cross-sectional data each have their own special problems, although I only highlighted the time series problems in Chap. 7. Panel data have these problems and more. You can view them as reflecting *additivity*. This means the number of problems, and the number of ways to analyze your data, are more than the problems individually. Basically, the sum of the problems is more than the individual problems.

Panel data analysis has to consider the joint behavior of the spatial and temporal dimensions as well as their individual behaviors and effects. It is the joint behavior that causes the sum to be greater than its parts. For example, with transactions data, you have variations in orders by customers and variations in orders by time periods, say months. But you also have variations by customers by time periods. You thus have variation within and between customers. The within variation is within each time period over customers (e.g., months) and the between variation is between customers (e.g., cross-sectional units). If you just have cross-sectional data, then you have between-variation, but not with-in. The same holds for time series although it could become complicated because you could have between years and within years. For panel data, there is a total variation over all units. There are, thus, three possible sources of variations in panels:

1. an overall variation by customers and time periods;
2. a within-customer variation; and
3. a between-customer variation.

The additivity effect is due to the inclusion of the overall variation.

There is a very large and complex econometric literature on modeling panel data sets. See, for example, Baltagi (1995), Hsiao (1986), and Wooldridge (2002). For now, it is just important to recognize that the problems associated with the entire Data Cube are complex to say the least. I will return to these in Chap. 10. My focus here is on simplifying the Data Cube.

9.3 The Data Cube and DataFrame Indexing

The Pandas DataFrame mimics the Data Cube using its indexing functionality. There are two forms of indexing: one for rows and one for columns. Column indexing just makes it easier to organize the features, to put them in a more intuitive order which does not affect analysis or modeling. It is just a convenience feature. The columns mimic the measure dimension of the Data Cube. The row index, however, has definite implications for not just organizing data but also for subsetting, querying, sorting, and sampling a DataFrame. It is the general form of the row index that mimics the spatial and temporal dimensions of the Data Cube. But remember that these two dimensions do not have to be space and time *per se*. They could be any complex categorization of the measures in the columns.

A row index is an identifier of each row. It does not have to be unique, but it is much better if it is unique. The method *Index.is_unique* returns True if the index is unique, False otherwise. I show some options in Fig. 9.1. The method *duplicated* has some options for a *keep* parameter to indicate how the duplicates are marked:

- "first": mark duplicates as True except for the first occurrence;
- "last": mark duplicates as True except for the last occurrence; and
- False (no quotation marks): mark all duplicates as True.

Notice how I use the *set_index* method in the top panel of Fig. 9.1. Initially, the index is just a series of integers beginning with zero. You use this *set_index* method to set it to a specific column. The *inplace = True* argument overwrites the DataFrame; otherwise a copy is made that must be named. If *inplace = False* (the default) is used, the method returns None. You can use any column, a list of columns,

```
##
## Create DataFrame with duplicate index value
##
data = { 'idx':[ 'a', 'b', 'c', 'a' ], 'x':[ 1, 2, 3, 4 ] }
df = pd.DataFrame( data )
df.set_index( [ 'idx' ], inplace = True )
df
```

idx	x
a	1
b	2
c	3
a	4

```
##
## Check for duplicate index
##
df.index.is_unique
```

```
False
```

```
##
## Identify the duplicates
##
df.index.duplicated( keep = False )
```

```
array([ True, False, False,  True])
```

Fig. 9.1 There are several options for identifying duplicate index values shown here

or a list of other objects such as other integers or strings. For example, if you are working with state-level data, a logical index is the two-character state code.

The new index in Fig. 9.1 is a series of string objects. Previously, it was integers beginning with zero. You could use, as another option, a *DatetimeIndex*, an index that uses datetime values. The *DatetimeIndex* is an array of datetime values that can be created in several ways. You can use the Pandas *to_datetime* function with a list of dates. For example, you could use *pd.to_datetime(df.date)* where "date" is a date variable in the DataFrame *df*. These dates are automatically converted to datetime values and placed into an array which is the *DatetimeIndex*. You can then set the DataFrame index to the *DatetimeIndex* using the *set_index* method. Notice that the *DatetimeIndex* is not an index *per se*; you must still set it to be the DataFrame index. Another way to create a *DatetimeIndex* is to use the *date_range* function which has parameters for when the date range is to begin and end, the number of periods to include, and the type of period (e.g., month, quarter). The "start" parameter is required; the "end" is not necessary but the number of periods must then be specified. You specify the end of the range or the number of periods, not both. The type of period is optional with "day" as the default, otherwise you must specify the type. I provide a short list of date types in Table 9.2 which I originally showed in Chap. 7. An example of the *date_range* function is *pd.date_range(start = '1/1/2021', end = '8/1/2021', freq = 'M')*. Another is *pd.date_range(start = '1/1/2021', periods = 8, freq = 'M')*.

Frequency alias	Description
B	Business day
D	Calendar day
W	Weekly
M	Month end
Q	Quarter end
A or Y	Year end

Table 9.2 This is a short list of available frequencies and aliases for use with the "freq" parameter of the date_range function. A complete list is available in McKinney (2018, p. 331)

Attribute	Description
day	Days of the period
dayofweek	Day of the week with Monday = 0, Sunday = 6
dayofyear	Ordinal day of the year
days_in_month	Number of days in the month
daysinmonth	Number of days in the month
hour	Hour of the period
is_leap_year	Logical indicator if the date belongs to a leap year
minute	Minute of the period
month	Month as January = 1, December = 12
quarter	Quarter of the date
second	Second of the period
week	Week ordinal of the year
weekday	Day of the week with Monday = 0, Sunday = 6
weekofyear	Week ordinal of the year
year	Year of the period

Table 9.3 This is a list of the attributes for the PeriodIndex. A complete list is available in McKinney (2018, p. 331)

You can also create a *PeriodIndex* from a *DatetimeIndex*. A *PeriodIndex* is like a *DatatimeIndex* but for periods. A *period* is a span of time such as a day, month, or quarter. A *DateTimeIndex* is a specific instant in time, but this is sometimes fuzzy. The values in a *PeriodIndex* are ordinal indicating regular periods in time. I provide a list of attributes to specify a period in Table 9.3.[3] You can create a *PeriodIndex* using a statement such as *pd.period_range('2021–02', periods = 8, freq = 'M')* which will create eight monthly periods beginning with February, 2021. You can also convert a *DatatimeIndex* to a *PeriodIndex* using the *to_period* method. I show an example in Fig. 9.2 so you can see how this is done.

[3] See https://pandas.pydata.org/pandas-docs/stable/reference/api/pandas.PeriodIndex.html. for more detail. Last accessed Januray 21, 2021.

An important functionality of an index allows you to query a DataFrame based on the index using the *query* method. For example, you can query the DataFrame in Fig. 9.1 using *df.query("index == 'a"')*. Notice the use of the single and double quotation marks. If the index is a *DatetimeIndex*, you could use *df.query("index == '2020-11-04"')* or *df.query("index == 'November 4, 2020"')*. Pandas will internally translate different date representations into a single datetime representation.

Figure 9.1 has a single index. You could create a multi-level index called a *MultiIndex*. You create it using the same *set_index* method I described above but the argument is a list of several DataFrame columns or other lists. Each element in the list becomes a level in the overall index in the order they appear in the list. This is a very important and useful feature of DataFrames because it more closely mimics the Data Cube concept. Recall that a Data Cube has a $measure \times space \times time$ configuration. You can view the space and time dimensions as indexes of the Data Cube: they define different views of the measure. For example, if the Data Cube represents quantities of a product manufactured at five plants for four calendar quarters, then plants and quarter index the product production. The corresponding DataFrame has one column for the amount produced and a two-level index: one for the plant identifier and one for the calendar quarter.

```
##
## Create DataFrame with DatetimeIndex
##
df = pd.DataFrame( { 'X': [ 1, 2, 3, 4 ]},
                    index = pd.to_datetime( [ '01/01/2021', '02/02/2021', '03/01/2021', '04/05/2021' ] ) )
display( df.style.set_caption( 'Using DatetimeIndex' ).\
    set_table_styles( tbl_styles ) )
##
## Convert DatetimeIndex to PeriodIndex
##
df.index = df.index.to_period( "M" )
display( df.style.set_caption( 'Using PeriodIndex' ).\
    set_table_styles( tbl_styles ) )
```

Using DatetimeIndex

	X
2021-01-01 00:00:00	1
2021-02-02 00:00:00	2
2021-03-01 00:00:00	3
2021-04-05 00:00:00	4

Using PeriodIndex

	X
2021-01	1
2021-02	2
2021-03	3
2021-04	4

Fig. 9.2 This illustrate how to convert a *DatetimeIndex* to a *PeriodIndex*

A benefit of a MultiIndex is the hierarchical structure you can impose on your data. For example, you may have sales data by state and localities within each state. Clearly, the localities are nested within the corresponding states. A MultiIndex reflects this hierarchical structure with state as the first (outer) level and locality as the second (inner) level. You can think of the second level as nested within the first. This capability comes with a cost: added complexity for accessing and manipulating a MultiIndex structure. The online Pandas documentation provides a thorough guide for handling and manipulating MultiIndexes.

There is another issue with a MultiIndex. The overall index is actually an array composed of tuples. The number of tuples equals the number of rows of the DataFrame and the number of elements in each tuple equals the number of levels in the MultiIndex. Each element of a tuple is the name or identifier of a level. For instance, if a MultiIndex has two levels "Product" and "Period", then each tuple has two elements in this order. The issue with this structure is that if you want to change the values of one of the levels, you have to recreate the tuples and then the MultiIndex. You cannot change thetuples directly since tuples are immutable. As an example, suppose you first create a MultiIndex with the two levels "Product" and "Period" with "Period" as a DatetimeIndex. Now you want to change "Period" to a PeriodIndex. You must first recreate the tuples by changing each DatetimeIndex

```
##
## Generate some data
##
np.random.seed( 42 )
data = { 'Product':list( "".join( [ i*3 for i in 'ABCDE' ] ) ), \
      'Period':[ '10/01/2020', '11/01/2020', '12/01/2020' ] * 5,\
     'Discount':np.random.uniform( size = 15 ) }
df_pan = pd.DataFrame( data, index = range( 15 ) )
##
## Create DatetimeIndex
##
df_pan.Period = pd.to_datetime( df_pan.Period )
##
## Set MultiIndex as Product and Period
##
df_pan.set_index( [ 'Product', 'Period' ], inplace = True )
##
## Convert 'Period' level of MultiIndex to PeriodIndex
## Based on: https://stackoverflow.com/questions/45243291/parse-pandas-multiindex-to-datetime
##
new_tup = df_pan.index.map( lambda x: ( x[ 0 ], x[ 1 ].to_period( 'M' ) ) )
df_pan.index = pd.MultiIndex.from_tuples( new_tup, names=[ "Product", "Period" ] )
##
## Display DataFrame
##
df_pan.head( n = 6 ).style.set_caption( 'MultiIndexed DataFrame' ).set_table_styles( tbl_styles )
```

MultiIndexed DataFrame

		Discount
Product	Period	
A	2020-10	0.374540
	2020-11	0.950714
	2020-12	0.731994
B	2020-10	0.598658
	2020-11	0.156019
	2020-12	0.155995

Fig. 9.3 Changing a MultiIndex to a new MultiIndex

```
##
## Query a MultiIndex DataFrame
##
x = pd.Period('2020-11', freq='M')
df_pan.query( "Period == @x" )
```

Product	Period	Discount
A	2020-11	0.472282
B	2020-11	0.576673
C	2020-11	0.903837
D	2020-11	0.090752
E	2020-11	0.593090

Fig. 9.4 This is one way to query a PeriodIndex in a MultiIndex. Notice the @. this is used then the variable is in the environment, not in the DataFrame. This is the case with “x”

value to a PeriodIndex value and then overwrite the DataFrame index. I show you how you can do this in Fig. 9.3.[4]

You can query a DataFrame based on the MultiIndex using the *query* method. For example, You could query on “Product” in Fig. 9.3 using *df_pan.query(“Product == ‘A”’)* to get all the records for product “A”. You could also query on a period, but this is done slightly differently because “Period” in Fig. 9.3 is a *PeriodIndex*. You need another *PeriodIndex* value for the comparison. I show one solution in Fig. 9.4. You can query both components of the MultiIndex using two expressions joined with either the symbol and or the word *and* for a “logical and” statement or the symbol | or the word *or* for a “logical or” statement.

9.4 Sampling From a DataFrame

I developed basic data analytic methodologies in the previous chapters and illustrated how you use them on whole data sets. The reason I used this approach is two-fold. First, this is how these methodologies are typically developed and presented. This will give you consistency with other books. My goal was to develop tools that you could quickly use in your work. Second, you may not always have a very large data set so using all the data you have is more the norm than the exception. Both reasons must now be dropped. More complex methodologies warrant advanced data handling and the sheer size of many data sets for modern business problems is large, to say the least.

[4] The steps to change the index are based on https://stackoverflow.com/questions/45243291/parse-pandas-multiindex-to-datetime. Last accessed January 21, 2021.

It is often impractical to use all the data available in many business data sets. You may have to use a subset. The question is how do you subset your data. The answer depends on whether or not you collapsed the Data Cube. If you did collapse it on space so you are working with a time series, the chances are that you also collapsed the time series to manageable levels by resampling. If you collapsed on time to produce a cross-sectional data set, you may also have to further collapse these data to more manageable levels by summing or averaging. But what if you didn't do either and you want to remain at the granularity you have? You have to sample.

There are three ways to sample a DataFrame:

1. simple random sampling (*SRS*);
2. stratified random sampling; and
3. cluster random sampling.

9.4.1 Simple Random Sampling (**SRS**)

When you use a panel data set (i.e., in a Data Cube format), your immediate inclination might be to draw a simple random sample of size n from the entire data set. The problem with this approach is that you may, and probably will, produce a smaller data set that has breaks in the continuity of the time dimension of each cross-sectional unit. For example, if you have monthly data from January to December for 1 year for a cross-sectional unit, the resulting sampled data could have data for March, May, June, and September of that year. There is no reason for continuity to be preserved when sampling, but continuity is a desired characteristic of time series; time series continuity must be preserved. There is no continuity in cross-sectional units, however. Technically, you could take a purely cross-sectional data set, estimate a regression model, shuffle the data, and then re-estimate the model. The estimated parameters will be identical. This suggests that you should draw a random sample of the cross-sectional units, keeping all the time series elements for each sampled unit.

Drawing a random sample of cross-sectional units and keeping the time dimensional data intact for each sampled unit is tantamount to selecting a random cluster sample with each cross-sectional unit viewed as a cluster. In sampling methodology, cluster sampling involves drawing a random sample of groups (*a.k.a.*, clusters) and then using all the objects within each sampled cluster. The sampling is at the cluster level.[5] There is a definite implication for modeling such as regression modeling. See any textbook on sampling methodology such as Cochrane (1963), Levy and Lemeshow (2008), and Thompson (1992) for a discussion of cluster sampling. See Wooldridge (2003) for a discussion of the impact of cluster sampling in regression analysis.

[5] You could use a two-stage clustering approach in which the clusters are randomly sampled for the first stage but then the objects are randomly sampled with each cluster for the second stage.

You can randomly sample your DataFrame using the Pandas method *sample*. The parameters for this are the sample size (*n* as an integer) or the fraction of the entire DataFrame to sample (*frac* as a float between 0 and 1), although you cannot use both at the same time. You also need to indicate if sampling is with or without replacement (*replace* with False as the default). Sampling is random so, in order to get the same sample each time you use this method, you should specify the *random seed* (*random_state* as an integer such as 42). See the Appendix for some background on random numbers and the random number seed. You can also specify weights for a stratified random sample. I used the *sample* method in Chap. 4 when I discussed visualizing *Large-N* data set, so you should review those examples.

9.4.2 Stratified Random Sampling

To select a stratified random sample, you need a strata identifier in your DataFrame. This could be, for example, product or marketing region. You would use the *groupby* method with this strata identifier to group your data. You should include the parameter *group_keys = False* to avoid having the grouping levels appear as an index in the returned DataFrame. For each group created, use the *sample* method to draw the sample. However, this is applied to each group using the *apply* function with the *sample* method in a *lambda function* since *apply* requires a function. This could all be chained into one expression. For example, if "Product" is the strata identifier, then you would use the set-up I illustrate in Fig. 9.5.[6]

```
##
## Stratified random sampling
##
df.groupby( 'Product', group_keys = False ).apply( lambda x: x.sample( n = 2, random_state = 42 ) )
```

	Product	Period	Discount
0	A	2020-10-01	0.374540
1	A	2020-11-01	0.950714
3	B	2020-10-01	0.598658
4	B	2020-11-01	0.156019
6	C	2020-10-01	0.058084
7	C	2020-11-01	0.866176
9	D	2020-10-01	0.708073
10	D	2020-11-01	0.020584
12	E	2020-10-01	0.832443
13	E	2020-11-01	0.212339

Fig. 9.5 This illustrates how to draw a stratified random sample from a DataFrame

[6] Based on https://stackoverflow.com/questions/44114463/stratified-sampling-in-pandas. Last accessed Januray 23, 2021.

9.4.3 Cluster Random Sampling

Cluster sampling is a two-step process. You first have to select the clusters and then select a sample within the selected clusters. This differs from the stratified random sampling where a random sample was selected from each stratum. For the first stage, you can use the Numpy *choice* function in the *random* package. This function randomly selects a set of values (i.e., choices) from an array. The selection is returned as an array. This array is then used in the second stage to subset the DataFrame using the *query* method. I illustrate how to do this in Fig. 9.6.[7]

```
##
## Cluster random sampling
##
## Select clusters using Numpy choice function
##
np.random.seed( 42 )
clusters = np.random.choice( df.Product, size = 3, replace = False )
##
## Get only unique clusters
##
clusters = np.unique( clusters )
display( clusters )
##
## Select clusters
##
df.query( 'Product in @clusters' )
```

```
array(['A', 'D'], dtype=object)
```

	Product	Period	Discount
0	A	2020-10-01	0.374540
1	A	2020-11-01	0.950714
2	A	2020-12-01	0.731994
9	D	2020-10-01	0.708073
10	D	2020-11-01	0.020584
11	D	2020-12-01	0.969910

Fig. 9.6 This illustrates how to draw a cluster random sample from a DataFrame. Notice that the Numpy *unique* function is used in case duplicate cluster labels are selected

9.5 Index Sorting of a DataFrame

You can sort a DataFrame by values in one or more columns or by the index. To sort by values, use the Pandas *sort_values* method. The parameters are *by*, *axis* (the default is 0), *ascending* (the default is True), and *inplace* (the default is False).

[7] Based on https://www.statology.org/cluster-sampling-in-pandas/. Last accessed Januray 23, 2021.

You can sort by an index level if you use $axis = 0$ or $axis = index$ in which case the *by* parameter may contain the level names of a MultiIndex.

You could also use *sort_index* to sort an index. Use $axis = 0$ (the default) to sort the rows. You can specify $level$ using an integer, a level name, or a list of level names. If you have a MultiIndex, then you can sort on the levels. The sort order is specified by the $level$ parameter.

9.6 Splitting a DataFrame: The Train-Test Splits

I have, so far, illustrated how to conduct statistical and econometric analyses to extract Rich Information from a data set. These methods allow you to get sufficiently Rich Information to answer most practical business questions. There are times, however, when you need to dig deeper into your data to extract even richer information because the implications of decisions based on that information are so critical for your business that any lesser information is unacceptable. Some decisions are "mission critical"; others are not. The "mission critical" ones require more sophisticated methods.

Part of these sophisticated methods involve checking the accuracy of your analysis. For example, if you train a logistic regression model[8] of customers' credit default risk with the intention of classifying customers by their credit worthiness, then you should want to know how accurately the model classifies them before you implement it. Your goal is to predict a customer's credit worthiness. If you predict incorrectly, you could either lose a valuable customer so you incur a loss or retain one who defaults so you also incur a loss. The stakes of being wrong are too high.

Your inclination might be to check "goodness-of-fit" statistics comparable to those I discussed in Chap. 6. This is incorrect because these statistics only tell you how well your model was trained by the data. Training a model and predicting with it are two different and separate tasks. The obvious solution might seem to be to use the trained model to check the model's residuals. A score, such as *Mean Square Error* (*MSE*), can be calculated for a prediction error and used to judge if the model is satisfactory or not. This score is a *performance error rate*. When the residuals are used in this fashion, the performance error rate is sometimes referred to as *resubstitution error rate* because the training data are substituted into the prediction calculations and then substituted again into the prediction error calculations. See Witten et al. (2011, p. 148). Using them to check predictions is equally incorrect because it involves using the same data twice: once to train the model and then again to test it via the residuals. The trained model, in a sense, already knows this data. So, you are attempting to use the same data to do double duty. Since the trained model already knows the data, using them again leads to a prediction bias.

---

[8] I discuss logistic regression modeling in Chap. 11.

Best Practice, whether for model training or not, is to use one data set for model training and another separate and independently drawn data set for performance testing beyond calculating goodness-of-fit statistics. Performance testing involves examining how well you can predict with your trained model. This is *Predictive Error Analysis* (*PEA*) which I introduced in Chap. 1.

Best Practice involves using at least two data sets: one strictly for training a model and the other for testing the performance of the trained model. These two sets, in addition to being independent, should also be mutually exclusive and completely exhaustive. Mutually exclusive means that any one record appears in only one of the two sets. Completely exhaustive means that all the records in the master data set are allocated; no record is ignored.

The source of all this data is an issue. One possibility is to collect new data that are independent of the "old" data. This independence is important because it means you have two separate random samples. The predictions based on the testing random sample data will then be unbiased estimates of the true values.

In most situations, the collection of a separate, independent random sample is impractical or infeasible. A possibility is to reuse the full or master data set in an iterative fashion through what is called *cross-validation*. This means you split off a piece of your data set and keep this on the side as a *hold-out sample*, train a model with the remaining pieces and then test the trained model with the hold-out piece. You do this repeatedly by splitting off a new piece each time. This is a good strategy if your data set is small in terms of the number of observations. If your data set is large this procedure could become computationally intensive depending on the size of the piece split off.

A simplified alternative is to split your data set just once into two parts: one called the*training data set* and the other the *testing data set*. The testing data are placed on the side (locked in a safe, if you wish) so that the model never sees it until it is ready to be tested. A final variation is available if you have a very large data set. This involves dividing your data into three parts: one for training, one for validation, and one for testing. The testing data set is still locked in a safe. The validation set is used iteratively to check a model during its training, but it is not used for testing. Validation and testing are two different operations. The validation data set plays a specialized, separate role in fine tuning the model which I describe in the next section. See Reitermanov (2010) for a good summary of the process. Most *PEA* use only two data sets. I illustrate the possible splitting in Fig. 9.7 and the divisions in Fig. 9.8.

9.6.1 Model Tuning of Hyperparameters

I discussed one modeling framework, *OLS* for a continuous target, in the previous chapters. This was presented from a simple perspective: one feature. There can certainly be many features, hence you could have a more complex *OLS* model. I address some of these complexities in the next chapter. There could be other

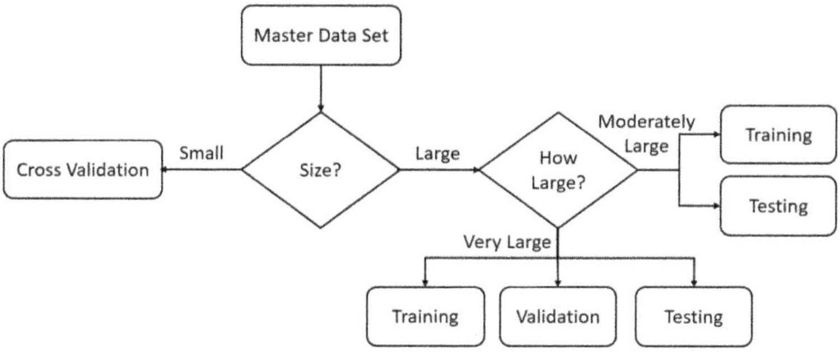

Fig. 9.7 This schematic illustrates how to split a master data set

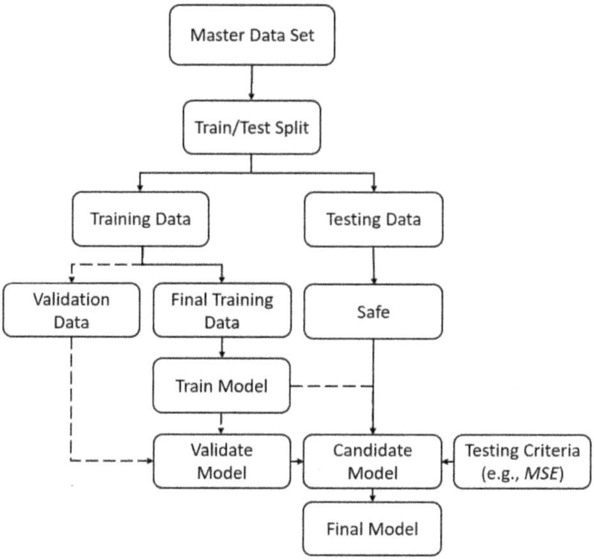

Fig. 9.8 This illustrates a general correct scheme for developing a model. A master data set is split into training and testing data sets for basic model development but the training data set is split again for validation. If the training data set itself is not split, perhaps because it is too small, then the trained model is directly tested with the testing data set. This accounts for the dashed arrows

types of models aside from *OLS* for different types of targets and objectives such as identifying impacts or developing continuous measures (*OLS*), estimating probabilities, or classifying objects. In fact, as I discussed above, there may not even be a target and the objective may be just to group objects. All of these models or schemes open up great possibilities. But for each one, there are many subordinate possibilities determined by parameters that are not part of the model but are extra to it and not estimated from data. These are *hyperparameters*.

As examples of hyperparameters, consider the models I will introduce in Chaps. 11 and 12. I will introduce the *K-Nearest Neighbor* (*KNN*) classification model in Chap. 11. The "K" determines how many objects (e.g., customers) have to be near or close to a specific point to be considered a "neighbor". This "K" is a hyperparameter: you specify what is acceptable. I will also introduce *Decision Trees* which are grown to a specific depth. The depth is a hyperparameter. Finally, in Chap. 12 I will introduce *K-Means Clustering* where "K" is the number of clusters to create. The "K" is a hyperparameter. There are many more examples in advanced learning.

Each setting of a hyperparameter specifies a new model, or variant of a basic model. What is the best or optimal hyperparameter setting? This is a very complex question to answer. Suffice it to say that one answer is to try different but reasonable values for them and assess the predictive performance of each resulting model. Basically, you establish a range of reasonable values for the hyperparameters and then iteratively move through the range trying each value to identify the optimal one. This can be refined to a grid search. This is *model tuning* because the search for an optimal hyperparameter is the search for a finely tuned predictive model. This is where the validation set is used. The word "validation" is thus misleading. It leads people to believe that this is another testing data set. This is incorrect. It is a tuning data set and should be referred to as such. See Ripley (1996, p. 354) who provides a very unambiguous definition of the validation data set for this specific purpose.

The bottom-line implication is that you have one candidate model scheme with variations determined by the hyperparameters. For each variation, you use the tuning data set to make a prediction and calculate a prediction error score. The variations can be ranked with the best variation (i.e., the best hyperparameter setting) becoming the final model. You may want to test the statistical differences among the scores rather than just raking them. Witten et al. (2011), for example, recommend a paired t-test, but this is incorrect because of the multiple comparison problem which results in an inflated significance level (i.e., α). See Paczkowski (2021b) for a discussion of the issue and how to deal with the problem. Regardless how you analyze the scores, you will have a final model that is then used with the test data set to calculate a final error prediction rate. I will discuss error rate calculations in Chaps. 10 and 11 in the context of actual prediction examples.

9.6.2 *Incorrect Use of Testing Data*

The testing data are frequently used incorrectly as I illustrate in Fig. 9.9. In this use, the training data set is used to train a model, but then the model is tested with the testing data. This part is okay. But if the test results are unacceptable, the analyst circles back to the model training, retrains the model and then retests it with the same testing data set. The reason for this is the mistaken belief that the testing data set is supposed to be used this way. In this process, the testing data are part of the model training thus negating its independent status. The testing data set is to be used just once as I illustrate in Fig. 9.8.

9.6.3 Creating the Training/Testing Data Sets

Saying a data set has to be split begs a question: "*How is a master data set split into two parts*?" This is not simple to answer. First, the relative size of training and testing data must be addressed. There is clearly a trade-off between a large and small assignment to training and testing. The more you assign to testing, then obviously the less you can assign to training. I illustrate this trade-off in Fig. 9.10. The implication is that a smaller training data set jeopardizes training because you have fewer degrees-of-freedom. Also, if you have $n > p$ before a training-testing split, then you could have an issue if you assign too much data to testing because you could produce a situation where you now have $n < p$ in the training data set. Now training a linear model becomes an issue—you cannot estimate the model's parameters. Dimension reduction methods, such as principal components, become important.

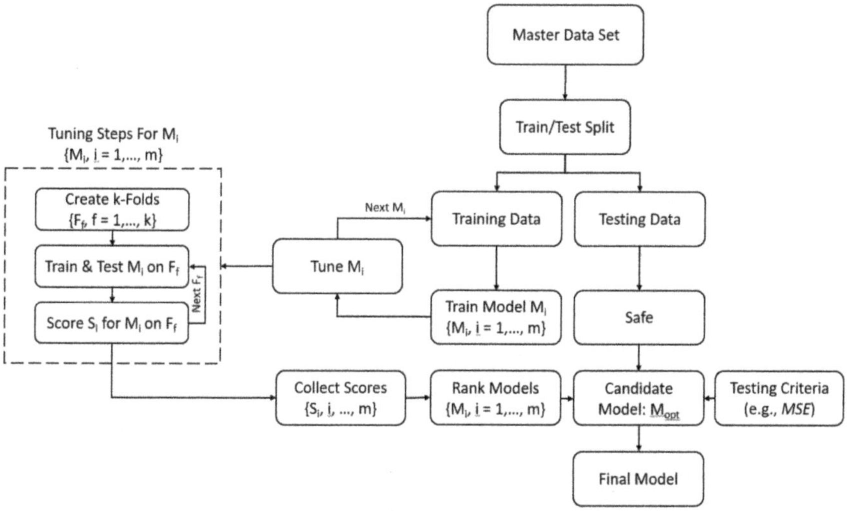

Fig. 9.9 This illustrates a general incorrect scheme for developing a model. The test data are used with the trained model and if the model fails the test, it is retrained and tested again. The test data are used as part of the training process

You will, of course, have a different issue if you assign too much to the training data set. Then you jeopardize testing because you may have patterns in the smaller testing set that are either unrepresentative of overall patterns or unrealistic. If the testing data are bigger, you might find that the trained model does better than it would with a smaller testing data set. See Faraway (2016) and Picard and Berk (1990).

There are no definitive rules for the relative proportions in each subset. There are, instead, rules-of-thumb (*ROT*s). One is to use three-fourths of the master data set for

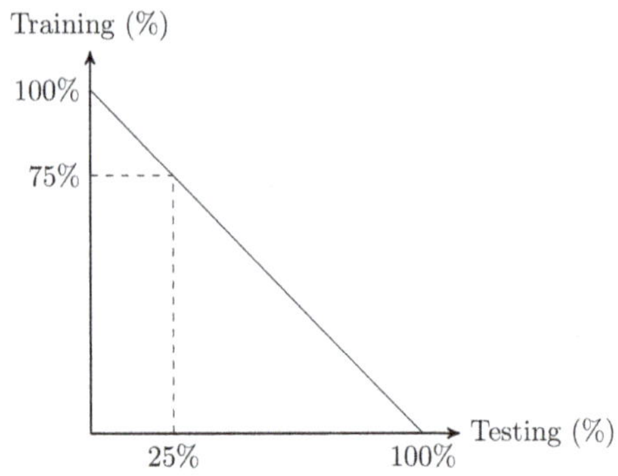

Fig. 9.10 There is a linear trade-off between allocating data to the training data set and the testing data set. The more you allocate to the testing, the less is available for training

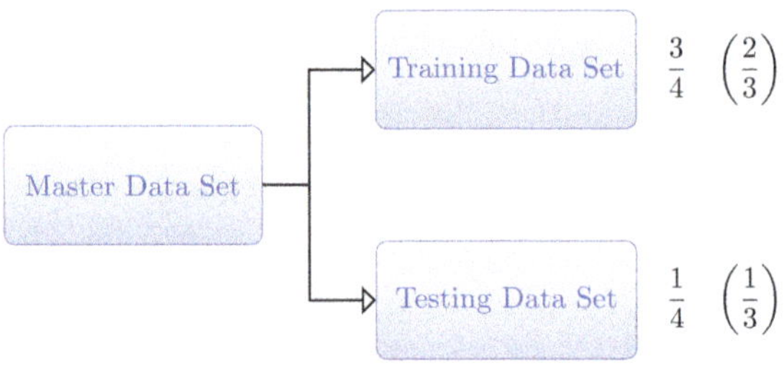

Fig. 9.11 As a rule-of-thumb, split your data into three-fourths training and one-fourth testing. Another is two-thirds training and one-third testing

training and the remaining one-fourth for testing. A second is two-thirds training and one-third testing. Regardless of the relative proportions, the training data set must be larger since more data are typically needed for training than testing. I show you such a splitting in Fig. 9.11. See Picard and Berk (1990) who show for small data sets that for optimal splits, the proportion of data allocated for testing should be less than 1/2 and preferably in the range 1/4–1/3 which are the proportions I quoted above. Also see Dobbin and Simon (2011) who found that the *ROT* strategy of using 1/3 of the data for testing is near optimal.

A second question is begged: "*How is the assignment made?*" The answer depends on the type of your data. Think about the Data Cube once more. One axis is time, another is cases or objects, and the third is measures. Cross-sectional data are the Cube collapsed on time; time series is the Cube collapsed on the cases. If your data are cross-sectional, then a random assignment is appropriate because, as I noted in Sect. 9.4.1, you could shuffle cross-sectional data and get the same answer. If your data are time series, the time continuity must be preserved so you would pick a time period such that anything before it is training and after is testing. By the way, the testing data are used to determine how well a time series model forecasts which is a special case of forecasting; this is a different issue I will address later. If your data is a panel (i.e., the whole Cube), then you have three options:

1. collapse the time dimension and randomly allocate cross-sectional units;
2. collapse cross-sectional units and choose a time period; or
3. use a random allocation of cross-sectional units to training and testing to preserve the integrity of time.

I consider each situation in the following subsections.

9.6.3.1 Comment on Strategy

The use of a splitting strategy is neither uncommon nor illogical. Yet there are criticisms. For example, Romano and DiCiccio (2019) state that a random allocation may still produce conflicting results merely because of the luck of the draw of a sample. Two analysts (or even the same one) could randomly allocate data to training and testing and get different data which may produce different results. Results should be independent of the allocation but this may not be the case because of the draws. Of course, if the same random number seed is used by both, the results will be identical, but then what is the point of doing a second study? See the Appendix about random numbers and the random number seed. Another criticism is that analysts feel the acceptability of results is reduced if only a portion of the available data is used in training. Trying to explain why a split was done may be more onerous than they feel is warranted. See Romano and DiCiccio (2019) for a comment about this. Also, there is an issue about a potential loss of statistical power when the data are split and a smaller set is used for inference. This is an open issue. See Romano and DiCiccio (2019) for comments. Finally, the proportions used for the split will have an impact on results. If one analyst uses the one-third *ROT* and another uses a one-fourth *ROT*, then their results will differ merely because of different proportions.

9.6.3.2 Handling Cross-Sectional Data

For the first sampling option involving cross-sectional data, a function in the Python library *scikit-learn* (in the submodule *model-selection*) named *train_test_split* can be used to split the master data set. You import this important function using *from sklearn.model_selection import train_test_split*.

Splitting is a simple mechanical procedure. There are three main parameters:

- The data to split: lists, Numpy arrays, scipy-sparse matrices, or Pandas DataFrames. I recommend the DataFrame.
- The proportion for the training or testing data set: specify one; the second is automatically calculated.
- A random state or seed.

The random state is needed if you want to reproduce your splits and, therefore, your work. A random number sequence is generated in agreement with the number of records in the master data set. The master data set is then randomly shuffled based on this sequence before it is split. This sequence will be different each time you split the master data set which will result in a completely different split each time. The difference is a consequence of the starting point for generating the sequence. Specifying a random state or seed, however, will start the generation from the same point (i.e., the seed) each time. Consequently, the split will be exactly the same each time. Hence, the reproducibility.

I illustrate the process in Fig. 9.12. In this example, the master data set, *df_cs*, is a Pandas DataFrame with 150 rows and 2 columns. This is used as the data argument in the function *train_test_split*. I set the training size proportion to 0.75. I could have set the testing size to 0.25 using *test_size = 0.25*. I will get the same results. The random seed was set to 42, an arbitrary number. This ensures that the split will be exactly the same each time I run this example. The function returns a list which is unpacked into two DataFrames, training and testing, in that order. The resulting split has 112 (= 0.75 × 150) records in the training data set and the remaining 38 in the

```
##
## Split the example cross-sectional data into two DataFrames
##
train_cs, test_cs = train_test_split( df_cs, train_size = 0.75, random_state = 42 )
##
display( f'Shape of the cross-sectional training data: {train_cs.shape}' )
display( f'Shape of the cross-sectional testing data: {test_cs.shape}' )
##
display( train_cs.head().style.set_caption( 'Cross-Sectional Training Data: Post-Split' ).\
         set_table_styles( tbl_styles ) )
display( test_cs.head().style.set_caption( 'Cross-Sectional Testing Data: Post-Split' ).\
         set_table_styles( tbl_styles ) )
```

```
'Shape of the cross-sectional training data: (112, 2)'

'Shape of the cross-sectional testing data: (38, 2)'
```

Cross-Sectional Training Data: Post-Split

	X1	X2
4	0.156019	1.985650
32	0.065052	1.924694
142	0.497249	1.822601
85	0.325183	1.722452
86	0.729606	1.280772

Cross-Sectional Testing Data: Post-Split

	X1	X2
73	0.815461	1.367716
18	0.431945	1.040775
118	0.892559	1.359491
78	0.358466	1.892047
76	0.771270	1.973011

Fig. 9.12 This is an example of a train-test split on simulated cross-sectional data

testing data set. Notice that the index numbers in the displayed sets are randomized reflecting the random shuffling I described above. You can now use the training data set to train a model and the testing data set to test the predictions from the trained model.

You could override the shuffling using the argument *shuffle = False* (True is the default). This would result in a stratified allocation, but you have to specify the stratifying labels.

9.6.3.3 Handling Time Series Data

Splitting a master time series into training and testing series is deceptively easy. The trick is to pick a point in time to divide the master series. All time series are naturally ordered by time so this should not be an issue. Refer to the definition of times series that I provided earlier. I show one way to do this in Fig. 9.13. See Picard and Berk (1990) who also address this issue of splitting on time.

As you will learn in Chap. 10, the training data may have to be further split for *cross-validation* of a trained supervised model before that model is subjected to a "final" test with the testing data set. In other words, there are multiple tests a model may undergo before you can say it is a "good" one and its predictions should be followed or used. The added complication from cross-validation is selecting subsets of the training data such that in each subset the chronological aspect of the data is preserved. Cross-validation works by withholding a subset of the training data set, making a prediction using the learned model and the remaining data, and then comparing the predictions against the withheld data. This is repeated a number of times or *folds*, each fold being a validation of the learned model's predictive ability. You could have a 5-fold validation, a 10-fold validation, or a *k-fold* validation. Regardless of the number of folds, the implication is obvious: you cannot select subsets that violate a natural ordering in time; the continuity must be preserved. So, the last date value in one fold must be before the first date value in the next fold. For now, just the simple time series splitting I illustrate in Fig. 9.13 will suffice for training a model, but it will be the basis for the k-fold validation that I will discuss later.

9.6.3.4 Handling Panel Data

Splitting a panel data set (i.e., the whole Data Cube) into training and testing parts is more complicated. Splitting a cross-sectional data set is done by random assignment since the order of objects (i.e., observations) is irrelevant. Splitting a time series is done by choice, not at random, since the time sequence order of objects (i.e., observations) must be preserved. But a panel is a combination of both, so which splitting procedure should you use? Preserving the time aspect of the data is a constraint on your options. Your sole degree of freedom is the cross-sectional aspect.

```
##
## Split time series into traing/testing data sets
##
train_ts, test_ts = ts_split( df_ts, plot = True )
##
## Plot the two series with a summary table
##
tbl_dict = { 'Master Series Length': len( df_ts ), 'Training Series Length': len( train_ts ), \
            'Testing Series Length':len( test_ts ) }
tbl = pd.DataFrame( tbl_dict, index  = [ None ] )
##
fig, ax = plt.subplots()
train_ts.plot( ax = ax, color = 'black' )
test_ts.plot( ax = ax, color = 'red' )
##
plt.title( 'Training and Testing Data Sets', pad = 35, fontsize = font_title )
##
plt.axvline( train_ts.index[ -1 ] )
##
plt.legend( [ 'Training', 'Testing' ], title = 'Data Sets', loc = 'upper right',\
          frameon = False, bbox_to_anchor = ( 1.25, 1.0 ) )
##
plt.table( cellText = tbl.values,
          rowLabels = tbl.index,
          colLabels = tbl.columns,
          cellLoc = 'center', rowLoc = 'center',
          loc = 'top' );
```

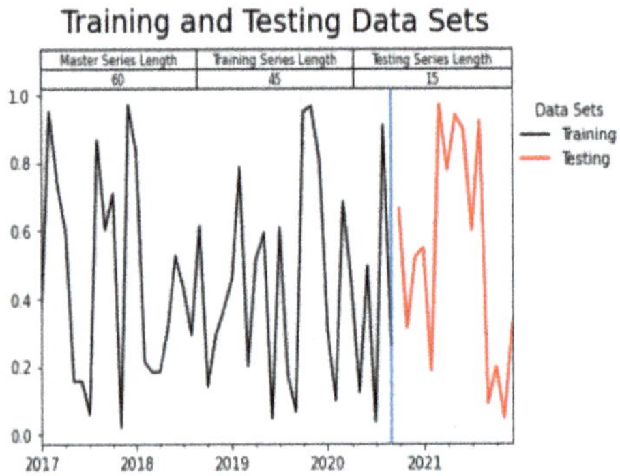

Fig. 9.13 This is an example of a train-test split on simulated time series data. Sixty monthly observations were randomly generated and then divided into one-fourth testing and three-fourths training. A time series plot shows the split and a table summarizes the split sizes

The way to approach splitting a master panel data set into training and testing data is to randomly assign the cross-sectional units, carrying along with each of these units all its time series data. I show this schematically in Fig. 9.14.

To illustrate how to implement the procedure, I created a small example DataFrame that is MultiIndexed and with only one data column. See Fig. 9.3. There are two levels to the index: *Product* and *Period*. The one data column is the average *Discount* for the product in the designated period. Figure 9.15 shows how to split this DataFrame into training and testing subsets based on the *Product* index level. The unique labels for the *Product* index are first extracted into a list and then the list is split into the training and testing subsets using the *train_test_split* function

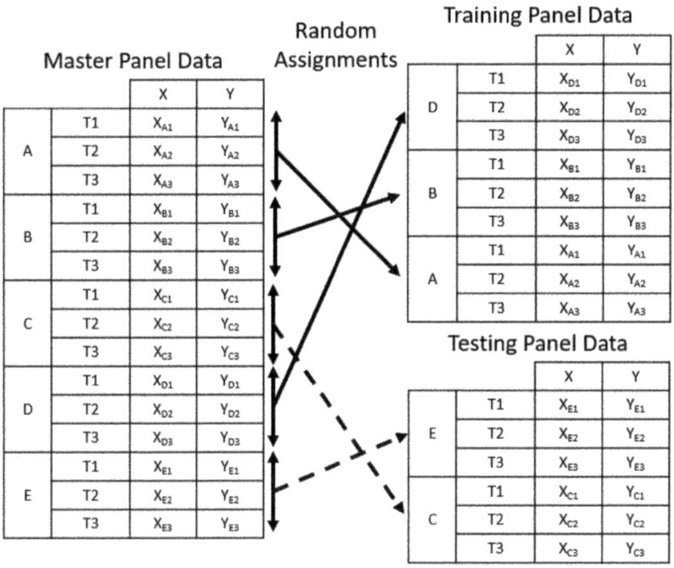

Fig. 9.14 This illustrates a master panel data set consisting of five cross-sectional units, each with three time periods and two measures (X and Y) for each combination. A random assignment of the cross-sectional units is shown. Notice that each unit is assigned with its entire set of time periods

I described above. These are training and testing product labels. The master panel DataFrame is then queried once using the training product labels and then again using the testing product labels. Each query returns the appropriate complete part of the master panel DataFrame. You now have the training and testing data sets for your analysis.

9.6.4 Recombining the Data Sets

Since data are often scarce, you may want to recombine your data after the final model test and then retrain your final model one last time. It is this final-final model that is deployed for its intended purpose. The estimated parameters will differ slightly from the penultimate model before the recombining, but this should not be dramatic. See Witten et al. (2011, p. 149).

```
##
## Select cross-sectional labels to split
## Selection based on https://stackoverflow.com/questions/24495695/pandas-get-unique-multiindex-level-values-by-label
##
lbl = list( df_pan.index.unique( level = 'Product' ) )
lbl_train, lbl_test = train_test_split( lbl, train_size = 0.60, random_state = 42 )
##
## Get training data
## Query statement based on https://stackoverflow.com/questions/53927460/select-rows-in-pandas-multiindex-dataframe
##
train_pan = df_pan.query( 'Product in @lbl_train' )
test_pan = df_pan.query( 'Product in @lbl_test' )
##
display( train_pan.style.set_caption( 'Training Subset' ).set_table_styles( tbl_styles ),
        test_pan.style.set_caption( 'Testing Subset' ).set_table_styles( tbl_styles ) )
```

Training Subset

Product	Period	Discount
	2020-10	0.374540
A	2020-11	0.950714
	2020-12	0.731994
	2020-10	0.058084
C	2020-11	0.866176
	2020-12	0.601115
	2020-10	0.708073
D	2020-11	0.020584
	2020-12	0.969910

Testing Subset

Product	Period	Discount
	2020-10	0.598658
B	2020-11	0.156019
	2020-12	0.155995
	2020-10	0.832443
E	2020-11	0.212339
	2020-12	0.181825

Fig. 9.15 This illustrates how the master panel data set of Fig. 9.3 is split into the two required pieces. Notice that I set the training size parameter to 0.60

9.7 Appendix

I will briefly describe how random numbers are generated in general and in Python.

9.7.1 Primer on Random Numbers

It is a misconception held by many people, analysts included, that a computer can actually generate random numbers. The numbers called "random" are actually generated by an algorithm. The produced numbers have properties that allow us to call them "random", but they are not random. They are called *pseudo-random numbers* because they are generated.

An old algorithm, now used in only a few legacy systems, is the *linear congruential generator* based on a deterministic formula

$$X_{n+1} = (c + a \times X_n) \bmod \mathrm{m}$$

where n is the iteration number, $0 \leq c < m$ is a base value, $0 < a < m$ is a multiplier, and *mod* is the modulo operator with m as the modulo divider usually as a power of 2 such as $m = 2^{32}$ or $m = 2^{64}$. This is a recursive formula meaning that it repeats as many times as necessary to give you the number of pseudo-random numbers you need, but the prior value generated by it is used with each repeat of the formula. The starting point, X_0, is the *seed*, which is usually the current clock time on the computer. You can, however, set the seed to whatever value you wish. I use 42 in my examples in this chapter. Actually, 42 is used by many geeks. It comes from Douglas Adams' novel *A Hitchhikers Guide to the Galaxy*. It also comes from a math problem: *The Sum of Three Cubes Problem*. Simply stated, what is the solution to $x^3 + y^3 + z^3 = k$ for k between 1 and 100? All numbers were found to either have a solution or not—except 33 and 42. The 33 was eventually solved and 42 was just recently solved. The solution is $42 = (-80538738812075974)^3 + 80435758145817515^3 + 12602123297335631^3$. See "MIT News" for a discussion.[9] Incidentally, in early 2021, a second solution for $k = 3$ was found.

```
##
## Import random package
##
import random
```

```
##
## Random number based on current clock time
##
## Get current clock time using the datetime package
##
##from datetime import datetime
now = datetime.now()
##
current_time = now.strftime( "%H:%M:%S" )    ## strftime is a string format function
print( f'Current clock time: {current_time}' )
##
## The random number is uniform
##
print( f'Random number using current clock time: {random.random()}' )
```

```
Current clock time: 11:43:55
Random number using current clock time: 0.19131531223425502
```

Fig. 9.16 This shows how to generate a random number based on the computer's clock time. The *random* package is used

---

[9] Available at https://news.mit.edu/2019/answer-life-universe-and-everything-sum-three-cubes-mathematics-0910. Last accessed February 12, 2021.

```
##
## Set the seed to be other than the current cloack time
##
random.seed( )
print( f'Random number without seed: {random.random()}' )
##
random.seed( 42 )
print( f'Random number with seed: {random.random()}' )
```

```
Random number without seed: 0.6565031317691028
Random number with seed: 0.6394267984578837
```

Fig. 9.17 This shows how to generate a random number based on a seed. I used 42 The *random* package is used

Python, through the *random* package, uses the *Mersenne Twister* which is based on prime numbers. The technicalities of this are beyond the scope of this book. Suffice it to say that this generator produces numbers such that a repeat occurs after $2^{19937} - 1$ runs, so a repeat is very unlikely; hence, the random number. See the Wikipedia articles on random number generation (https://en.wikipedia.org/wiki/Random_number_generation), pseudo-random number generation (https://en.wikipedia.org/wiki/Pseudorandom_number_generator), and the Mersenne Twister (https://en.wikipedia.org/wiki/Mersenne_Twister#Advantages).

The Python *random* package will generate random numbers. To use it, you first import it using *import random*. I show how to generate a random number using this package and the computer's clock time for the seed in Fig. 9.16. I also show how to use a seed value in Fig. 9.17.

You can also use Numpy's *random* module to generate random numbers. The advantage of this module is that you can generate an array of numbers. I provide an example in Fig. 9.18. The cluster sampling example in Fig. 9.6 uses the *choice* function in the Numpy *random* module to randomly select from an array of values. It basically selects a random sample from the array.

Fig. 9.18 This shows how to generate a random number based on seed and using the Numpy *random* package

```
##
## Set Numpy's random seed
##
np.random.seed( 42 )
##
## Specify the array size
##
n = 4
##
## Generate the random numbers
##
random_numbers = np.random.random( size = n )
random_numbers
```

```
array([0.37454012, 0.95071431, 0.73199394, 0.59865848])
```

Chapter 10
Advanced *OLS* for Business Data Analytics

I introduced the twin concepts of supervised and unsupervised learning in the previous chapter as a segue into a discussion about advanced manipulations of the Data Cube. The reason for the segue is that Business Data Analytics involves more than just modeling. It also involves classification. Both are methods for learning from your data, the former in a supervised fashion and the latter in an unsupervised fashion. More importantly, both involve more "steps" than just modeling or classifying. They also involve testing, prediction, and validation. This is where the train-test split discussion enters the picture.

I will begin to delve into an advanced discussion of supervised learning in this chapter. I will focus on a more extensive discussion of regression modeling, primarily *OLS* which I first introduced in Chap. 6. I will follow this chapter with a discussion of logistic regression. Finally, I will present material on classification, an unsupervised learning set of procedures, in the last chapter.

10.1 Link Functions: An Introduction

The single most important supervised analytical tool is the *regression model*, primarily *OLS*. It is the oldest formal technique for estimating models, but it is also just one member of a family of methods: the *Generalized Linear Model* (*GLM*) family. The *GLM* family is large, going beyond basic *OLS*. There is a different family member depending on the nature of the target. Each one has a function of the mean of the target linked to a linear function of the features. The mean is the expected value of the target. The function of the mean is called a *link function*. The link function transforms the mean of a random variable so that it equals a linear combination of the features. That is, ifμ_{tr} is the mean (expected value) of the target, then the transformation is $g(\mu_{tr}) = \beta_0 = \sum_{i=1}^{p} \beta_i \times X_i$ where $g(\ \cdot\ )$ is the link function. There is a large number of link functions, hence the large family. See

W. R. Paczkowski, *Business Analytics*,
https://doi.org/10.1007/978-3-030-87023-2_10

McCullagh and Nelder (1989) for the main reference on link functions and the *GLM* family. Also see Dobson (2002) for a detailed discussion of the *GLM* family.

The family of models is a general family because there is really only one model with variations connecting the target and features. The link function identifies the variations (i.e., family cousins). I list several link functions in Table 10.1. I will only consider two link functions in common use in *BDA*:

1. Identity Link; and
2. Logit Link.

Method	Dependent variable	Link
OLS	Continuous	Identity
Logistic	Binary	Logit
Poisson	Count	Log

Table 10.1 This is a list of the most commonly used link functions

The link for *OLS* is called the Identity Link because the expected value of the target (i.e., its mean) is already identically equal to a linear combination of the features; no further transformations are needed. Recall from my discussion in Chap. 6 that a basic regression model is $Y_i = \beta_0 + \sum_{j=1}^{p} \beta_j \times X_{ij} + \epsilon_i$ where $\epsilon_i \sim \mathcal{N}(0, \sigma^2), \forall i$. Since the disturbance term, ϵ_i, is a random variable, then so is the target, Y_i. The expected value of the target is $E(Y) = \beta_0 + \sum_{j=1}^{p} \beta_j \times X_{ij}$ so it is identically a linear combination of the features. Hence, the name. I will discuss the Logit Link, which has a different mean, in Chap. 11.

10.2 Data Preprocessing

I have to say something about data preprocessing which I introduced in Chap. 5. In particular, I have to cover:

1. standardization;
2. encoding of categorical data; and
3. data reduction primarily to handle multicollinearity.

I will cover the first two in this section and multicollinearity in a later section.

10.2.1 Data Standardization for Regression Analysis

Data standardization is an important first step in any data analysis. The objective of standardization, you will recall, is to place all the variables on the same base so that any one variable does not unduly influence analysis results. This undue influence could happen, for instance, because of widely differing measurement scales where one variable dominates another solely because of its scale. So, it seems it should also be a first step in regression analysis. This is, however, controversial. There are two camps. One, of course, says you should standardize while the other says you do not need to because the regression coefficients automatically adjust for scale differences with the p-values, R^2, and F-statistic unchanged. So, the substantive results are unchanged: what is statistically significant and how much variance is explained are unchanged. In addition, an elasticity calculated with the parameter estimates is also unchanged so the measure of the degree of impact of a change in the feature is unchanged. The issue is really just interpretation of the parameter estimates.

Recall from my discussion in Sect. 5.1.1 that one linear transformation is the Z-transform which linearly maps a variable to a scale that has zero mean and unit variance. There are two parts to the transformation equation: a numerator that centers the data and a denominator that scales the data. There are, therefore, two simultaneous transformations: centering and scaling. You could actually apply just one or both. In other words, if X is your feature variable, you could do $Z_i = X_i - \bar{X}$, $Z_i = X_i/s_X$, or $Z_i = X_i - \bar{X}/s_X$.

Regardless which one you use, you have to *fit* a statistical function to the data and then use that fitted statistic to *transform* the feature variable. The mean and/or the standard deviation are fit to the data and then they are used to do the Z-transformation. There is a *fit* step and a *transform* step. In Python's *SciKit's* libraries, these are *fit( )*, *transform( )*, and *fit_transform( )* which does both with one call. You will see examples for these later, although you already saw the use of the *fit( )* in Chap. 6.

For *OLS*, you do not need to scale your feature variables. To see this, assume you have a simple one variable model: $Y_i = \beta_0 + \beta_1 \times X_i + \epsilon_i$. Suppose you adjust X by a scaling factor f_X so the feature is now $X_i^* = f_X \times X_i$. This factor could be $f_X = 1$ for no scaling or $f_X = 1/s_X$ for scaling by the standard deviation. The new model is $Y_i = \beta_0 + \beta_1^* \times X_i^* + \epsilon_i$. To see the impact on the estimated slope parameter, first notice that the sample mean for X_i^* is simply $\bar{X}$ adjusted by f_X: $\bar{X}^* = f_X \times \bar{X}$. Then, $\bar{X}^* = \bar{X}$ for no adjustment; $\bar{X}^* = 1/s_X \times \bar{X}$ for scaling. Next, recall the formula for calculating the slope estimate from Chap. 6 and substitute the new X:

$$\begin{aligned}
\hat{\beta}_1^* &= \frac{\sum (Y_i - \bar{Y}) \times (X_i^* - \bar{X}^*)}{\sum (X_i^* - \bar{X}^*)^2} \\
&= \frac{f_X \times \sum (Y_i - \bar{Y}) \times (X_i - \bar{X})}{f_X^2 \times \sum (X_i - \bar{X})^2} \\
&= \frac{1}{f_X} \times \hat{\beta}_1.
\end{aligned}$$

The estimated slope is the true estimate scaled by the inverse of the factor. Kmenta (1971) shows that the standard error is also scaled by the inverse of the factor. This implies that the t-ratio is unchanged which further means that the p-value for significance is unchanged. Since the F-statistic for a simple model is the t-statistic squared, then the F-statistic is also unchanged. Finally, the R^2 is unchanged as also shown by Kmenta (1971). We can go the one extra step to note that the elasticity is also unchanged since

$$\begin{aligned}\eta^{Y}_{X*} &= \hat{\beta}^{*}_{1} \times \frac{\bar{X^{*}}}{\bar{Y}} \\ &= \hat{\beta}_{1} \times \frac{\bar{X}}{\bar{Y}}\end{aligned}$$

which is the elasticity without scaling.

What about the intercept? It is easy to show that the intercept is unchanged. So, basically, nothing is accomplished by scaling.

If the target is similarly scaled by a factor f_Y, then you can show that the estimated slope is $f_Y \times \hat{\beta}_1$. The intercept is $\hat{\beta}^{*}_{0} = f_Y \times \bar{Y} - \hat{\beta}_1 \times \bar{X}$ so it is adjusted by the factor. Generally, for standardizing Y and X the estimated slope is $f_Y/f_X \times \hat{\beta}_1$ and the intercept is appropriately adjusted.

Now consider centering by subtracting the mean. It is easy to show that the slope estimator is unchanged if either the mean of X or the mean of Y, or both, is subtracted. Simply recognize that the mean of the deviations from the mean is zero. The conclusion is that centering has no effect.

The final implication is that standardization is not necessary for *OLS*.

10.2.2 One-Hot and Effects (or Sum) Encoding

It is not unusual to include a categorical variable in a regression model to estimate the effect of a non-quantitative factor on the target. It may be known through experience or suspected because of anecdotal evidence that a factor has an impact and so should not only be included in the regression model but also tested for its effect. Does the non-quantitative factor have an effect beyond mere suspicion? Examples include marketing regions and store types for marketing studies; education level or degree for *HR* salary studies; robotic/non-robotic manufacturing plants for cost efficiency studies; and so forth.

Since these factors are non-quantitative, they cannot be directly used in a regression model whose parameters are estimated using only numeric data. The estimation formulas rely on means and deviations from these means; categorical variables do not have means. They have to be encoded, or converted, to numeric measures. I examined different encoding methods in Chap. 5. The most commonly used encoding in econometrics and the machine learning space is *dummy encoding* or *one-hot encoding*; both names refer to the same encoding scheme and are used interchangeably. I use "dummy encoding."

There is another form of encoding called *effects coding* which uses a -1 and $+1$ coding scheme rather than a 0 and 1 scheme. There is one effects coded variable created for each level of the categorical variable just as for dummy encoding. As for dummy coding, one of these effects coded variables must be excluded from a linear model[1] to avoid the dummy variable trap. The advantage of this encoding is that the sum of the estimated coefficients for the included effects coded variables equals the negative of the one for the excluded effects coded variable. The sum of the included and one excluded effects coded estimates must sum to 0.0. You could, therefore, always "retrieve" the omitted coefficient if you need it. Effects coding is often used in market research and statistical design of experiments. I will not use it in this book, instead restricting my discussions only to dummy encoding. See Paczkowski (2018) for a discussion of effects coding.

I mentioned that for dummy encoding, one of the dummy variables is dropped to avoid a numeric problem. The problem is referred to as the *dummy variable trap* which I mentioned several times. If all possible dummy variables for a single categorical variable are included in a linear model, then the sum of those variables equals the constant term. The constant term is actually 1.0 for all observations and the sum of the dummy variables is also 1.0. I illustrate this in Table 10.2. Therefore, there is a perfect linear relationship between the constant and the dummies. This is

Observation	Region	Constant term	D1	D2	D3	D4	Sum
1	Midwest	1	1	0	0	0	1
2	Northeast	1	0	1	0	0	1
3	South	1	0	0	1	0	1
4	West	1	0	0	0	1	1

Table 10.2 This table illustrates the dummy variable trap. The constant term is 1.0 be definition. So, no matter which Region an observation is in, the constant has the same value: 1.0. The dummy variables' values, however, vary by region as shown. The sum of the dummy values for each observation is 1.0. This sum and the Constant Term are equal. This is perfect multicollinearity. The trap is not recognizing this equality

perfect multicollinearity. Numerically, the unknown parameters cannot be estimated in this case, so all estimation breaks down and stops. Some software will detect this problem and drop one of the dummy variables, but not always.

The *Statsmodels* package provides an alternative solution. Recall that you to specify a formula for your model as a character string using the *Patsy* package formula paradigm.[2] I provided an example in Chap. 6. There is a Patsy function,

[1] As long as there is a constant in the model, which there should be.

[2] See https://patsy.readthedocs.io/en/latest/ for a full description of this formula paradigm. Last accessed October 27, 2020.

C( ), that takes a categorical variable as a parameter and creates the required encoded variables. Note the upper case "C". There are several encoding schemes available, the default being the dummy encoding. Dummy encoding in Patsy is referred to as *Treatment* encoding, represented by an upper case *T* in output, just to make naming conventions confusing. This function detects the levels of the categorical variable and creates the correct number of dummy variables. It also drops one of these dummies; this level is referred to as the *reference* level in the Patsy documentation. The reference level is the first level in alphanumeric order for the categorical levels. For example, referring to the furniture Case Study, there is a Region variable that has four levels: Midwest, Northeast, South, and West. Adding "C(Region)" to the Patsy formula string results in three dummy variables: one for each of the Northeast, South, and West; the Midwest in the reference level and is omitted.

10.3 Case Study Application

Consider the furniture regression model I discussed in Chap. 6. That model formula had only one explanatory variable: log of the pocket price for the furniture. The model can be expanded to include the discount rates offered by the sales force. There are four: dealer (*Ddisc*), competitive (*Cdisc*), order size (*Odisc*), and pick-up (*Pdisc*). The last one is offered if the customer drives to the manufacture's warehouse to pick up the order. The model should also be expanded to include the marketing region since, in this example, the sales force is regionally, not centrally, managed so each region basically has its own discount policy; only the list price is centrally determined. There are four marketing regions that coincide with the U.S. Census Regions: Midwest, Northeast, South, and West.

The first step is to collapse the Data Cube which is a panel data set: time periods by customers by orders. For this example, I collapsed the time periods to create a data set of customer IDs (*CID*) with the total orders, mean price, and mean discounts per *CID*. Since each customer is in only one marketing region, that region was included. I show the code snippet for this aggregation in Fig. 10.1. This is now cross-sectional data.

I then split the aggregated cross-sectional data into training and testing data sets using the *train_test_split* function. I set the random allocation to three-fourths training and one-fourth testing. A random seed was set at 42 so that the same split is produced each time I run the splitting function. I show the code snippet for this in Fig. 10.2.

Once I had the training data, I then set-up the regression. I show the set-up in Fig. 10.3. This follows the same four steps I outlined in Chap. 6. The formula is the key part. I wrote this as a character string: *formula = 'log_totalUsales ~ log_meanPprice + meanDdisc + meanOdisc + meanCdisc + meanPdisc + C(Region)'*. Notice that there is a term for the marketing region: *C(Region)*. As I stated above, the *Region* variable is categorical with four levels. The *C( )* function assesses the number of levels and creates a dummy variable for

```
##
## Aggregate panel data to CID level for modeling
##
## Identify variables for modeling and aggregation
##
cols = [ 'CID', 'Region', 'loyaltyProgram', 'buyerRating', 'buyerSatisfaction',
        'Usales', 'Pprice', 'Ddisc', 'Odisc', 'Cdisc', 'Pdisc' ]
##
## Identify variables for grouping
##
grp = [ 'CID', 'Region', 'loyaltyProgram', 'buyerRating', 'buyerSatisfaction' ]
##
## Specify aggregations
##
aggregations = { 'Usales':'sum', 'Pprice':'mean', 'Ddisc':'mean', 'Odisc':'mean',
                 'Cdisc':'mean', 'Pdisc':'mean'}
##
## Use groupby with agg function to aggregate
##
tmp = df[ cols ].copy()
df_agg = tmp.groupby( grp ).agg( aggregations )
##
## Rename columns and reset index
##
df_agg.rename( columns = { 'Usales':'totalUsales', 'Pprice':'meanPprice', 'Ddisc':'meanDdisc',
                           'Odisc':'meanOdisc', 'Cdisc':'meanCdisc',
                           'Pdisc':'meanPdisc'}, inplace = True )
df_agg = df_agg.reset_index()
##
## Print head
##
df_agg.head().style.set_caption( 'Aggregated Data' ).set_table_styles( tbl_styles )
```

Aggregated Data

	CID	Region	loyaltyProgram	buyerRating	buyerSatisfaction	totalUsales	meanPprice	meanDdisc	meanOdisc	meanCdisc	meanPdisc
0	14	Northeast	No	Good	1	3461.000000	5.400400	0.131820	0.051113	0.068248	0.038241
1	17	West	Yes	Good	4	1001.000000	5.797762	0.128930	0.050093	0.071814	0.036907
2	26	West	Yes	Good	5	787.000000	5.744514	0.136357	0.046714	0.071179	0.040036
3	28	West	Yes	Good	4	1873.000000	5.702487	0.136744	0.050641	0.071500	0.040564
4	38	West	No	Excellent	3	2350.000000	5.722586	0.135367	0.049990	0.070184	0.041439

Fig. 10.1 This is the code to aggregate the orders data. I had previously created a DataFrame with all the orders, customer-specific data, and marketing data

```
##
## Create the training/testing data.  Use the aggregated data.
##
cols = [ 'CID', 'Region', 'loyaltyProgram', 'buyerRating', 'buyerSatisfaction', 'totalUsales', 'meanPprice',
        'meanDdisc', 'meanOdisc', 'meanCdisc', 'meanPdisc' ]
X = df_agg[ cols ].copy()
##
## Add logs
##
X[ 'log_totalUsales' ] = np.log1p( X.totalUsales )
X[ 'log_meanPprice' ] = np.log1p( X.meanPprice )
##
## Split the data: 1/4 test and 3/4 train.
##
ols_train, ols_test = train_test_split( X, train_size = 0.75, random_state = 42 )
##
## Add log terms
##
##ols_train[ 'log_totalUsales' ] = np.log1p( ols_train.totalUsales )
##ols_train[ 'log_meanPprice' ] = np.log1p( ols_train.meanPprice )
##
display( ols_train.head().style.set_caption( 'Training Data Set' ).set_table_styles( tbl_styles ) )
df_size( ols_train )
##
display( ols_test.head().style.set_caption( 'Testing Data Set' ).set_table_styles( tbl_styles ) )
df_size( ols_test )
```

Training Data Set

	CID	Region	loyaltyProgram	buyerRating	buyerSatisfaction	totalUsales	meanPprice	meanDdisc	meanOdisc	meanCdisc	meanPdisc	log_totalUsales
164	685	Midwest	Yes	Excellent	3	3660.000000	6.112898	0.136045	0.049871	0.067748	0.040458	8.205492
28	218	West	No	Excellent	4	572.000000	5.192861	0.127714	0.048952	0.072000	0.033095	6.350886
708	2454	South	Yes	Excellent	4	966.000000	5.747111	0.081488	0.049463	0.070488	0.041366	6.874198
193	777	West	Yes	Good	1	804.000000	6.185009	0.129231	0.053974	0.069949	0.040795	6.690842
218	855	Midwest	Yes	Excellent	5	1396.000000	5.714906	0.138788	0.051173	0.066519	0.041442	7.242082

Fig. 10.2 This is the code to split the aggregate orders data into training and testing data sets. I used three-fourths testing and a random see of 42. Only the head of the training data are shown

```
##
## OLS
##
## There are four steps for estimatng a model:
##
##    1. define a formula (i.e., the specific model to estimate)
##    2. instantiate the model (i.e., specify it)
##    3. fit the model
##    4. summarize the fitted model
##
## ===> Step 1: Define a formula <===
##
## The formula uses a "~" to separate the left-hand side from the right-hand side
## of a model and a "+" to add columns to the right-hand side.  A "-" sign (not
## used here) can be used to remove columns from the right-hand side (e.g.,
## remove or omit the constant term which is always included by default).
##
formula = 'log_totalUsales ~ log_meanPprice + meanDdisc + meanOdisc +\
    meanCdisc + meanPdisc + C( Region )'
##
## Since Region is categorical, you must create dummies for the regions. You
## do this using 'C( Region )' to indicate that Region is categorical.
##
## ===> Step 2: Instantiate the OLS model <===
##
mod = smf.ols( formula, data = ols_train )
##
## ===> Step 3: Fit the instantiated model <===
##        Recommendation: number your models
##
reg01 = mod.fit()
##
## ===> Step 4: Summarize the fitted model <===
##
display( reg01.summary() )
```

Fig. 10.3 This is the code to set up the regression for the aggregated orders data. Notice the form for the formula statement

each, omitting the first level in alphanumeric order, which is the Midwest. Except for the formula modification, all else is the same as I outlined in Chap. 6. I show the regression results in Fig. 10.4.

The R^2 indicates that only about 27% of the variation in the target variable is explained by the set of independent variables. Unfortunately, this measure is inflated by the number of independent variables. It is a property of the R^2 that it is inflated as more variables are added. This is because the error sum of squares (*SSE*) is reduced with each new variable added to the model. The regression sum of squares (*SSR*) is therefore increased since the total sum of squares (*SST*) is fixed. Since $R^2 = SSR/SST$, R^2 must increase. We can impose a penalty for adding more variables in the form of an adjustment reflecting the degrees-of-freedom. The new measure is called R^2*-adjusted* or *Adjusted* R^2 or $\bar{R}^2$; the name varies. It is true that *Adjusted* $R^2 \leq R^2$.

OLS Regression Results

Dep. Variable:	log_totalUsales	**R-squared:**	0.268
Model:	OLS	**Adj. R-squared:**	0.258
Method:	Least Squares	**F-statistic:**	26.37
Date:	Tue, 23 Feb 2021	**Prob (F-statistic):**	7.65e-35
Time:	13:28:33	**Log-Likelihood:**	-638.15
No. Observations:	584	**AIC:**	1294.
Df Residuals:	575	**BIC:**	1334.
Df Model:	8		
Covariance Type:	nonrobust		

	coef	std err	t	P>|t|	[0.025	0.975]
Intercept	13.8888	1.775	7.824	0.000	10.402	17.375
C(Region)[T.Northeast]	0.3569	0.119	3.003	0.003	0.124	0.590
C(Region)[T.South]	-0.1574	0.431	-0.365	0.715	-1.004	0.689
C(Region)[T.West]	0.3620	0.077	4.698	0.000	0.211	0.513
log_meanPprice	-2.8279	0.252	-11.242	0.000	-3.322	-2.334
meanDdisc	-0.9686	7.601	-0.127	0.899	-15.898	13.961
meanOdisc	-0.2430	14.733	-0.016	0.987	-29.180	28.694
meanCdisc	-24.3524	10.307	-2.363	0.018	-44.596	-4.109
meanPdisc	14.5734	19.645	0.742	0.458	-24.010	53.157

Omnibus:	23.466	**Durbin-Watson:**	1.995
Prob(Omnibus):	0.000	**Jarque-Bera (JB):**	12.238
Skew:	0.157	**Prob(JB):**	0.00220
Kurtosis:	2.364	**Cond. No.**	1.39e+03

Fig. 10.4 This is the results for the regression for the aggregated orders data

I described the F-statistic's use in testing the Null Hypothesis that the model is no different than what I referred to as the Stat 101 model. The latter is a model with only the constant term. This is said to be *restricted* since all the parameters, except the constant, are zero. The model we are considering is *unrestricted* since all the parameters are included. The Null Hypothesis is, therefore $H_O : \beta_1 = \beta_2 = \ldots = \beta_p = 0$ and the Alternative Hypothesis is that at least one of these parameters is not zero. It does not matter if the one non-zero parameter is greater than or less than zero; it must be just non-zero. The F-test is a test of the restricted vs the unrestricted models. The p-value for the F-statistic tells you the probability of getting an F-value greater than the calculated value. A p-value less than 0.05 tells you to reject the Null Hypothesis. The p-value in Fig. 10.4 is 7.65e-35 which is definitely zero.[3] The Null Hypothesis should be rejected.

[3] The "e" notation is scientific notation. The statement "e-35" tells you to shift the decimal point 35 places to the left. A positive sign, not shown here, tells you to shift to the right.

The log price variable, *log_meanPprice*, has a negative estimated coefficient (−2.8279) with a p-value of 0.000. The Null Hypothesis is that this variable's parameter is zero and this Null must be rejected. The implication is that the (log) price has a (negative) effect on (log) sales. This estimated coefficient is the price elasticity as I showed in Chap. 6. Living room blinds are still highly price elastic.

The discounts are a slight mix. The competitive discount, *Cdisc*, is significant while the other three are insignificant. You could run a second regression with only *Cdisc* to see what happens.

The region dummies need some explanation. The dummies are indicated by the labels *C(Region)[T.XXX]* where *XXX* is Northeast, South, or West. The *C(Region)* part tells you that the variable was created by the *C( )* function and the "Region" part tells you that this is for the marketing region variable. The "T" stands for "Treatment" which is the *Statsmodels*' term for dummy variable. Therefore, the statement *C(Region)[T.Northeast]* represents the dummy variable for the Northeast marketing region.

You will notice that the estimated dummy coefficients are a mix of positive and negative values which may lead you to conclude that some regions have a positive effect and others do not. You must, however, also look at their p-values. They are also a mix. Unlike the discounts, this leads to a problem. For the discounts, you could drop the insignificant ones as I noted above. For the dummies, however, you must either drop them all or keep them all; it is a "none-or-all" decision. The reason for this is that region is a concept and the dummies are just the encoding of this concept. You cannot drop part of a concept. Suppose, however, that you drop the insignificant dummies. What happens? You redefine what all of them represent. You have to test the significance of all the dummies to determine whether or not you keep the concept of marketing region. The test is another F-test.

The general form for the F-statistic is

$$F_C = \frac{(SSR_U - SSR_R)/(df_U - df_R)}{SSE_U/(n-p-1)} \tag{10.3.1}$$

$$= \frac{(SSE_R - SSE_U)/(df_U - df_R)}{SSE_U/(n-p-1)} \tag{10.3.2}$$

where "U" indicates the unrestricted model "R" indicates the restricted model. If the restricted model is the Stat 101 model with just a constant term, then $SSR_R = 0$ (and $df_R = 0$) since there are no independent variables.

You can now run two regressions: one with and one without the dummy variables. The Null Hypothesis is that all the coefficients for the dummies are zero and the Alternative Hypothesis is that at least one is not. The results can be compared by examining the difference in the residual mean squares for both models. As an example, consider a DataFrame that has simulated data on two variables and 15 observations. One variable is a quantitative measure and the other is a categorical variable with three levels. This latter variable has to be dummified in a regression model. I ran two separate regressions, one with and one without the dummy

```
##
## Create a variable to hold the model names; this is a list
##
model_names = [ 'Unrestricted\nModel', 'Restricted\nModel' ]
##
## create the portfolio summary table
##
summary_table = summary_col( [ reg01, reg02 ],
            float_format = '%0.2f',
            model_names = model_names,
            stars = True,
            info_dict = info_dict
)
summary_table.add_title( 'Summary Table for Simulated Data Models' )
display( summary_table )
```

	Unrestricted Model	Restricted Model
C(X)[T.B]	-12.00	
	(17.32)	
C(X)[T.C]	-48.00**	
	(17.32)	
Intercept	258.00***	7.58***
	(12.25)	(0.03)
R-squared	0.31	0.00
	0.41	0.00
n	15	584
R2	0.409	0.000
R2Adjust	0.311	0.000
AIC	144.522	1460.805

Fig. 10.5 These are the regression results for simulated data. The two lines for the R^2 are the R^2 itself and the adjusted version

variables. I succinctly summarize the results in Fig. 10.5 rather than present the entirety of the regression output. I also created the relevant *ANOVA* tables which I show in Fig. 10.6. Using the data in Fig. 10.6, I manually calculated the F-statistic using (10.3.2) and show this in Fig. 10.7. Notice that the manually calculated F-Statistic agrees with the one in Fig. 10.5. You could just do an F-test comparing the two models as I show in Fig. 10.8. Notice that the results agree with what I showed in the other figures.

10.4 Heteroskedasticity Issues and Tests

A key Classical Assumption is that the variance of the disturbance term is constant, $\sigma^2_{\epsilon_i} = \sigma^2, \forall i$. This is called *homoskedasticity*. If the variance is not constant, then you have *heteroskedasticity*: $\sigma^2_{\epsilon_i} = \sigma^2_i, \forall i$. This is typical for cross-sectional data

```
##
## ANOVA for unstricted model
## Simulated data
##
aov01 = anova_lm( reg01 )
aov01
```

	df	sum_sq	mean_sq	F	PR(>F)
C(X)	2.0	6240.0	3120.0	4.16	0.042418
Residual	12.0	9000.0	750.0	NaN	NaN

(a)

```
##
## ANOVA for restricted model
## Simulated data
##
aov02 = anova_lm( reg02 )
aov02
```

	df	sum_sq	mean_sq	F	PR(>F)
Residual	14.0	15240.0	1088.571429	NaN	NaN

(b)

Fig. 10.6 Panel (**a**) is the unrestricted *ANOVA* table for simulated data and Panel (**b**) is the restricted version

```
##
## Manual calculation of F-Statistic
##
num = ( 15240.0 - 9000.0 )/(14 - 12)
den = 750
F = num/den
F
```

```
4.16
```

Fig. 10.7 This is the manual calculation of the F-Statistic using the data in Fig. 10.6. The F-statistic here agrees with the one in Fig. 10.5

```
##
## Specify the joint hypothesis -- note the form
##
hypothesis = '( C(X)[T.B] = 0, C(X)[T.C] = 0 )'
## This is interpreted as dummy for Level B = 0 AND dummy for Level C = 0
##
## Run and print the test
##
f_test = reg01.f_test( hypothesis )
##
print( f'F = {round( f_test.fvalue[0][0], 2 )}\np-Value = {np.around( f_test.pvalue, 4)} \
        \nTest: {f_test.pvalue < 0.05 }' )
```

```
F = 4.16
p-Value = 0.0424
Test: True
```

Fig. 10.8 This is the F-test of the two regressions I summarized in Fig. 10.5

which are concerned with different units, such as households, firms, industries, states, or countries. They vary by some measured characteristic at a point in time and the differences in those characteristics across the units result in variations in the disturbance terms. In time series data, however, the same unit is measured at different points in time and it is the relationship between the measurements at different points that is at issue.

What is the impact of heteroskedasticity on the properties of *OLS* estimators? Recall from Chap. 6 that there are four desirable estimator properties. See Kmenta

(1971) for these properties. They define the estimators to be *Best Linear Unbiased* estimators (*BLU*). Linearity is still met since this has nothing to do with the variance. Similarly for unbiasedness and consistency of the estimators. The minimum variance property, however, is not met because this is connected to the variance of the disturbance term. The result is that the non-constant variance makes *OLS* estimators differ from *BLU* estimators. For the *BLU* estimator, the desirable properties are imposed at the estimator's derivation so it, by definition, has minimum variance even under heteroskedasticity.

For a single feature model, the *BLU* estimator is

$$\hat{\beta}_1^* = \frac{\sum w_i \left(X_i - \overline{X^*}\right)\left(Y_i - \overline{Y^*}\right)}{\sum w_i \left(X_i - \overline{X^*}\right)^2} \tag{10.4.1}$$

$$\overline{X^*} = \sum \frac{w_i X_i}{\sum w_i} \text{ ; similarly for } \overline{Y^*} \tag{10.4.2}$$

$$w_i = \frac{1}{\sigma_i^2}. \tag{10.4.3}$$

See Kmenta (1971) for the derivation of (10.4.1). Clearly, with the Classical Assumptions the *OLS* estimator is

$$\widehat{\beta_1} = \frac{\sum(X_i - \overline{X})(Y_i - \overline{Y})}{\sum(X_i - \overline{X})^2}. \tag{10.4.4}$$

because the weight is constant and so cancels from the numerator and denominator. This suggests that there is a family of estimators. In fact, there is. It is called *Generalized Least Squares* (*GLS*). *OLS* is a special case of this much larger family. So, yes, another family inside a family member.

The weights, w_i, are important. They are the inverses of the variances. An issue is the source of the weights to implement a *BLU* estimator. Several possibilities are:

- construct them from information from other studies;
- make an assumption about the *Variance Generating Process* (*VGP*) based on the data analysis;
- replicate observations during the data collection phase of a study; or
- use other estimation techniques.

I will explore the second and fourth options below.

10.4.1 Heteroskedasticity Problem

What problem does heteroskedasticity produce in an *OLS* estimation? Kmenta (1971) clearly shows the implication. He shows that if you persist in using *OLS* and ignore heteroskedasticity, you would calculate an estimate that is not efficient in the class of linear, unbiased estimators since *BLU* has a smaller variance as I noted above. The correct minimum variance under heteroskedasticity is

$$\sigma^2_{\widehat{\beta_1^*}} = \frac{\sum w_i}{\sum w_i \sum w_i X_i^2 - (\sum w_i X_i)^2} \tag{10.4.5}$$

This obviously differs from the *OLS* variance under the Classical Assumptions, but simplifies to it under those Assumptions which is easy to show. We can show that $\sigma^2_{\widehat{\beta_1^*}} \leq \sigma^2_{\widehat{\beta_1}}$. You need an estimator of the variance of the estimator. Under homoskedasticity, this is $s^2 = {SSE}/{n-2}$ for the one variable model. If you use this under heteroskedasticity, you would be using a biased estimator. Unfortunately, we do not know the direction of the bias. As noted by Kmenta (1971), if the bias is negative, i.e., too small, then the t-statistics would be too large and you would reject the Null Hypothesis when you should not reject it. You would then make the wrong decision which could be very damaging to your business decisions. You would not provide the Rich Information that is needed. This is the problem due to heteroskedasticity. See Kmenta (1971), Gujarati (2003), and Hill et al. (2008) for extensive discussions of heteroskedasticity.

10.4.2 Heteroskedasticity Detection

You can detect a heteroskedasticity problem using either graphical displays or formal statistical tests. I will comment on both.

The graphical approach is the simplest. You merely plot the *OLS* residuals and look for a signature pattern for heteroskedasticity. Use the residuals against the estimated dependent variable which captures the effect of the features. If there is only one feature, then using it will suffice; if there are several, then use the estimated dependent variable. As with all graphs, you must look for patterns as I emphasized in Chap. 4. The desired pattern is a constant random spread around a line with zero intercept and zero slope. The reason for the zero intercept is that zero is the mean of the residuals. The zero slope is due to lack of trend in the residuals. I illustrate this desirable pattern in Panel (a) of Fig. 10.9. Panel (b) of Fig. 10.9, on the other hand, shows an undesirable pattern. The residuals fan out with increasing values of the x-axis variable. This fan pattern is the signature indicating heteroskedasticity.

A residual graph is easy to produce. Recall that the orders data were split into training and testing data sets. The training data were used to train the model shown in Fig. 10.4 above. The residuals and predicted (i.e., fitted) values were retrieved and plotted. I show the plot in Fig. 10.10. Notice that the pattern is similar to the one in Fig. 10.9, Panel (a) suggesting no heteroskedasticity.

Although these graphs are powerful for highlighting a problem when one exists, they are also subject to interpretation. Different people can and will see different

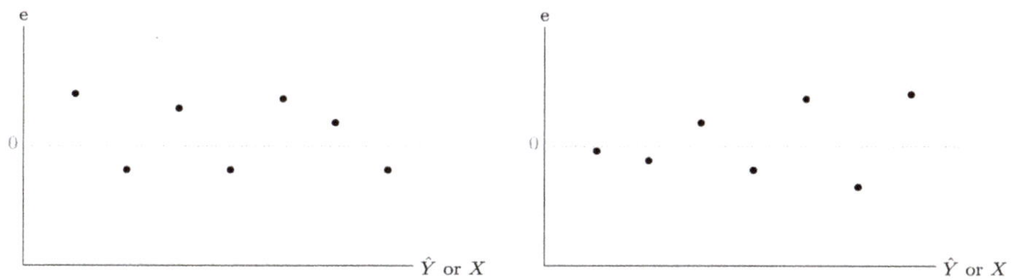

(a) Homoskedasticity Signature (b) Heteroskedasticity Signature

Fig. 10.9 These are the signature patterns for heteroskedasticity. The residuals are randomly distributed around their mean in Panel (**a**); this indicates homoskedasticity. They fan out in Panel (**b**) as the X-axis variable increases; this indicates heteroskedasticity

```
##
## Create DataFrame of residuals and predicted values
##
data = { 'residual':reg01.resid, 'predicted':reg01.fittedvalues }
df_residuals = pd.DataFrame( data )
##
## Create residual plot for reg01: orders data
##
base = 'Data from orders training data; n = ' + str( ols_train.shape[ 0 ])
ax = df_residuals.plot.scatter( x = 'predicted', y = 'residual' )
ax.set_title( 'Residual Plot', fontsize = 20 )
ax.set( ylabel = 'Residual', xlabel = 'Predicted Values' )
plt.axhline( y = 0 )
footer();
```

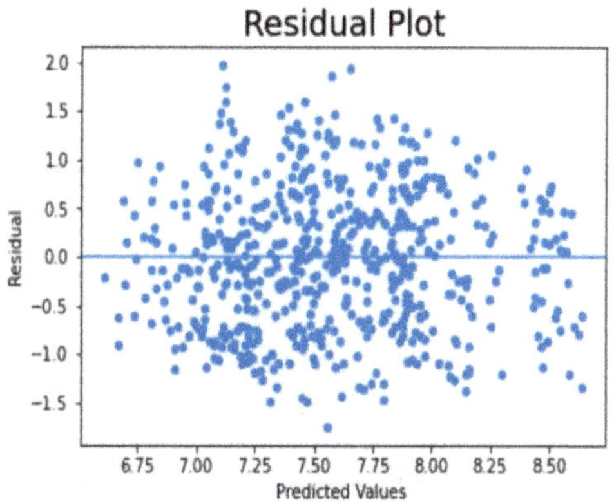

Fig. 10.10 This is the residual plot for the residuals in Fig. 10.4

patterns in a graph. A better approach is to use a formal statistical test, although there is controversy regarding the interpretation of p-values. Nonetheless, statistical tests should always be done in conjunction with graph examinations.

There are a number of statistical tests for heteroskedasticity. The one I will describe is the *White Test*. See Hill et al. (2008) for a good discussion of this test as well as Greene (2003) for an advanced treatment. For this explanation, first assume there is only one feature so $p = 1$. The Null Hypothesis is $H_0 : Homoskedasticity$ and the Alternative Hypothesis is $H_A : Heteroskedasticity$. The test involves several steps:

1. Estimate the model and save the residuals.
2. Square the residuals and estimate a second, auxiliary model:

$$e_i^2 = \gamma_0 + \gamma_1 X_i + \gamma_2 X_i^2 + u_i$$
$$u_i \sim \mathcal{N}(0, \sigma_u^2)$$

   Squared residuals are proxies for the disturbance terms' variance, so you are modeling the variances.
3. Save the R^2.
4. Calculate $\chi_C^2 = n \times R^2 \sim \chi_2^2$ where χ_C^2 is the calculated chi-square and χ_2^2 is its theoretical value with two degrees-of-freedom. The "2" results from the X_i and X_i^2 terms.
5. Reject H_0 of homoskedasticity if p-value < 0.05.

If more than one feature is involved, use the individual terms, their squared values, and all interactions (i.e., cross-products). For example, for $p = 2$, you have

$$e_i^2 = \gamma_0 + \gamma_1 X_{i1} + \gamma_2 X_{i1}^2 + \gamma_3 X_{i2} + \gamma_4 X_{i2}^2 + \gamma_5 X_{i1} X_{i2} + u_i$$

There are now five degrees of freedom for the chi-square test. In general, the test statistic is

$$nR^2 \sim \chi_{2p+p-1}^2 \tag{10.4.6}$$

where the degrees-of-freedom are the number of parameters in the auxiliary model less 1 for the constant. I show the set-up and results for the White Test in Fig. 10.11. I used *statsmodels*' *het_white* function in the *stats.diagnostics* submodule. This has two parameters: the regression residuals and the features. I retrieved the features from the estimated model as I show in Fig. 10.11. These results clearly provide evidence for rejecting the Null Hypothesis of homoskedasticity which differs from the conclusion based on Fig. 10.10. I will discuss in the next section how to remedy this issue.

10.4.3 Heteroskedasticity Remedy

If heteroskedasticity is suspected or shown to be an issue based on the White Test, then you must address it; that is, it must be fixed. MacKinnon and White

```
##
## White Test for Heteroskedasticity
##
## Retrieve X variables from the regression
##
X = reg01.model.data.orig_exog
##
## Use the residuals and the exogenous variables
##
white = sm.stats.diagnostic.het_white( reg01.resid, X )
##
## Display the results
##
returns = [ 'Chi Square Test Statistic', 'p-Value', 'F-Statistic', 'p-Value' ]
data = { 'Stats': returns, 'Values':white }
df_white = pd.DataFrame( data )
display( df_white.style.set_caption( "White's Test for Heteroskedasticity" ).\
        hide_index().set_table_styles( tbl_styles ) )

```

White's Test for Heteroskedasticity

Stats	Values
Chi Square Test Statistic	60.767839
p-Value	0.010918
F-Statistic	1.665683
p-Value	0.008604

Fig. 10.11 These are the White Test results

(1985) developed procedures to correct the standard errors which are the factors impacted by heteroskedasticity. There are four versions of the remedy which I list in Table 10.3. See Hausman and Palmery (2012) for descriptions and discussions. Also, see White (1980) and MacKinnon and White (1985). Hausman and Palmery (2012, p. 232) notes that *HC0_se* is "prevalent in econometrics. However, it has long been known that t-tests based on White standard errors over-reject when the null hypothesis is true and the sample is not large. Indeed, it is not uncommon for the actual size of the test to be 0.15 when the nominal size is the usual 0.05." Nonetheless, they later state that *HC1_se* is most commonly used, so this is the recommended version. I show how this is implemented in Fig. 10.12.

Test command	Description
HC0_se	White (1980)'s original robust standard errors correction
HC1_se	MacKinnon and White (1985)'s alternative; adjusts for degrees of freedom and is the most commonly used
HC2_se	MacKinnon and White (1985)'s alternative; adjusts for the leverage values
HC3_se	MacKinnon and White (1985)'s alternative; a slight modification to #2

Table 10.3 These are the four White and MacKinnon correction methods available in *statsmodels*. The test command notation is the *statsmodels* notation. The descriptions are based on Hausman and Palmery (2012)

```
##
## Fixing standard errors using HC1 for covariance type
## Specify HC1 in the fit() function
##
reg02 = smf.ols( formula, data = ols_train ).fit( cov_type = 'HC1' )
display( reg02.summary() )
```

OLS Regression Results

Dep. Variable:	log_totalUsales	R-squared:	0.268
Model:	OLS	Adj. R-squared:	0.258
Method:	Least Squares	F-statistic:	31.54
Date:	Wed, 25 Nov 2020	Prob (F-statistic):	4.44e-41
Time:	15:40:37	Log-Likelihood:	-638.15
No. Observations:	584	AIC:	1294.
Df Residuals:	575	BIC:	1334.
Df Model:	8		
Covariance Type:	HC1		

	coef	std err	z	P>|z|	[0.025	0.975]
Intercept	13.8888	1.779	7.805	0.000	10.401	17.377
C(Region)[T.Northeast]	0.3569	0.122	2.923	0.003	0.118	0.596
C(Region)[T.South]	-0.1574	0.449	-0.350	0.726	-1.038	0.723
C(Region)[T.West]	0.3620	0.075	4.816	0.000	0.215	0.509
log_meanPprice	-2.8279	0.248	-11.408	0.000	-3.314	-2.342
meanDdisc	-0.9686	7.985	-0.121	0.903	-16.620	14.682
meanOdisc	-0.2430	16.443	-0.015	0.988	-32.471	31.985
meanCdisc	-24.3524	11.488	-2.120	0.034	-46.868	-1.837
meanPdisc	14.5734	20.905	0.697	0.486	-26.399	55.546

Omnibus:	23.466	Durbin-Watson:	1.995
Prob(Omnibus):	0.000	Jarque-Bera (JB):	12.238
Skew:	0.157	Prob(JB):	0.00220
Kurtosis:	2.364	Cond. No.	1.39e+03

Fig. 10.12 This is the standard error correction based on *HC1_se* from MacKinnon and White (1985)

10.5 Multicollinearity

Multicollinearity is a potential major issue with business data when high dimensional data sets are used. These are data sets with many variables that could be used in a linear model. This could be a problem because the probability of any two or more variables being related grows as more variables are considered. See, for example, Zhao et al. (2020) and Fan et al. (2009). There are some interchangeable terms for this situation:

- multicollinearity;
- collinearity; and
- ill-conditioning (primarily used by numerical analysts).

10.5.1 Digression on Multicollinearity

Multicollinearity is an important topic in *BDA*, more so than what is believed. It has a major impact when the number of features is large because the probability that several of them will be linearly related becomes high. I want to digress for a discussion of multicollinearity in this section. I mentioned it a number of times in previous chapters but never fully discussed it.

Recall that *OLS* estimators are given by (6.4.2), which I repeat here for convenience:

$$\hat{\beta} = \left(\mathbf{X}^\top \mathbf{X}\right)^{-1} \mathbf{X}^\top \mathbf{Y} \tag{10.5.1}$$

The key component of this is the inverse of the sum of squares and cross-products matrix, $(\mathbf{X}^\top \mathbf{X})^{-1}$, which is a function of all the features. Calculating it is not trivial and is impossible if two or more features are linearly related. In this case, you are unable to estimate the unknown parameters of a linear model.

Suppose three features are $\mathbf{X_1}$, $\mathbf{X_2}$, and $\mathbf{X_3}$. They are linearly related if, say,

$$\mathbf{X_1} = \alpha_1 \mathbf{X_2} + \alpha_2 \mathbf{X_3}$$

This implies that the information contained in $\mathbf{X_2}$ and $\mathbf{X_3}$ is also contained in $\mathbf{X_1}$ so nothing is gained by including all three in a model. As an example, recall from basic economics that Real GDP is defined as $C + I + G + NX$ where C is real personal consumption expenditures, I is real gross private domestic investment, G is real net government expenditures and investments, and NX is real net exports of goods and services. All the information about Real GDP is contained in $C + I + G + NX$. Including Real GDP, C, I, G, and NX in the $\mathbf{X}$ matrix is tantamount to including information redundancy; this is multicollinearity.

There are three multicollinearity cases to consider:

1. Perfect Multicollinearity
   - Perfect linear relationship among the variables.
   - One or more variables are redundant.
   - Holds for all observations in the dataset.
   - Usually introduced into a problem by accident. Example: the dummy variable trap.
2. Some Multicollinearity
   - Some linear relationship.
   - Typical of business data.
3. No Multicollinearity
   - No linear relationship at all.
   - Not typical of business data.

Some degree of multicollinearity will usually be present in your Business Analytics data because of the size of the data sets you will use[4] as well as the fact that many business data are related. For example, production and sales. The case of no multicollinearity is not an issue because it will rarely occur with business and economic data.

When you have Perfect Multicollinearity, you are unable to invert the $(\mathbf{X}'\mathbf{X})$ matrix. This is strictly a computational problem. A non-invertible matrix is called *Singular* or *Ill-conditioned.* Some packages give a warning such as *MATRIX IS SINGULAR*. If the inverse cannot be found, then you cannot find parameter estimates. The whole estimation process breaks down.

When you have some multicollinearity, you can invert the matrix and calculate estimates. The problem, however, is that the estimates are unstable. In this case, the inverted matrix is "inflated" so that results may be counterintuitive with reversed signs and magnitudes that may change with a slight change in the data. A far more important problem than just unstable estimates is that the variances of the estimators "blow-up." Recall from Sect. 6.4 that

$$\sigma^2_{\hat{\boldsymbol{\beta}}} = \sigma^2 \left(\mathbf{X}^\top \mathbf{X}\right)^{-1}. \tag{10.5.2}$$

If $\left(\mathbf{X}^\top \mathbf{X}\right)^{-1}$ is inflated, then so are the variances.

Suppose you have just two features, X_1 and X_2. Then the variance for $\hat{\beta}_1$ for X_1 is

$$\sigma^2_{\hat{\beta}_1} = \frac{1}{1 - r^2_{12}} \times \frac{\sigma^2}{\sum (X_1 - \overline{X_1})^2} \tag{10.5.3}$$

where $0 \leq r^2_{12} \leq 1$ is the squared correlation between X_1 and X_2 so $0 \leq (1 - r^2_{12}) \leq 1$. In essence, the multiplicative factor in the denominator, $(1 - r^2_{12})$, deflates the other factor in the denominator, $\sum (X_1 - \bar{X}_1)^2$, and thus inflates $1/\sum (X_1 - \overline{X_1})^2$. If $r^2_{12} = 0$, then $\sigma^2_{\hat{\beta}_1}$ is as before. If $r^2_{12} = 1$, then $\sigma^2_{\hat{\beta}_1}$ is undefined, so it "blows up." For large values of r^2_{12}, $\sigma^2_{\hat{\beta}_1}$ is inflated.

More generally, let VIF_j be the *Variance Inflation Factor* for parameter j, $j = 1, 2, \ldots, p$. This is the amount by which the variance is inflated due to multicollinearity. It is a generalization of the factor $1/(1 - r^2_{12})$ in (10.5.3). You can write this variance for j as

$$\sigma^2_{\hat{\beta}_j} = \sigma^2 \times f\left[\left(\mathbf{X}^\top \mathbf{X}\right)^{-1}\right] \times VIF_j. \tag{10.5.4}$$

[4] Refer to my discussion about the size and complexity of data sets in Chap. 1.

The VIF_j is related to the correlation between variable j and all other variables. If $VIF_j = 1$ (i.e., no correlation) then there is no variance inflation; there is no multicollinearity. You should expect $VIF_j \geq 1$. Basically,

$$\sigma^2_{\hat{\beta}_j} = \sigma^2 \times \frac{1}{\sum(x_{ij} - \bar{x_j})^2(1 - R^2_{j,x_1,x_2,x_{j-1},x_{j+1},\ldots,x_p})}. \quad (10.5.5)$$

If the variances are inflated by the *VIF*, then the t-statistics are too small. Recall that

$$t_{C,\hat{\beta}_j} = \frac{\hat{\beta}_j}{s_{\hat{\beta}_j}}.$$

Consequently, you will not reject the Null Hypothesis as often as you should. This will lead you to believe that the coefficient is statistically zero when it is not and thus you will make an incorrect decision.

10.5.2 Detection with VIF *and the Condition Index*

How do you detect multicollinearity? It is not a present/absent problem, but one of degree. You must check for the severity of multicollinearity, not its presence or absence. Always assume it is present in your business data, especially high-dimensional data.

One way to check for it is to look for a contradiction between the t-statistics and R^2. The t-statistics are deflated because of inflated standard errors, but the R^2 is inflated since it is a function of inflated parameter vector. It can be shown that $SSR = \hat{\beta}_1^2 S_{XX}$ so $R^2 = SSR/SST$ is too large.

You can also check the *Pearson correlation matrix*. Correlations are near $|1|$ if a linear relationship exists between two features. The "two" is important because correlations are pair-wise. If more than two features are involved in a linear relationship, then correlations may be low despite a high degree of multicollinearity. But if correlations are high, then you have multicollinearity between two features; if correlations are low, then do another check.

An alternative check is to look at linear relationships between each feature and all the other features. Let R^2_j be the squared multiple correlation between X_j and all other features in the model. This is $R^2_{j,X_1,X_2,X_{j-1},X_{j+1},\ldots,X_p}$ that we saw before. That is, regress X_j on the other features and save R^2. These are *auxiliary regressions*. If you have five features, then you have five auxiliary regressions.

Recall the variance inflation factor, VIF, I introduced above. Based on that definition,

$$VIF_j = \frac{1}{1 - R_j^2}. \tag{10.5.6}$$

No linear relationship implies that $VIF_j = 1$. Large values of VIF_j indicate a high degree of multicollinearity. This suggests a rule-of-thumb (*ROT*) for assessing multicollinearity: the maximum value of VIF_j greater than 10 indicates the presence of a high degree of multicollinearity.

Another indicator of multicollinearity is the *condition number*. This is the square root of the ratio of the largest to smallest eigenvalues of the $\mathbf{X}^\top\mathbf{X}$ matrix. The eignevalues are found from the Singular Value Decomposition (*SVD*) of this matrix. A *ROT* is that multicollinearity is an issue if the condition number is greater than 20. See Belsley et al. (1980) and Greene (2003) for a discussion of the condition number, especially the former reference.

The regression output from Fig. 10.4 provides the base for checking for multicollinearity. First, I checked the correlation matrix which I display in Fig. 10.13. You can see that the correlations are all very low suggesting that multicollinearity is not an issue. Figure 10.14, which shows the *VIF*'s for this problem indicate otherwise. The *VIF*s are the diagonal elements of the inverse of the correlation matrix. As such, they "measure the effect of the (independent) variables on the variances of the estimated coefficients." See Snee (1977, p, 417). For a high *VIF*, the associated feature is likely to be "poorly estimated because of the correlation among the" features. See Snee (1977, p, 417). From Fig. 10.14, you can see that there is multicollinearity revealed by the highlighted features. There is a high degree of collinearity affecting the estimated coefficients for the Southern Region and dealer discount (*Ddisc*).

10.5.3 Principal Component Regression and High-Dimensional Data

You have several remedies or fixes for multicollinearity if the *VIF*s or the condition numbers indicate a problem. One recommended remedy is to drop one feature from the model. There is multicollinearity if you have redundancy, so simply drop the redundant feature. The *VIF* measures are the guide for which one(s) to drop. Another remedy is to use Principal Components analysis (*PCA*) from Sect. 5.3. The extracted components are linearly independent by design and are fewer in number than the original set of features. The components can be used in a regression which is sometimes called a *Principal Components Regression* (*PCR*). See Johnston (1972) for some discussion.

```
##
## Subset the design matrix to eliminate the first column of 1s
##
cols = reg01.model.data.orig_exog.columns[ 4: ]
##
## Create the correlation matrix
##
x = reg01.model.data.orig_exog[ cols ]
corr_matrix = x.corr()
corr_matrix.style.format( '{:,.2f}' ).set_caption( 'Correlation Matrix' ).\
    set_table_styles( tbl_styles )
```

Correlation Matrix

	log_meanPprice	meanDdisc	meanOdisc	meanCdisc	meanPdisc
log_meanPprice	1.00	-0.27	0.08	-0.01	0.03
meanDdisc	-0.27	1.00	-0.10	0.06	-0.08
meanOdisc	0.08	-0.10	1.00	-0.09	-0.03
meanCdisc	-0.01	0.06	-0.09	1.00	0.05
meanPdisc	0.03	-0.08	-0.03	0.05	1.00

Fig. 10.13 This is the correlation matrix to check for multicollinearity in Fig. 10.4

10.6 Predictions and Scenario Analysis

The objective for estimating a linear model is not only to produce estimated effects of key driver features, but also to use the model to predict. Recall from my discussion in Chap. 1 that *BDA* is concerned with what will happen, unlike Business Intelligence which is concerned with what did happen. What will happen is predicting.

There are two ways to predict. The first is to determine how well the model functions with a data set it has not seen and the second to make predictions for totally new situations. The first uses the testing data set which, until now, has been unused. The second involves scenario analysis in which values are specified for the key drivers in the linear model. The testing data set is not needed for this. I will first describe the use of the testing data set and then show how to specify a scenario. It is scenario analysis that is the true reason for making predictions. I will continue to use the aggregated orders data and the associated regression model to illustrate making predictions.

10.6.1 Making Predictions

When a regression model is estimated and stored in an object variable, such as *reg01* in Fig. 10.4, a *predict* method is automatically created and attached to this object.

```
##
## Calculate VIFs
## The VIFs are the diagonal elements of the inverted correlation
## matrix of the independent variables.
##
## Subset the design matrix to eliminate the first column of 1s.
##
## Create the correlation matrix
##
cols = reg01.model.data.orig_exog.columns[ 1: ]
x = reg01.model.data.orig_exog[ cols ]
corr_matrix = x.corr()
##
## Invert the correlation matrix and extract the main diagonal
##
vif = np.diag( np.linalg.inv( corr_matrix ) )
##
## Create a DataFrame using a dictionary
##
indepvars = [ i for i in x.columns ]
data = { 'Variable':indepvars, 'VIF':vif }
df_vif = pd.DataFrame( data )
df_vif.set_index( 'Variable', inplace = True )
##
df_vif.style.applymap( lambda x: 'background-color : yellow' if x > 10 else '' ).\
    set_caption( 'VIF Values' ).\
    set_table_styles( tbl_styles )
```

VIF Values

Variable	VIF
C(Region)[T.Northeast]	1.286874
C(Region)[T.South]	41.746375
C(Region)[T.West]	1.338442
log_meanPprice	1.085393
meanDdisc	40.161801
meanOdisc	1.025272
meanCdisc	1.020673
meanPdisc	1.012210

Fig. 10.14 These are the *VIF*s to check for multicollinearity in Fig. 10.4

The parameter for this method is simply the testing data set. The model applies the features in this testing data set to the estimated parameters of the linear model and produces predicted or estimated values for the linear model's target. Measures can then be used to compare the predicted values to the actual values which are also in the testing data set. I illustrate this approach in Fig. 10.15. For this example, unit sales are predicted but the predictions are in (natural) log terms since the training set had log sales. The predicted log sales are converted back to unit sales in "normal" terms by exponentiation. The predictions can be compared to the actual values in the testing data set using an R^2 measure.

10.6.2 Scenario Analysis

A common business question is in a *what-if* form in which a hypothetical case is specified and a prediction is needed for that case. For example, for the orders

```
##
## Calculate predicted log of unit sales, the dependent variable.
##
## Note: the inverse of the log is needed; use np.expm1( x )
## since log1p was used: np.expm1 = exp(x) - 1.
##
log_pred = reg01.predict( ols_test )
y_pred = np.expm1( log_pred )
##
## Combine into one temporary DataFrame for convenience
##
tmp = pd.DataFrame( { 'y_test':ols_test.totalUsales, 'y_logPred':log_pred, 'y_pred':y_pred } )
tmp.info()
```

```
<class 'pandas.core.frame.DataFrame'>
Int64Index: 195 entries, 595 to 165
Data columns (total 3 columns):
 #   Column     Non-Null Count  Dtype
---  ------     --------------  -----
 0   y_test     195 non-null    int64
 1   y_logPred  195 non-null    float64
 2   y_pred     195 non-null    float64
dtypes: float64(2), int64(1)
memory usage: 6.1 KB
```

Fig. 10.15 This illustrates making a prediction using the *predict* method attached to the regression object. The testing data set, *ols_test* is used

data model, a business manager could ask: "*What would sales be in the Western marketing region if the pocket price is \$2.50, the order size discount is set at 5%, the other discounts are all set at 3% each*?" This is a specific scenario that can be answered using the estimated model. I illustrate the set-up for this scenario in Fig. 10.16.

10.6.3 Prediction Error Analysis (PEA)

I described above one simple measure of prediction accuracy. This is just an R^2 statistic relating the actual and predicted values. There are more complex methods for determining prediction accuracy and it is these methods that I will describe in this section.

These methods, referred to as *cross-validation methods*, involve successively dividing the entire data set into two parts: training and validation. This is comparable to the division of the entire data set into training and testing data sets, but there is a slight difference here. It is how the data are divided and where they fit within the context of the entire modeling process. The two validation approaches I will shortly describe are on the left in one generic descriptor box labeled "Validation Data" in Fig. 9.8.

```
##
## Specify scenario values to use for prediction
##
## Create a dictionary
##
data = {'meanPprice': [ 2.50 ], 'meanDdisc': [ 0.03 ], 'meanOdisc': [ 0.05 ],
        'meanCdisc': [ 0.03 ], 'meanPdisc': [ 0.03 ], 'Region': [ 'West' ]
       }
##
## Create a DataFrame using the dictionary
##
df_scenario = pd.DataFrame.from_dict( data )
##
## Insert a log price column after the Pprice variable i.e., insert into position 1
##
df_scenario.insert( loc = 1, column = 'log_meanPprice',
                    value = np.log1p( df_scenario.meanPprice ) )
##
## Display the settings and the predicted unit sales
##
display( df_scenario.style.set_caption( 'Scenario Settings' ).set_table_styles( tbl_styles ).hide_index() )
##
## Create a prediction
##
log_pred = reg01.predict( df_scenario )
y_pred = np.expm1( log_pred )
##
boldprt( '\nPredicted Unit Sales:' )
print( f'\t{round( y_pred[ 0 ], 0 )}' )  ## Position 0 is used since y_pred is a Series object
```

Scenario Settings

meanPprice	log_meanPprice	meanDdisc	meanOdisc	meanCdisc	meanPdisc	Region
2.500000	1.252763	0.030000	0.050000	0.030000	0.030000	West

Predicted Unit Sales:
32002.0

Fig. 10.16 This illustrates doing a scenario what-if prediction using the *predict* method attached to the regression object. The scenario is put into a DataFrame and then used with the *predict* method

One approach is to remove one observation at a time from the master data set and treat that one observation as the test data. The other is to remove k observations at a time and treat them as the test data. The first is referred to as *leave-one-out cross validation* (*LOOCV*) and the other as *k-fold cross validation*. I will describe each in the next subsections. One reference is Paczkowski (2018) for some comments.

10.6.3.1 *LOOCV* Approach

LOOCV is an iterative procedure. For my description, I will refer to the entire data set as the master data set to distinguish the entire data set from the training and validation sets. As described by Paczkowski (2018), there are four steps:

1. Remove the first observation from the master data set and set it aside as the validation set.
2. Use the remaining $n - 1$ observations to train the model.

3. Repeat Steps 1 and 2 for the entire data set iterating over all n. Calculate the *mean square error* (*MSE*) with the validation set for each iteration. There should be n *MSE* values.
4. Estimate the validation error score as $Score_{(n)} = 1/\sum_{i=1}^{n} e_i^2 = MSE$ where $e_i = Y_i - \hat{Y}_i$ is the prediction residual comparable to the residual I defined in Chap. 6 of *OLS*. This error score is the *Mean Square Error* (*MSE*).

The advantage of *LOOCV* is that it avoids biases. The disadvantage is that it is computationally expensive if n is large since you have to iterate through all cases. It has been estimated that *LOOCV* requires $O(n^2)$ computational time. If the sample size is n, then on each iteration it will use $n - 1$ values for the training and 1 for testing. Also, if n is large, then there is no difference between using all n samples and using $n - 1$ samples. The estimated results should be almost, if not exactly, the same. This means that the added computational costs will yield little to no pay-back.

10.6.3.2 *k-Fold* Approach

A second approach is *k-fold* which is an extension of *LOOCV* to the situation in which not one sample is left but several are left out. This involves five steps that are similar to the *LOOCV*:

1. Divide the data set into k groups (i.e., *folds*) of approximately equal size. Usually $k = 5$ or $k = 10$. If n is the total sample size, then $\sum_{i=1}^{k} n_i = n$.
2. Use the first fold as a validation set.
3. Train a model with the remaining $k - 1$ folds.
4. Repeat Steps 2 and 3 for the remaining folds, calculating *MSE* for each iteration using the validation sets. There should be k of these. If e_{ij} is residual $j, j = 1, \ldots, n_i$ for fold $i, i = 1, \ldots, k$, then $MSE_i = 1/n_i \sum_{j=1}^{n_i} e_{ij}^2$.
5. Estimate the validation error score as $Score_{(k)} = \dfrac{1}{k} \times \sum_{i=1}^{k} MSE_i$.

If n is the sample size and k is the number of folds, then each training will use $n - n/k = (k-1) \times n/k$ samples. Clearly, this method approaches the *LOOCV* as k becomes large. It is computationally less expensive than the *LOOCV* since you are working with chunks of data, not each individual case. Nonetheless, some estimates of computational time are $O(kn^2)$ and $O(n^3)$. A *ROT* is that 5 or 10 folds should be used.

10.6.3.3 Score Measures

I mentioned the *MSE* as a score to compare models or assess one model for either the *LOOCV* or *k-fold* method. This is probably the most commonly used one. There are others available such as

- Mean Absolute Error;
- Mean Squared Logarithmic Error;

- Median Absolute Error;
- Explained Variance Score; and
- Maximum Error.

These measures are applicable for a regression model. For classification model, there are different measures that I will discuss in the next chapter. See the *scikit-learn* documentation for discussions of each of these and several others.[5]

10.6.3.4 Variations on Validation Methods

There are variations on the *k-fold* approach:[6]

Repeated k-fold: Do the *k-fold* method m times whereas the *k-fold* itself does it $m = 1$ times. This is an iterative process with m iterations. For each iteration, the master data set is split into k-folds. This means you will have different folds for each iteration.
LPOCV: Like *LOOCV* but with $p > 1$ left out. If $p = 1$, then this is just *LOOCV*.
Shuffle and Split: Shuffle the master data set rather than do just a random split.
Stratified k-fold: Use a class grouping.
Repeated Stratified k-fold: A combination of Repeated k-fold and Stratified k-fold.
Stratified Shuffle and Split: a combination of Shuffle and Split and Stratified k-fold.
Group k-fold: Do k-fold validation using groups.
LOOGCV: Like *LOOCV* but based on groups of data.
LOPGCV: Like *LOPCV* based on groups.
Group Shuffle and Split: Like Shuffle and Split but using groups.

I summarize these cross-validation functions in Table 10.4.

There is a caveat for implementing a cross-validation using any of the above methods. Whatever preprocessing is done using any of the methods I discussed in Chap. 5, then that same preprocessing should be done for the training and testing data sets. Recall that preprocessing is actually done in two steps: fitting (i.e., calculating) necessary statistics and then transforming the data using those statistics. As an example, standardization involves first calculating the mean and standard deviation for the data followed by applying them to each observation. The first step is the fit, the second is the transformation. You can do them separately using the *fit()* method followed by the *transform()* method, or you can do them both with one command: *fit_transform()*. For cross-validation, you should fit the training data with the *fit()* method; transform the training data with the *transform()* method; and then

[5] At https://scikit-learn.org/stable/modules/model_evaluation.html#mean-squared-error. Last accessed November 24, 2020.

[6] See the scikit-learn documentation at https://scikit-learn.org/stable/modules/cross_validation.html for more thorough descriptions. Web site last accessed November 24, 2020.

Function	Description
GroupKFold	k-fold iterator variant with non-overlapping groups
GroupShuffleSplit	Shuffle-Group(s)-Out
KFold	K-Folds
LeaveOneGroupOut	Leave One Group Out (*LOOGCV*)
LeavePGroupsOut	Leave P Group(s) Out (*LOPGCV*)
LeaveOneOut	Leave-One-Out (*LOOCV*)
LeavePOut(p)	Leave-P-Out (*LPOCV*)
PredefinedSplit	Predefined split
RepeatedKFold	Repeated K-Fold
RepeatedStratifiedKFold	Repeated Stratified K-Fold
ShuffleSplit	Random permutation
StratifiedKFold	Stratified K-Folds
StratifiedShuffleSplit	Stratified Shuffle Split
TimeSeriesSplit	Time Series

Table 10.4 These are the available cross-validation functions. See https://scikit-learn.org/stable/modules/classes.html for complete descriptions. Web site last accessed November 27, 2020

use the same *transform()* method on the testing data. Using this approach preserves continuity between the training and testing data sets.

10.6.3.5 Complexity of Testing

A new level of complexity is introduced by the family of cross-validation methods. The way I originally presented testing was simply to split the master DataFrame into two parts, training and testing: train the model with the training data set and then test it with the testing data set. This simple process still holds but there are more intermediate steps. First, more than one model usually is under consideration, each defined by a different hyperparameter or set of them. Let there be m models with model i indicated as $M_i, i = 1, 2, \ldots, m$. The master DataFrame is still split into two parts with the testing data set put into a "safe" until all training and cross-validation is complete. Then for each model, M_i, do a *k-fold* split of the training data, calculate a score for each fold using the validation data, and then calculate and store the score. When all folds for model M_i have been used, average the scores for that model. Let this average be $S_i, i = 1, 2, \ldots, m$. Repeat this for all models. You should have m scores. Rank the m scores and select the optimal one which is tantamount to selecting the optimal model. Then test this model one more time using the testing data set from the "safe." If you are satisfied based on some criteria for evaluating the tested model, you then have your final model. The final criteria could be another score such as the *Mean Square Error*. I illustrate this extended, more complex process in Fig. 10.17.

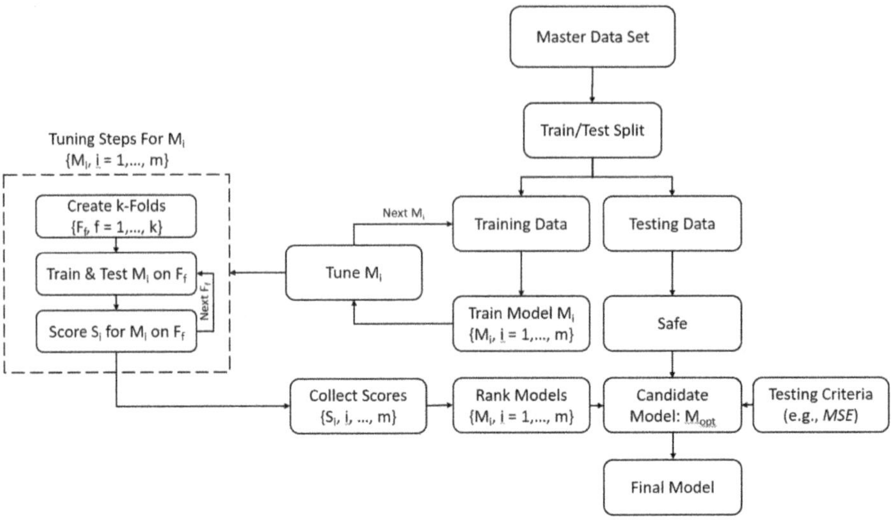

Fig. 10.17 This is the extended, more complex train-validate-test process I outlined in the text

10.6.3.6 Examples of k-Fold Split

I provide a code snippet in Fig. 10.18 to illustrate how k-fold splitting is done. This snippet is rather long, but basically it uses a DataFrame that has three variables and four samples, does a 2-fold split of the DataFrame, and saves the row indexes for each fold for the training and testing components. Then it prints the training DataFrame based on the training indexes and the testing DataFrame for the respective testing indexes. It does this printing for each of the two folds. I then show the first fold DataFrames in Fig. 10.19. You can see from this example how the folds work and how the master DataFrame is "folded."

I repeat this DataFrame splitting example in Figs. 10.20 and 10.21, but with a grouping variable added to the master DataFrame. The groups could represent marketing regions, customer segments, manufacturing plants, and so forth. I display the distribution of the groups in the output in Fig. 10.21. In this example, I treated the groups separately so that all the rows of the DataFrame associated with a group are extracted as a unit by the function and then that unit is split into training and testing components. The requirement for the function is that the number of groups must at least equal the number of folds. In this case, there are three groups and three folds; only fold 1 results are shown. The grouping function maintains one group for the testing data set. Although I do not show all the folds, the testing data set in each one does have only one group represented. The training data sets have several groups represented. You can see in Fig. 10.21 that the training data set has groups 2 and 3 while the testing data set has only group 1.

```
##
## k-fold split example
## Based on: https://scikit-learn.org/stable/modules/cross_validation.html#cross-validation-iterators
##
## Define X master DataFrame
##
data = { 'y':[ 0, 1, 0, 1 ], 'X1':[ 0, 1, -1, 2 ], 'X2':[ 0, 1, -1, 2 ]  }
X = pd.DataFrame( data )
display( X.style.set_caption( 'Master DataFrame' ).set_table_styles( tbl_styles ) )
print( '='*20 )
##
## Instantiate k-fold splitter
##
n_splits = 2
kf = KFold( n_splits = n_splits )
##
## Print the splits
##
print( 'k-fold splits indexes\n' )
for train, test in kf.split( X ):
    print( 'Train  Test' )
    print( '=====  ====' )
    print( f'{train} {test}\n' )
print( '='*20 )
##
## Print
##
fold = 1
for train, test in kf.split( X.index ):
    print( f'Fold ' + str( fold ) + f' --- train: {train} test: {test}\n' )
    ##
    display( X.query( 'index in @train' ).style.set_caption( 'Training DataFrame' ).set_table_styles( tbl_styles ) )
    ##
    display( X.query( 'index in @test' ).style.set_caption( 'Testing DataFrame' ).set_table_styles( tbl_styles ) )
    print( '='*20 )
    fold += 1
```

Fig. 10.18 This is the code snippet for the example k-fold splitting of a DataFrame. See Fig. 10.19 for the results

10.7 Panel Data Models

This literature discusses several types of models, in particular, they are

1. Pooled Model;
2. Fixed Effects Model; and
3. Random Effects Model.

The first, *Pooled Model*, is based on pooling the data in the entire panel data set so that the model is

$$Y_{it} = \alpha + \boldsymbol{\beta} \times \mathbf{X_{it}} + \boldsymbol{\epsilon_{it}} \tag{10.7.1}$$

where Y_{it} is the dependent variable measure for object i in period t; α is a constant term that does not vary by object or time (hence, this is why it is a "constant": it is a constant for all objects and time periods); $\mathbf{X_{it}}$ is a matrix of features with values that vary by objects and time periods; $\boldsymbol{\beta}$ is a vector of weights that are constant for each object and each time period; and $\boldsymbol{\epsilon_{it}}$ is a random disturbance term that varies by objects and time periods. The "objects" can be consumers buying a product through an online ordering system; retailers ordering products for resale; employees in a business unit; production by a robotic system at different plants; different suppliers of raw material; and so forth.

Master
DataFrame

	y	X1	X2
0	0	0	0
1	1	1	1
2	0	-1	-1
3	1	2	2

```
====================
k-fold splits indexes

Train  Test
=====  ====
[2 3] [0 1]

Train  Test
=====  ====
[0 1] [2 3]

====================
Fold 1 --- train: [2 3] test: [0 1]
```

Training
DataFrame

	y	X1	X2
2	0	-1	-1
3	1	2	2

Testing
DataFrame

	y	X1	X2
0	0	0	0
1	1	1	1

Fig. 10.19 This is result for fold 1 for the code snippet in Fig. 10.18. Fold 2 would be the same but for different indexes

The *Pooled Model* is based on putting all the data into one estimation. There is no allowance for variation in the parameters by objects or time. The parameters are constant. This is unrealistic because there is variations in both dimensions as I already noted, which is why there is a Data Cube. If there really is no variation, a Cube is not needed. This model is very rarely used in practice because of this unrealistic assumption of a lack of variation.

The second model, *Fixed Effects Model*, allows for variation from one group to another so that between group variation is recognized. The model is

$$Y_{it} = \alpha_i + \boldsymbol{\beta} \times \mathbf{X_{it}} + \boldsymbol{\epsilon_{it}} \tag{10.7.2}$$

where α_i is a constant for object i that varies from one object to another, so it allows for group effects. The groups could be individual consumers, firms, manufacturing plants, suppliers, and so forth. This term is important because it allows for heterogeneity across the objects. For example, it allows you to consider differences in customers whereas the Pooled Model does not allow for any heterogeneity.

```
##
## k-fold split example with groups
## Based on: https://scikit-learn.org/stable/modules/cross_validation.html#cross-validation-iterators
##
## Define X master DataFrame
##
data = { 'y':[ 0, 1, 0, 1, 1, 1, 0, 0 ], 'X1':[ 0, 1, -1, 2, 1, 2, 3, 1 ], \
         'X2':[ 0, 1, -1, 2, 2, 1, 3, 1 ], \
         'grp':[ 'grp1', 'grp2', 'grp1', 'grp3', 'grp3', 'grp1', 'grp2', 'grp1' ]  }
X = pd.DataFrame( data )
display( X.style.set_caption( 'Master DataFrame' ).set_table_styles( tbl_styles ) )
print( 'Grouping Variable Distribution:' )
display( X.grp.value_counts( ) )
print( '='*20 )
##
## Instantiate k-fold splitter
##
n_splits = 3
gkf = GroupKFold( n_splits = n_splits )
##
## Print the splits
##
print( 'k-fold splits indexes\n' )

for train, test in gkf.split( X, groups = X.grp ):
    print( 'Train  Test' )
    print( '=====  ====' )
    print( f'{train} {test}\n' )
print( '='*20 )
##
## Print
##
fold = 1
for train, test in gkf.split( X, groups = X.grp ):
    print( f'Fold ' + str( fold ) + f' --- train: {train} test: {test}\n' )
    ##
    display( X.query( 'index in @train' ).style.set_caption( 'Training DataFrame' ).set_table_styles( tbl_styles ) )
    ##
    display( X.query( 'index in @test' ).style.set_caption( 'Testing DataFrame' ).set_table_styles( tbl_styles ) )
    print( '='*20 )
    fold += 1
```

Fig. 10.20 This is the code snippet for the example k-fold splitting of a DataFrame with three groups. See Fig. 10.21 for the results

A question is the relationship between this heterogeneity factor and the features. If there is a correlation between the group effects and the features, then the model is a *Fixed Effects Model*. If there is no correlation between the group effect, α_i, and the features, then you have a *Random Effects Model*. This intercept term varies independently of the features, following its own random generating process. This is the same assumption as for the disturbance term, ϵ_{it}. Recall that this term is also assumed to be independent of the features. Consequently, there is a second random variation term added to this random disturbance to produce a composite random disturbance: $u_{it} = \alpha_i + \epsilon_{it}$. The model is now

$$Y_{it} = \boldsymbol{\beta} \times \mathbf{X_{it}} + \mathbf{u_{it}} \tag{10.7.3}$$

where $u_{it} = \alpha_i + \epsilon_{it}$ with $\sigma_u^2 = \sigma_\alpha^2 + \sigma_\epsilon^2$ and $cov(u_{it}, u_{is}) = \sigma_\alpha^2$. We often use a correlation rather than a covariance so the correlation is $\rho = cor(u_{it}, u_{is}) = \sigma_\alpha^2/\sigma_\alpha^2+\sigma_\epsilon^2$.

These panel models are more complex to estimate as should be evident from the model specifications. There is a Python package to handle them. The package, *linearmodels*, is installed using *pip install linearmodels* or *conda install -c conda-forge linearmodels*.

Master
DataFrame

	y	X1	X2	grp
0	0	0	0	grp1
1	1	1	1	grp2
2	0	-1	-1	grp1
3	1	2	2	grp3
4	1	1	2	grp3
5	1	2	1	grp1
6	0	3	3	grp2
7	0	1	1	grp1

```
Grouping Variable Distribution:

grp1    4
grp2    2
grp3    2
Name: grp, dtype: int64

====================
k-fold splits indexes

Train  Test
=====  ====
[1 3 4 6] [0 2 5 7]

Train  Test
=====  ====
[0 1 2 5 6 7] [3 4]

Train  Test
=====  ====
[0 2 3 4 5 7] [1 6]

====================
Fold 1 --- train: [1 3 4 6] test: [0 2 5 7
```

Training
DataFrame

	y	X1	X2	grp
1	1	1	1	grp2
3	1	2	2	grp3
4	1	1	2	grp3
6	0	3	3	grp2

Testing
DataFrame

	y	X1	X2	grp
0	0	0	0	grp1
2	0	-1	-1	grp1
5	1	2	1	grp1
7	0	1	1	grp1

```
====================
```

Fig. 10.21 This is result for fold 1 for the code snippet in Fig. 10.20. Folds 2 and 3 would be the same but for different indexes and groups

Chapter 11
Classification with Supervised Learning Methods

I covered modeling as a way to predict either future events (i.e., forecasting) or the outcome of decisions (i.e., predicting) in the previous chapters. Recall that "prediction" is a broad label that includes forecasting as a subset: all forecasts are predictions but not all predictions are forecasts. Predicting in general is a very important function of Business Data Analytics which is why I spent so much time on it. There is, however, another important function: classifying or assigning new cases to groups. The cases could be new or existing customers, applicants for loans or business lines of credit, new products in a development pipeline, and so forth. For potential new customers, the issue is their likelihood to actually buy your product. For existing customers, the issue is segmenting them so you could develop differential marketing strategies. For loan applicants, the issue is their likelihood to default. For new products, the issue is which ones are likely to succeed and, therefore, should continue in the pipeline.

For each case, you can imagine that a collection of objects, such as customers, are split or divided into groups, the minimum, of course, is two. If a 2-D graph is created that displays measures on the objects with the objects as the points on the graph, then the objects are classified by drawing a line through the graph. Objects on one side of the line are classified into one group, and those objects on the other side are classified into the other group. The issue is how to draw the line. This is a high-level conceptualization of the classification process, perhaps overly simplistic, but it nonetheless exemplifies the issue.

There are several methods for classifying objects. The most commonly used are:

- Logitistic regression based on a Logit Link;
- K-Nearest Neighbor (*KNN*);
- Naive Bayes;
- Decision Trees; and
- Support Vector Machines.

W. R. Paczkowski, *Business Analytics*,
https://doi.org/10.1007/978-3-030-87023-2_11

Logistic regression is sometimes used as a classifier and sometimes as a regression modeling tool to estimate effects. I chose to incorporate it into this chapter because it is used as a classifier in business applications more often than not. Therefore, I will begin this chapter with the logit link to continue the flow from the previous one which dealt with the Identity link.

11.1 Case Study: Background

I will use the furniture manufacturer Case Study. In particular, I will focus on the customer satisfaction scores provided by the customers. These are on a five-point Likert Scale with 5 being Very Satisfied. Although there are many ways to analyze these ratings, the most common is the *top-two box* score. I frequently refer to the top-two box as $T2B$ and the bottom-three box as $B3B$. This is a transformation from the five points to two. Basically, it is a binary recoding or dummifying of the scores. An indicator function defines the encoding as $\mathbb{I}(score \geq 4)$ so that any score greater than or equal to 4 is coded as 1; 0 otherwise. The "1" is referred to as the top-two box. Customers in the top-two box are considered satisfied; all others dissatisfied. This is done with a list comprehension to create the new variable, *sat_t2b*, that is added to the furniture DataFrame: *df_agg['sat_t2b'] = [1 if x >= 4 else 0 for x in df_agg.buyerSatisfaction]* where *df_agg* is the aggregated DataFrame.

The problem is to predict if a randomly selected retail boutique customer is top-two box satisfied or not.

11.2 Logistic Regression

I introduced the concept of a link function in Chap. 10 and illustrated one link function—the Identity link for *OLS* regression. This link function is applicable when the target is continuous. If it is binary, however, then this link is inappropriate. Examples of binary targets are:

- Will someone buy a product? Yes or No
- Is someone satisfied with service? Yes or No
- Will someone attend a conference? Yes or No
- Will a current association member renew? Yes or No

The targets do not have to be binary. They could be, say, *Yes/No/Maybe* in which case they are multinomial, a case I will not consider in this book. They could also be ordinal as in ranking preference for a product: first, second, third. *OLS* models, although very powerful for a host of problems, are not applicable for this class of problems for several reasons:

- They predict any range of values. This class of problems has only two, such as *Yes* and *No*.
- They have a disturbance term that is normally distributed. This class of problems has a binomial distribution.
- They have a constant variance. This class of problems has a non-constant variance.

11.2.1 A Choice Interpretation

The binary target can be viewed as a *choice set* consisting of $J = 2$ items. Let $\Omega = \{item_1, item_2\}$ be this choice set. The minimum size is two; anything less implies no choice. The choices could be simply *Yes* or *No*. A restriction on the items is that they are mutually exclusive and completely exhaustive. Mutually exclusive because you can select only one item and completely exhaustive because they cover or span all possibilities. Only one item from the set can be selected, there is no in-between, making this is a binomial problem. For a multinomial problem, $\Omega = \{item_1, \ldots, item_J\}$, $J > 2$.

You can express the coding of the target or choice variable for the binary case as

$$Y_{i1} = \begin{cases} 1 & \text{if object } i \text{ chooses } item_1 \text{ from } \Omega \\ 0 & \text{otherwise} \end{cases}$$

for $i = 1, \ldots, n$ objects such as customers. This is just a dummy variable definition applied to the target variable.

11.2.2 Properties of this Problem

Features to explain choice can be any type of variable. They are called *attributes* in marketing. Examples are price, weight, color, and time. The task is to measure the importance of a feature on the choice of the item. For example, measure the effect of price on the purchase of a product. In a classification problem, knowing these effects allows you to classify objects (e.g., customers, loan applicants, employees). A possible model relating the target to a feature is

$$Y_i = \beta_0 + \beta_1 X_i + \epsilon_i, i = 1, \ldots, n \tag{11.2.1}$$

where X_i is a feature. This is called a *Linear Probability Model* (*LPM*). It assumes that $E(\epsilon_i) = 0$. Under this assumption $E(Y_i) = \beta_0 + \beta_1 X_i$. To see the binomial nature of the problem, let $Pr(Y_i = 1) = p_i$ and $Pr(Y_i = 0) = 1 - p_i$. Therefore, $E(Y_i) = 1 \times Pr(Y_i = 1) + 0 \times Pr(Y_i = 0) = p_i$. This implies that

$$E(Y_i) = \beta_0 + \beta_1 X_i \qquad \text{(a linear model)} \tag{11.2.2}$$

$$= p_i \qquad \text{(a probability)} \tag{11.2.3}$$

so the mean (a linear function of X_i) must lie between 0 and 1. Hence the name: *linear probability model*. Clearly, $p_i = \beta_0 + \beta_1 \times X_i$ which changes as X_i changes. This is why there is a subscript on p.

Suppose the model is $Y_i = \beta_0 + \beta_1 X_i + \epsilon_i$. If $Y_i = 1$, you have $1 = \beta_0 + \beta_1 X_i + \epsilon_i$ so $\epsilon_i = 1 - \beta_0 - \beta_1 X_i$ with probability p_i which is the probability the 1 occurs. If $Y_i = 0$, then $\epsilon_i = -\beta_0 - \beta_1 X_i$ with probability $1 - p_i$. The disturbance term can have only two values: $1 - \beta_0 - \beta_1 X_i$ with probability p_i and $-\beta_0 - \beta_1 X_i$ with probability $1 - p_i$ so it is binomial, not normal. This is not an issue since for large samples a binomial random variable approaches a normally distributed random variable because of the *Central Limit Theorem*. See Gujarati (2003).

Now consider the variance of the disturbance:

$$V(\epsilon_i) = E\left[\epsilon_i - E(\epsilon_i)\right]^2 \tag{11.2.4}$$

$$= E(\epsilon_i^2) \quad \text{Since } E(\epsilon_i) = 0 \tag{11.2.5}$$

$$= (1 - \beta_0 - \beta_1 X_i)^2 \cdot p_i + (-\beta_0 - \beta_1 X_i)^2 \cdot (1 - p_i) \tag{11.2.6}$$

$$= (\beta_0 + \beta_1 X_i)(1 - \beta_0 - \beta_1 X_i) \tag{11.2.7}$$

$$= E(Y_i)\left[1 - E(Y_i)\right] \tag{11.2.8}$$

$$= p_i(1 - p_i) \tag{11.2.9}$$

which clearly changes as X_i changes so the disturbance is heteroskedastic. This is not too bad since you can use weighted least squares for estimation.

11.2.3 A Model for the Binary Problem

A big issue is that the estimated value of Y, $\hat{Y}_i$, may not lie in the range [0, 1]. So, you may predict something that cannot physically happen. It has been suggested that this is also not troublesome because the Y_i can be scaled, perhaps using the MinMax standardization of Chap. 5. Application to this problem, however, only places a veil over the problem. The issue remains: predictions are potentially impossible. A proper solution is a transformation of the target before training that ensures the right magnitudes.

You need a probability model with $0 \leq p_i \leq 1$. A likely candidate is a *cumulative distribution function* (*CDF*), defined as $Pr(X_i < x_i)$, such as

$$Pr(Y_i = 1) = p_i \tag{11.2.10}$$

$$= \frac{e^{Z_i}}{1 + e^{Z_i}} \tag{11.2.11}$$

$$= \frac{e^{\beta_0 + \beta_1 X_i}}{1 + e^{\beta_0 + \beta_1 X_i}} \tag{11.2.12}$$

$$Z_i = \beta_0 + \beta_1 X_i. \tag{11.2.13}$$

This is a *logistic CDF* which I illustrate in Fig. 11.1. For a more complete definition of Z_i, you can write $Z_i = \beta_0 + \sum_{j=1}^{p} \beta_j \times X_{ij}$. What happens when Z_i becomes large or small? Note that

$$Pr(Y_i = 1) = p_i \tag{11.2.14}$$

$$= \frac{e^{Z_i}}{1 + e^{Z_i}} \tag{11.2.15}$$

$$= \frac{1}{1 + e^{-Z_i}} \tag{11.2.16}$$

If $Z_i \rightarrow +\infty$, then $e^{-Z_i} = 0$ and $p_i = 1$. Similarly, if $Z_i \rightarrow -\infty$, then $p_i = 0$. Finally, note that

$$\frac{e^{Z_i}}{1 + e^{Z_i}} + \left(1 - \frac{e^{Z_i}}{1 + e^{Z_i}}\right) = 1 \tag{11.2.17}$$

so the probabilities add correctly.

You can give the model in (11.2.11) a choice interpretation. The numerator represents the influence of the attribute and thus represents “choice.” The “1” in the denominator represents “no choice” (i.e., e^0) so the factor $1 + e^{Z_i}$ represents the total choice option. So, (11.2.11) is the probability of choosing $item_i$ from a choice set with two items. I will occasionally refer to p_i as a choice probability.

You can explore this model further for another interpretation by writing

$$\frac{p_i}{1 - p_i} = \frac{\frac{e^{Z_i}}{1 + e^{Z_i}}}{1 - \frac{e^{Z_i}}{1 + e^{Z_i}}} \tag{11.2.18}$$

$$= e^{Z_i} \tag{11.2.19}$$

```
##
## Set parameters and generate data
##
xmin = -6
xmax = 6
sp = 100
x = np.linspace( xmin, xmax, sp )  ## equally spaced values
y = [ np.exp(z)/(1 + np.exp(z)) for z in x  ]  ## list comprehension for logistic data
##
## Plot the curve
##
plt.plot( x, y )
plt.title( 'Logistic Function\n$Pr(Y_i = 1) = \dfrac{e^{Z_i}}{1 + e^{Z_i}}$', fontsize = 18 )
plt.xlabel( 'Quantiles' )
plt.ylabel( 'Probability' )
plt.vlines( 0, 0, 1, linestyles = 'dotted' )
plt.hlines( 0.5, xmin, xmax, linestyles = 'dotted' );
```

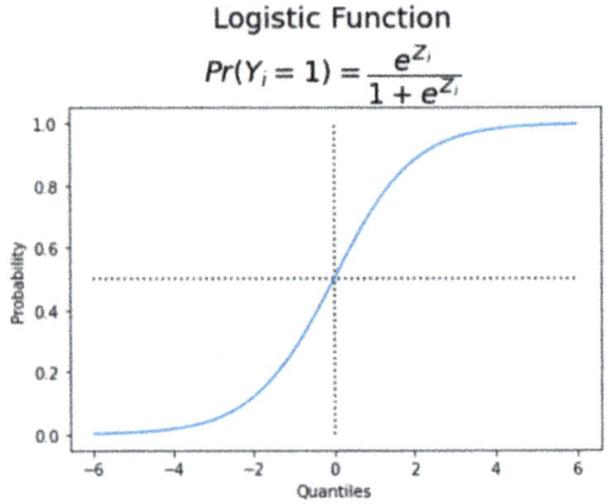

Fig. 11.1 This is an illustration of a logistic *CDF*. Notice the sigmoid appearance and that its height is bounded between 0 and 1. This is from Paczkowski (2021b). Permission to use from Springer

The ratio $p_i/1-p_i$ is the odds of choosing $item_1$ from the choice set. I introduced odds in Chap. 5. Taking the natural log of both sides of the odds, you get the "log odds", or

$$L = \ln\left(\frac{p_i}{1-p_i}\right) \tag{11.2.20}$$

$$= \ln\left(e^{Z_i}\right) \tag{11.2.21}$$

$$= Z_i \ln(e) \tag{11.2.22}$$

$$= Z_i \quad \text{Since } \ln \text{ and } e \text{ are inverses} \tag{11.2.23}$$

$$= \beta_0 + \sum_{j=1}^{p} \beta_j \times X_{ij} \tag{11.2.24}$$

where L is the *log odds* or *logit*. The word *logit* is short for "logarithmic transformation." The logistic regression model is, therefore, often called a *logit model*. I tend to use the two terms interchangeably. Maximum likelihood is used with the logit to estimate the unknown parameters, $\beta_k, k = 0, 1, \ldots, p$. Maximum likelihood has the form of *OLS* so it is clearly in the regression family. There are test statistics that tell you how good the estimates are just as in *OLS*.

Although you can estimate the logit parameters, you will have a hard time interpreting them, *per se*. Each estimated parameter value shows the change in the log-odds when the parameter's associated variable changes by one unit. This was not hard to understand for *OLS* because you can look at the change in Y for a change in X; that is, the *marginal effect*. Now you have the change in the log-odds. What does a change in the log-odds mean? To answer this question, suppose you have a simple one-variable model for buying a product where X is categorical. For example, let X represent the gender of a potential customer with the dummy coding: 0 = Females, 1 = Males. The log odds for Females is $L_0 = \ln\left(p_0/1-p_0\right) = \beta_0$ and the log odds for Males is $L_1 = \ln\left(\frac{p_1}{1-p_1}\right) = \beta_0 + \beta_1$. Exponentiating both log odds and forming the ratio of Males to Females, you get[1]

$$\frac{e^{\ln\left(\frac{p_1}{1-p_1}\right)}}{e^{\ln\left(\frac{p_0}{1-p_0}\right)}} = \frac{\frac{p_1}{1-p_1}}{\frac{p_0}{1-p_0}} \quad \text{Called the } \textit{odds ratio} \tag{11.2.25}$$

$$= \frac{e^{\beta_0+\beta_1}}{e^{\beta_0}} \tag{11.2.26}$$

$$= \frac{e^{\beta_0}e^{\beta_1}}{e^{\beta_0}} \tag{11.2.27}$$

$$= e^{\beta_1} \tag{11.2.28}$$

The exponentiation of β_1 is the odds of Males buying the product to the odds of Females buying it. If the odds ratio is, say, 3, then the likelihood of Males buying the product is 3x greater than the likelihood of Females buying it.

11.2.4 Case Study: Train-Test Data Split

I split the aggregated data for the Case Study into the training and testing data sets in the same manner as I did before. Each data subset's name is prefixed with "logit_" to distinguish them from other data sets. I provide the code snippet to do this in Fig. 11.2. I then used the training data set to train the logit model.

[1] Note: $e^{\ln x} = x$.

```
##
## Create the X data for splitting
##
cols = [ 'sat_t2b', 'Pprice', 'Ddisc', 'Odisc', 'Cdisc', 'Pdisc', 'Region' ]
X = df_agg[ cols ].copy()
##
## Split the data.  The default is 1/3 testing.
##
logit_train, logit_test = train_test_split( X, test_size = 0.33, random_state = 42 )
##
display( logit_train.head().style.set_caption( 'Training Data Set' ).set_table_styles( tbl_styles ) )
df_size( logit_train )
##
display( logit_test.head().style.set_caption( 'Testing Data Set' ).set_table_styles( tbl_styles ) )
df_size( logit_test )
```

Training Data Set

	sat_t2b	Pprice	Ddisc	Odisc	Cdisc	Pdisc	Region
587	0	4.385216	0.132289	0.048700	0.067578	0.040233	West
63	0	5.373925	0.131973	0.045865	0.073081	0.041054	Northeast
61	1	6.086248	0.133550	0.051300	0.071960	0.040400	West
361	1	3.944080	0.138375	0.051172	0.069422	0.037984	Midwest
425	1	4.680028	0.080937	0.049556	0.069873	0.042349	South

DataFrame Dimensions

	Count
Number of Rows	521
Number of Columns	7

Testing Data Set

	sat_t2b	Pprice	Ddisc	Odisc	Cdisc	Pdisc	Region
595	0	6.677310	0.079525	0.051075	0.067625	0.039325	South
587	0	5.145406	0.128512	0.048907	0.070326	0.040581	Northeast
543	0	5.056035	0.137036	0.050019	0.071783	0.041048	West
644	1	6.531215	0.080465	0.051212	0.074939	0.039162	South
487	1	3.977678	0.130104	0.051076	0.069620	0.039987	Northeast

DataFrame Dimensions

	Count
Number of Rows	258
Number of Columns	7

Fig. 11.2 This is the code snippet for the train-test split for the logit model. Each subset is prefixed with "logit_"

11.2.5 Case Study: Logit Model Training

You train a logit model using the same set-up as for *OLS*, but remember that maximum likelihood, not *OLS*, is used. The *statsmodels* function for training is *logit* which has two parameters: the formula and the DataFrame. I show the set-up for the customer satisfaction problem in Fig. 11.3. Customers are either satisfied or not, so this is a binary problem. The target is the top-two box satisfaction.

Since maximum likelihood is used, an R^2 is not defined since this requires sums of squares, in particular, the regression sum of squares which is undefined for maximum likelihood. There is, however, an alternative which is the *pseudo-*R^2, also called the *McFadden pseudo-*R^2, defined as

$$\text{pseudo-}R^2 = 1 - \frac{\text{Log-Likelihood}}{\text{LL-Null}} \tag{11.2.29}$$

```
##
## Train a logit model
##
## ===> Step 1: Define a formula <===
##
formula = 'sat_t2b ~ Pprice + Ddisc + Odisc + Cdisc + Pdisc + C( Region )'
##
## ===> Step 2: Instantiate the logit model <===
##
mod = smf.logit( formula, data = logit_train )
##
## ===> Step 3: Fit the instantiated model <===
##
logit01 = mod.fit()
##
## ===> Step 4: Summarize the fitted model <===
##
display( logit01.summary() )
```

```
Optimization terminated successfully.
         Current function value: 0.601296
         Iterations 6
```

Logit Regression Results

Dep. Variable:	sat_t2b	No. Observations:	521
Model:	Logit	Df Residuals:	512
Method:	MLE	Df Model:	8
Date:	Fri, 26 Mar 2021	Pseudo R-squ.:	0.02006
Time:	11:24:41	Log-Likelihood:	-313.28
converged:	True	LL-Null:	-319.69
Covariance Type:	nonrobust	LLR p-value:	0.1180

	coef	std err	z	P>|z|	[0.025	0.975]
Intercept	-8.1240	5.646	-1.439	0.150	-19.190	2.942
C(Region)[T.Northeast]	-0.1711	0.361	-0.474	0.636	-0.879	0.537
C(Region)[T.South]	-0.6648	1.365	-0.487	0.626	-3.340	2.010
C(Region)[T.West]	0.0195	0.246	0.079	0.937	-0.463	0.502
Pprice	-0.1773	0.135	-1.310	0.190	-0.443	0.088
Ddisc	-17.7325	23.976	-0.740	0.460	-64.725	29.260
Odisc	89.5921	47.354	1.892	0.058	-3.221	182.405
Cdisc	32.3431	33.146	0.976	0.329	-32.622	97.308
Pdisc	136.6637	64.435	2.121	0.034	10.373	262.954

Fig. 11.3 The customer satisfaction logit model estimation set-up and results

where *Log-Likelihood* is the maximum value of the log of the likelihood for the model and *LL-Null* is the value of the log of the likelihood for the model with only a constant. I discussed the log-likelihood in Chap. 6. The model with only a constant is called a *Null model*. Using the data in Fig. 11.3, you can see that pseudo-$R^2 = 1 - {-313.28}/{-319.69} = 0.02006$ as shown. The pseudo-R^2 is interpreted like the *OLS* R^2 in the sense that you want a value close to 1.0. The value for this example is 0.02 which indicates that this is not a very good model.

11.2.6 Making and Assessing Predictions

The *statsmodels*'s logit function has a *predict* method that requires one parameter: the test data. The predictions are the probability of belonging to a class. Respondents, however, are classified using two words ("B3B" and "T2B") so the probabilities are somewhat obscure. For example, is someone satisfied or not if their predicted probability is 0.55? The answer depends on a recoding of the probabilities to 0 and 1 for $B3B$ and $T2B$, respectively. A cut-off value, $\theta \in 0, 1$, is fixed so that an estimated probability greater than θ is recoded to 1, 0 otherwise. That is, $T2B = \mathbb{I}(p_i \geq \theta)$ defines top-two box satisfaction. The choice of θ is arbitrary, although most analysts use $\theta = 0.50$. If you set θ close to 1.0, then almost everyone is classified as dissatisfied because few predicted values will be greater than θ, but almost all will be below it. If you set it close to 0.0, however, then almost everyone will be classified as satisfied and few dissatisfied. There is no correct value for θ which is why $\theta = 0.50$ is typically used. It basically gives you a 50-50 chance of classifying people one way or the other. The parameter θ is a *hyperparameter*.

You can assess the predictive power of the model with a *confusion table* which uses the test data set's true classification of the respondents and the predicted class. The table basically shows how confused the model is in predicting classes; that is, how often it makes the correct classification or gets confused and makes the wrong classification. I show how to create a confusion table in Fig. 11.4. Since you know the true classification in the testing data set and the predicted classification from the model based on your choice of θ, you can determine if the predictions are correct (i.e., True), or incorrect (i.e., False). If a person is truly dissatisfied and you predict they are dissatisfied, then they are counted as a *True Negative*. The "Negative" is the dissatisfaction (i.e., the bottom end of the binary scale). This second word refers to the prediction. The first word, the adjective, refers to and clarifies the correctness of the prediction. If you predict someone is dissatisfied and they are dissatisfied, then the prediction is correct and they are counted as a *True Negative*. There are four possible labels since there are two possible states for predictions and two for true states: *True Negative*, *False Positive*, *False Negative*, and *True Positive*. A simple count of the number of respondents in each category is displayed in a 2×2 table as I show in Fig. 11.4. This is sometimes displayed as a heatmap matrix which I show in Fig. 11.5.

There are a number of summary diagnostic measures derived from a confusion table. I show the set-up to derive them and the resulting output, called an *accuracy report* in Fig. 11.6. One point to notice is my recoding of 0 and 1 to 'B3B' and 'T2B', respectively. The former is "Not Satisfied" and the latter is "Satisfied." The unrecoded report just has 0 and 1 which is not too readable; changing the labels fixes this issue. I used a regular expression to do this.

```
##
## Make predictions and check accuracy measures
##
predictions = logit01.predict( logit_test )
##
## Recode predictions with cut-off
##
theta = 0.5
predictions_recoded = [ 0 if x < theta else 1 for x in predictions ]
##
## Create a confusion matrix
##
cm = confusion_matrix( logit_test.sat_t2b, predictions_recoded ).ravel()
##
## Create a dictionary of cm and labels
##
lbl = [ 'True Negative', 'False Positive', 'False Negative', 'True Positive' ]
data = { 'Classification':lbl, 'Frequency': cm, 'Percent':cm/cm.sum() }
##
## Display the confusion matrix in a DataFrame
##
df_confusion = pd.DataFrame( data )
df_confusion.set_index( 'Classification', inplace = True )
df_confusion.style.set_caption( 'Confusion Table' ).\
    format( {'Percent':format} ).\
    bar( align = 'mid', color = 'lightblue' ).\
    set_table_styles( tbl_styles )
```

Confusion Table

Classification	Frequency	Percent
True Negative	1	0.4%
False Positive	81	31.4%
False Negative	3	1.2%
True Positive	173	67.1%

Fig. 11.4 The logit model confusion table is based on the testing data set. Notice the list comprehension to recode the predicted probabilities to 0 and 1

The accuracy report has two halves. The top half has key measures by the two levels of satisfaction while the bottom half has aggregate measures. Both have four columns: "*precision*", "*recall*", "*f1-score*", and "*support*." I show a stylized confusion matrix in Table 11.1 to help you understand these measures. The sample sizes are all denoted by n with a double subscript. The first element of the subscript denotes the row of the matrix and the second the column of the matrix. This is just standard row-column matrix notation. The dot in the subscript denotes summation. So, $n_{.P}$ denotes the sum of the values in the two rows of column "P" which is "Positive."; $n_{..}$ denotes the overall sum or sample size so $n_{..} = n$. For the confusion matrix in Fig. 11.5 $n_{..} = n = 258$. The total number of cases, $n_{..}$, is the *support*. Figure 11.2 has the stylized table cells populated with the values for the satisfaction study.

The first calculation most analysts make, and what clients want to know, is the *error rate* followed by the *accuracy rate* which is 1 minus the error rate. The error rate is the number of false predictions, both negative and positive. This answers the question: "*How wrong is the prediction, on average*?" A rate is a ratio of two related quantities, so the error rate is the error sum relative to the total number of cases. In

```
##
## Create labels
##
lbl = [ 'Not Satisfied', 'Satisfied' ]
##
## Create DataFrame
##
x = cm.reshape( 2, 2 )
data = pd.DataFrame( x, index = lbl, columns = lbl )
##
## Plot the confusion matrix
##
sns.set( font_scale = 1.4 )   #for label size
##
base = 'Base: Testing data; n = ' + str( x.sum() )
ax = sns.heatmap( data, annot = True, fmt = '.3g', annot_kws = { "size": 16 } )  # annot_kws = font size
ax.set_title( 'Confusion Matrix\nLogit Classifier', fontsize = font_title )
ax.set( xlabel = 'Predicted', ylabel = 'True' )
ax.set_xticklabels( lbl )
ax.set_yticklabels( lbl )
footer( );
```

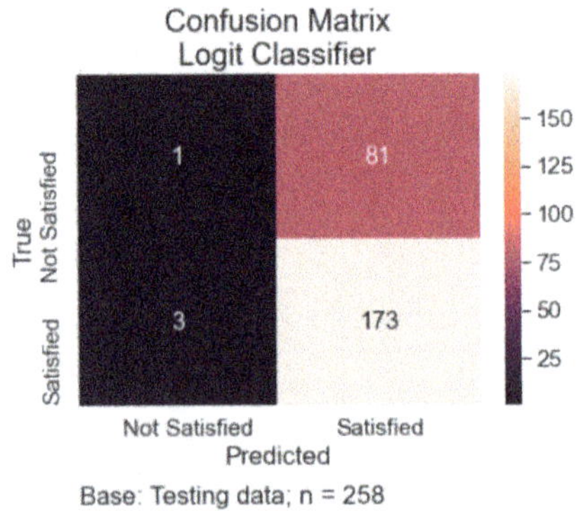

Fig. 11.5 The logit model confusion matrix is an alternative display of the confusion table in Fig. 11.4. The lower left cell has 3 people predicted as not satisfied (i.e., Negative), but are truly satisfied; these are False Negatives. The upper right cell has 81 False Positives. There are 173 True Positives and 1 True Negative

terms of Table 11.2, the error rate is

$$\begin{aligned} Error\ Rate &= \frac{FN + FP}{n_{..}} \\ &= \frac{84}{258} \\ &= 0.32558. \end{aligned}$$

The accuracy rate, the rate of correct predictions, answers the question: "*How correct is the prediction, on average*?" and is

$$Accuracy\ Rate = \frac{TN + TP}{n_{..}}$$

```
##
## Create a classification accuracy report
##
rpt = classification_report( logit_test.sat_t2b, predictions_recoded, digits = 3 )
##
## Use regex expression to recode 0 = Dissatisfied and 1 = Satisfied
##
rpt = re.sub( ' 0 ', 'B2B', rpt )
rpt = re.sub( ' 1 ', 'T2B', rpt )
##
## Display report
##
print( 'Classification Accuracy Report\n' )
print( f'{rpt}' )
print( f'Cut-off value: theta = {theta}' )
```

```
Classification Accuracy Report

              precision    recall  f1-score   support

         B2B      0.250     0.012     0.023        82
         T2B      0.681     0.983     0.805       176

    accuracy                          0.674       258
   macro avg      0.466     0.498     0.414       258
weighted avg      0.544     0.674     0.556       258

Cut-off value: theta = 0.5
```

Fig. 11.6 The customer satisfaction logit model accuracy report based on the testing data set

$$
\begin{aligned}
&= \frac{174}{258} \\
&= 0.67442 = 1 - 0.32558.
\end{aligned}
$$

A related concept is the *precision* of the prediction, which answers the question "*How correct are the predictions for a class, on average*?" For a two-class problem, there are two precision measures. A correct prediction is, of course, measured by the number of *True Positives* and the number of *True Negatives*. The precision for the satisfied class, which is "positive" in this problem, is

$$
\begin{aligned}
Precision_{T2B} &= \frac{TP}{n_{.P}} \\
&= \frac{173}{254} \\
&= 0.68110
\end{aligned}
$$

		Predicted state		
		Negative	Positive	Total
True state	Negative	True Negative (TN)	False Positive (FP)	$n_{N.}$
	Positive	False Negative (FN)	True Positive (TP)	$n_{P.}$
Total		$n_{.N}$	$n_{.P}$	$n_{..}$

Table 11.1 This illustrates a stylized confusion matrix. The n-symbols represent counts in the respective marginals of the table

		Predicted state		
		Negative	Positive	Total
True state	Negative	TN: 1	FP: 81	$n_{N.} = 82$
	Positive	FN: 3	TP: 173	$n_{P.} = 176$
Total		$n_{.N} = 4$	$n_{.P} = 254$	$n_{..} = 258$

Table 11.2 This is the stylized confusion matrix Table 11.1 with populated cells based on Fig. 11.5

and the precision for the dissatisfied class is

$$\begin{aligned} Precision_{B3B} &= \frac{TN}{n_{.N}} \\ &= \frac{1}{4} \\ &= 0.25 \end{aligned}$$

The interpretation of $Precision_{T2B}$ is that slightly more than two-thirds of the truly satisfied people were correctly predicted. For $Precision_{B3B}$, one-fourth of the truly dissatisfied people were correctly predicted. The logit model, therefore, was not very confused in predicting satisfied people.

The *recall* answers the question: "*How correct is the prediction in capturing those in the class, on average*?" This differs from precision which looks at how many of the predictions are correct while recall looks at how many were predicted correctly of all who are truly in a class. The base for precision and recall are different. You could view precision as conditional on the prediction class (i.e., column conditional) and recall as conditional on the true class (i.e., row conditional). The recall for the satisfied class is

$$\begin{aligned} Recall_{T2B} &= \frac{TP}{n_{P.}} \\ &= \frac{173}{176} \\ &= 0.98295 \end{aligned}$$

and for the dissatisfied class is

$$Recall_{B3B} = \frac{TN}{n_{N.}} \tag{11.2.30}$$

$$= \frac{1}{82} \tag{11.2.31}$$

$$= 0.01220. \tag{11.2.32}$$

.

The model correctly predicts 98.3% of those who are truly satisfied but only 1.2% of those who are truly dissatisfied. So again, the satisfied people did not confuse the model.

A final measure is the *f1-Score*. This is the average of the precision and recall for each class, but the average is not a simple arithmetic average because precision and recall are both rates. You must use the *harmonic average* when you average rates. If you did not and used the arithmetic average instead, then you run the risk of overstating the true average. This overstatement is due to the *Arithmetic Mean—Geometric Mean—Harmonic Mean Inequality*. Simply put, this inequality is $AR \geq GM \geq HM$ where AR is the arithmetic mean, GM is the geometric mean, and HM is the harmonic mean.[2] The arithmetic mean is

$$AR = \frac{1}{n \times \sum_{i=1}^{n} r_i};$$

the geometric mean is

$$GM = \left(\prod_{i=1}^{n} r_i \right)^{1/n};$$

and the harmonic mean is

$$HR = \frac{n}{\sum_{i=1}^{n} \frac{1}{r_i}}$$

where r_i is the ith rate. For a two-rate problem, HR simplifies to

$$HR = 2 \times \left[\frac{r_1 \times r_2}{r_1 + r_2} \right].$$

[2] See https://en.wikipedia.org/wiki/Average. Last accessed December 30, 2020.

The f1-Score for satisfied people, the harmonic mean, is

$$\begin{aligned}
\text{f1-Score}_{T2B} &= 2 \times \left[\frac{Precision_{T2B} \times Recall_{T2B}}{Precision_{T2B} + Recall_{T2B}} \right] \\
&= 2 \times \left[\frac{0.68110 \times 0.98295}{0.68110 + 0.98295} \right] \\
&= 0.805.
\end{aligned}$$

For f1-Score$_{B3B}$, it is 0.023.

These precision, recall, and f1-Score are summarized in the top portion of Fig. 11.6. The bottom portion shows the accuracy plus the measures averaged over both classes. The accuracy is aggregated over all classes. It is 0.674 on a support of 258 observations, the total sample size of the confusion matrix. The *macro avg* (short for *macro average*) is the simple arithmetic average of the two classes by the column categories: "precision", "recall", and "f1-score." The macro average for precision, for example, is

$$\begin{aligned}
macro\ avg &= \frac{0.250 + 0.681}{2} \\
&= 0.466.
\end{aligned}$$

The *weighted avg* is the weighted average of the classes for the respective column categories. The weights are true counts for the classes relative to the support. The true counts are used since the predicted counts are obviously dependent on the cut-off value, θ, for allocating objects to classes. You obviously change the weights if you change θ which is unacceptable since the very measure you want to use is dependent on your selection. The weighted average for precision is

$$\begin{aligned}
weighted\ avg &= \frac{82}{258} \times 0.250 + \frac{176}{258} \times 0.681 \\
&= 0.544.
\end{aligned}$$

11.2.7 Classification with a Logit Model

How is a logit model used for classification in practice? The confusion table and the accuracy report only tell you how the model performs based on the test data. Assume you are satisfied with these results. Now what? How do you implement the model? There are two possibilities:

1. case-by-case classification, and
2. in-mass classification.

Which one you use depends on your task, your objective. The first is applicable for classifying an individual, almost "on the spot." This is a field application. For example, if the problem is to classify a credit applicant at a field location, such as a bank branch office, as not risky or risky so they should be extended credit or not then a *typing tool* could be built to "type" or classify the applicant. As another example, a classification model could be built to predict if a prospective customer would place an order with your sales representative. A sales representative may not know enough about the potential customer prior to a sales visit and presentation, but once in front of that potential customer, they could learn enough to then use a typing tool to predict the probability of a sale and adjust the sales effort accordingly. The typing tool in both cases could be on a laptop.

The in-mass classification also types customers but not case-by-case in the field, on the spot, but by processing them all via a data processing application operating on a centralized data warehouse or other data mart. This application could potentially type thousands of prospects. For example, your task could be to classify customers in a database as potential buyers or not of a new product. A direct email campaign could be planned to promote the product to the most likely buyers. The typing tool would be built into a database processing system.

The central difference between the two typing tools is the level and extent of the features. For the case-by-case application, the features used in the model and implemented in the tool must be ones a sales representative could easily collect on-the-spot either through direct observation or by asking a few key questions. The typing tool should prompt the representative for these data points and then process the responses to produce a classification. For the in-mass application, the features should be in the data warehouse or could be added from an outside source.

I provide an example of predicting class assignments for a scenario in Fig. 11.7.

```
##
## Set scenario
##
data = { 'Pprice':[ 1, 2 ], 'Ddisc':[ 0.03, 0.03 ], 'Odisc':[ 0.03, 0.03 ], 'Cdisc':[ 0.03, 0.03
         'Pdisc':[ 0.03, 0.04 ], 'Region':[ 'West', 'Midwest' ] }
scenario = pd.DataFrame( data )
##
## Predict with scenario
##
scen_pred = logit01.predict( scenario )
##
## Label predictions
##
scenario[ 'Prediction' ] = [ 'Satisifed' if x >= 0.5 else 'Dissatisfied' for x in scen_pred ]
scenario.style.set_caption( 'Logit Scenario' ).set_table_styles( tbl_styles ).hide_index()
```

Logit Scenario

Pprice	Ddisc	Odisc	Cdisc	Pdisc	Region	Prediction
1	0.030000	0.030000	0.030000	0.030000	West	Dissatisfied
2	0.030000	0.030000	0.030000	0.040000	Midwest	Satisifed

Fig. 11.7 This illustrates how do a scenario classification analysis using a trained logit model

11.3 K-Nearest Neighbor (*KNN*)

The *K-Nearest Neighbor* (*KNN*) classification method is an intuitive and simple approach to classifying objects. As the name suggests, the approach is based on grouping objects that are close to their neighbors. The number of neighbors to consider is based on k which you specify. If $k = 1$, then only the nearest neighbor is used. If $k > 1$, then an object is grouped with those k objects. The k is a hyperparameter.

The group has to be labeled. Is it satisfied or not? The label associated with a group is based on a simple rule: majority wins. For example, suppose three objects are grouped together where two have one label and the third has a different one. The label for the group is based on the label for the two that share the same label. If more than two labels are involved, then the majority rule becomes a plurality rule.

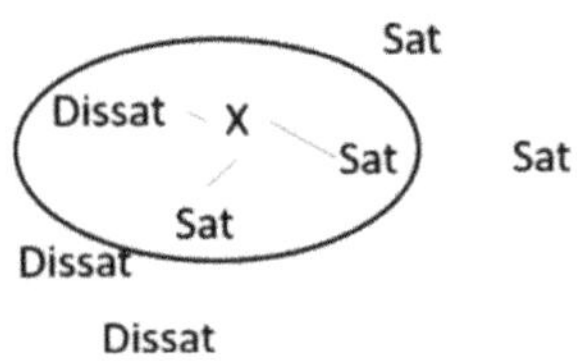

Fig. 11.8 This illustrates how the majority rule works for a *KNN* problem with $k = 3$

Since the label assignment rule is majority wins, this immediately suggests that the k for grouping the nearest objects should be odd; otherwise, there could be a tie in which case it is unclear what label is used. An odd k avoids this issue. You will see this same majority rule when I discuss decision trees later in this chapter. I illustrate the situation in Fig. 11.8 for $k = 3$ and seven objects: four labeled "Sat" for "Satisfied" and three labeled "Dissat" for "Dissatisfied." An eighth object, labeled X, must be classified as Satisfied or Dissatisfied. The three objects nearest X are shown in the circle. Notice that two are "Sat" and one is "Dissat." Therefore, by the majority rule, X is labeled Satisfied as is the whole group.

An issue with the majority rule centers on the distribution of the class labels. If the label distribution is skewed, then some labels will dominate an assignment merely because they occur so often. A fix for this situation is to weight the classes, perhaps based on the inverse of the distance from the point to be classified to each of its k nearest neighbors. Those neighbors closest to the point will have a large weight; those furthest away will have a small weight.

The distances between a point to be classified and all other points in a training data set are based on one of several *distance metrics*. Three out of many are the

- Euclidean Distance Metric;
- Manhattan Distance Metric (*a.k.a*, CityBlock); and
- Minkowski Distance Metric.

The *Euclidean Distance Metric* is the most frequently used. See Witten et al. (2011, p. 131). It is based on the square root of the sum of squared differences between all pairs of points:

$$d_{Euc}(x, y) = \left(\sum_{i=1}^{n} w_i \times \mid (x_i - y_i) \mid^2\right)^{1/2} \tag{11.3.1}$$

where x and y are two vectors (think columns of a DataFrame) each of length n and w is a weight vector. The weights are optional and are meant to address the label skewness issue I just mentioned. If they are not provided, then the default is $w_i = 1, \forall i$. The Euclidean measure is also scaled using the MinMax scaler from Chap. 5 because there is still a scale impact that may have to be accounted corrected. See Witten et al. (2011, p. 132) for a discussion. The distance metric is just an application of the Pythagorean Theorem.

The *Manhattan Distance Metric* (*a.k.a*, CityBlock) is the sum of the absolute values of differences between all pairs of points:

$$d_{Man}(x, y) = \sum_{i=1}^{n} w_i \times \mid (x_i - y_i) \mid . \tag{11.3.2}$$

The *Minkowski Distance Metric* is the sum of the absolute values of differences between all pairs of points, each absolute distance raised to a power, and the whole summation expression raised to the inverse of that power:

$$d_{Min}(x, y) = \left(\sum_{i=1}^{n} w_i \times \mid (x_i - y_i) \mid^p\right)^{1/p} \tag{11.3.3}$$

where p is an integer. Common values are $p = 1$ and $p = 2$. Notice that the Minkowski distance is a general case of the Euclidean and Manhattan distances: $p = 1$ is the Manhattan distance and $p = 2$ is the Euclidean distance. Also notice that for $p = 1$, the distance formula is equivalent to the *L1-norm Loss* I mentioned in Chap. 6 for the residual function where the deviations in that case are an actual value less its predicted value. You can now see the reason for the "1" in this name. Similarly, you can see that for $p = 2$, you have the *L2-norm Loss*. The parameter p is a hyperparameter.

You can use the *scipy.spatial.distance* package or the *sklearn.metrics.pairwise* package to calculate these distances although you would not do this in practice for a *KNN* classification problem. The former package allows for the weights and the latter will do all pairwise combinations. There are many more distance metrics available. See the *scipy* and *sklearn* documentation for a complete listing. I illustrate the *scipy* functions in Fig. 11.10 for the data I show graphically in Fig. 11.9.

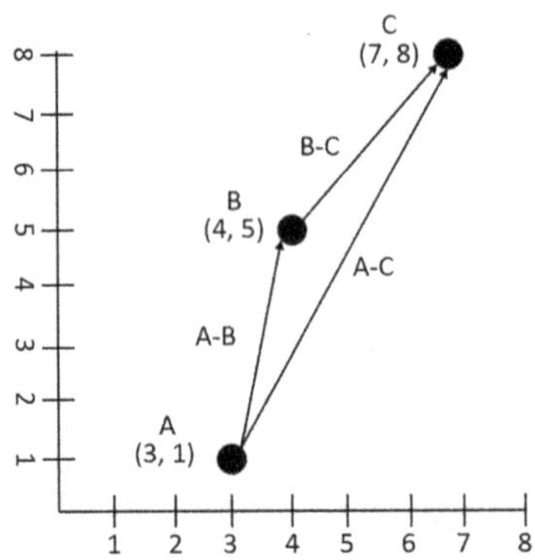

Fig. 11.9 This illustrates three points used in Fig. 11.10 for the distance calculations

```
from scipy.spatial import distance as scdst
##
## Create DataFrame
##
data = { 'x':[3, 4, 7], 'y':[1, 5, 8], 'labels':[ 'A', 'B', 'C'] }
df_dst = pd.DataFrame( data )
##
## All pairs of labels
##
x = list( itertools.combinations( df_dst.labels, 2) )
df_dst[ 'pairs' ] = [ ''.join(i) for i in x ]
##
## Euclidean Distance between rows of df_dst
##
cols = [ 'x', 'y' ]
pairs = list( itertools.combinations( df_dst.index, 2) )
x = [ scdst.euclidean( df_dst[ cols ].iloc[i], df_dst[ cols ].iloc[j] ) for i, j in pairs ]
df_dst[ 'Euclidean' ] = x
##
## Manhatten (Cityblock) Distance between rows of df_dst
##
x = [ scdst.cityblock( df_dst[ cols ].iloc[i], df_dst[ cols ].iloc[j] ) for i, j in pairs ]
df_dst[ 'Manhatten' ] = x
##
## Minkowski Distance between rows of df_dst
##
x = [ scdst.minkowski( df_dst[ cols ].iloc[i], df_dst[ cols ].iloc[j], p = 1 ) for i, j in pairs ]
df_dst[ 'Minkowski:1' ] = x
##
x = [ scdst.minkowski( df_dst[ cols ].iloc[i], df_dst[ cols ].iloc[j], p = 2 ) for i, j in pairs ]
df_dst[ 'Minkowski:2' ] = x
##
## Display distances
##
df_dst.style.set_caption( 'Distance Calculations Comparisons ').set_table_styles( tbl_styles ).hide_index()
```

Distance Calculations Comparisons

x	y	labels	pairs	Euclidean	Manhatten	Minkowski:1	Minkowski:2
3	1	A	AB	4.123106	5	5.000000	4.123106
4	5	B	AC	8.062258	11	11.000000	8.062258
7	8	C	BC	4.242641	6	6.000000	4.242641

Fig. 11.10 This illustrates the distance calculations using the *scipy* functions with the three points I show in Fig. 11.9

Once the nearest neighbors are determined based on the hyperparameter p, you can use them to classify a new object. This is where the test data can be used. The fitted model has a *predict* method just as the other methods do. I illustrate how you can do this in Fig. 11.11 with some other analysis displays in Figs. 11.12 and 11.13.

11.3.1 Case Study: Predicting

You can also create a scenario and predict specific outcomes. I show how you can do this in Fig. 11.14.

```
##
## Make predictions and check accuracy measures
##
cols = [ 'Pprice', 'Ddisc', 'Odisc', 'Cdisc', 'Pdisc', 'Midwest',
        'Northeast', 'South', 'West'  ]
pred = knn01.predict( knn_test[ cols ] )
##
cm = confusion_matrix( knn_test.sat_t2b, pred ).ravel()
##
## Create a dictionary of cm and labels
##
lbl = [ 'True Negative', 'False Positive', 'False Negative', 'True Positive' ]
data = { 'Classification':lbl, 'Frequency': cm, 'Percent':cm/cm.sum() }
##
## Display the confusion matrix in a DataFrame
##
df_confusion = pd.DataFrame( data )
df_confusion.set_index( 'Classification', inplace = True )
df_confusion.style.set_caption( 'Confusion Table' ).\
    format( {'Percent':format} ).\
    bar( align = 'mid', color = 'lightblue' ).\
    set_table_styles( tbl_styles )
```

Confusion Table

Classification	Frequency	Percent
True Negative	18	7.0%
False Positive	64	24.8%
False Negative	35	13.6%
True Positive	141	54.7%

Fig. 11.11 This illustrates how to create a confusion table for a *KNN* problem

11.4 Naive Bayes

Naive Bayes, despite its name, is a powerful, albeit simple, classifying methodology widely used in areas such as loan applications (e.g., risky or safe loan applicant), healthcare (e.g., needs assisted living or not), spam identification (e.g., emails and phone calls), to list a few. There are two operative parts to the method's name: *Naive* and *Bayes*. I will explain these in reverse order since the *Naive* part is the adjective modifying the second part.

11.4.1 Background: Bayes Theorem

The *Bayes* part of the name is due to the statistical theorem underlying the approach. This is *Bayes Theorem* taught in an introductory statistics course once conditional probabilities have been introduced. Recall that the probability of an event A conditioned on an event B is written as

```
##
## Create labels
##
lbl = [ 'Not Satisfied', 'Satisfied' ]
##
## Create DataFrame
##
x = cm.reshape( 2, 2 )
data = pd.DataFrame( x, index = lbl, columns = lbl )
##
## Plot the confusion matrix
##
sns.set( font_scale = 1.4 )   #for label size
##
base = 'Base: Testing data; n = ' + str( x.sum() )
ax = sns.heatmap( data, annot = True, fmt = '.3g', annot_kws = { "size": 16 } )  # annot_kws = font size
ax.set_title( 'Confusion Matrix\nKNN Classifier', fontsize = font_title )
ax.set( xlabel = 'Predicted', ylabel = 'True' )
ax.set_xticklabels( lbl )
ax.set_yticklabels( lbl )
footer( );
```

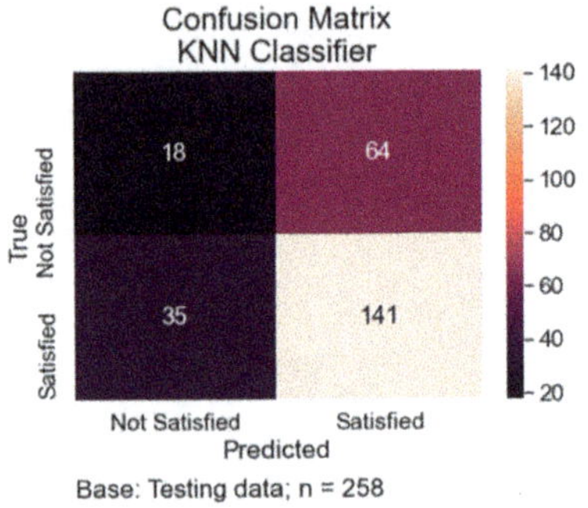

Fig. 11.12 This illustrates how to create a confusion matrix for a *KNN* problem

```
##
## Create a classification accuracy report
##
rpt = classification_report( logit_test.sat_t2b, predictions_recoded, digits = 3 )
##knn
## Use regex expression to recode 0 = Dissatisfied and 1 = Satisfied
##
rpt = re.sub( ' 0 ', 'B2B', rpt )
rpt = re.sub( ' 1 ', 'T2B', rpt )
##
## Display report
##
print( 'Classification Accuracy Report\n' )
print( f'{rpt}' )
print( f'Cut-off value: theta = {theta}' )
```

```
Classification Accuracy Report

              precision    recall  f1-score   support

         B2B      0.250     0.012     0.023        82
         T2B      0.681     0.983     0.805       176

    accuracy                          0.674       258
   macro avg      0.466     0.498     0.414       258
weighted avg      0.544     0.674     0.556       258

Cut-off value: theta = 0.5
```

Fig. 11.13 This illustrates how to create a classification accuracy report for a *KNN* problem

```
##
## Set scenario
##
data = { 'Pprice':[ 1, 2 ], 'Ddisc':[ 0.03, 0.03 ], 'Odisc':[ 0.03, 0.03 ], 'Cdisc':[ 0.03, 0.03 ],
           'Pdisc':[ 0.03, 0.04 ], 'Midwest':[ 0, 1 ], 'Northeast':[ 0, 0 ], 'South':[ 0, 0 ],
           'West':[ 1, 0 ]}
scenario = pd.DataFrame( data )
##
## Predict using scenario
##
cols = [ 'Pprice', 'Ddisc', 'Odisc', 'Cdisc', 'Pdisc', 'Midwest', 'Northeast', 'South', 'West' ]
scen_pred = knn01.predict( scenario )
##
## Label predictions
##
scenario[ 'Prediction' ] = [ 'Satisifed' if x == 1 else 'Dissatisfied' for x in list( scen_pred ) ]
scenario.style.set_caption( 'KNN Scenario' ).set_table_styles( tbl_styles ).hide_index()
```

KNN Scenario

Pprice	Ddisc	Odisc	Cdisc	Pdisc	Midwest	Northeast	South	West	Prediction
1	0.030000	0.030000	0.030000	0.030000	0	0	0	1	Dissatisfied
2	0.030000	0.030000	0.030000	0.040000	1	0	0	0	Satisifed

Fig. 11.14 This illustrates how to create a scenario analysis for a *KNN* problem

$$Pr(A \mid B) = \frac{Pr(A \cap B)}{Pr(B)} \tag{11.4.1}$$

where $\cap$ is the intersection operator. So, $Pr(A \cap B)$ is the probability of A AND B occurring. The denominator, $Pr(B)$, is the *marginal probability* of B occurring over all possibilities of A. You could also write

$$Pr(B \mid A) = \frac{Pr(A \cap B)}{Pr(A)}. \tag{11.4.2}$$

Since $Pr(A \cap B)$ is in both equations, you could solve for $Pr(A \cap B)$ in (11.4.2) and substitute the result in (11.4.1) to get

$$Pr(A \mid B) = \frac{Pr(A) \times Pr(B \mid A)}{Pr(B)}. \tag{11.4.3}$$

(11.4.3) is *Bayes Theorem*. It states that the probability of A conditioned on B is the probability of B given A adjusted by the marginal probability of A, all relative to the probability of B. $Pr(A)$ is the prior distribution, a term I introduced in Chap. 1. $Pr(A \mid B)$ is the posterior distribution. $Pr(B \mid A)$ is the likelihood of B given A. So, (11.4.3) says the prior is adjusted by the likelihood.

11.4.2 A General Statement

Suppose you want to classify objects (e.g., customers) into one of C_k segments or classes, $k = 1, \ldots, K$. Let $\mathbf{X}$ be a matrix of p factors, features, or independent variables, $\mathbf{X}_i, i = 1, 2, \ldots, p$, you will use for the classification. The problem is to select a class for the object given the classification data. You could equate the symbol A to the class and the symbol B to the features. Then (11.4.3) is[3]

$$Pr(C_k \mid \mathbf{X}) = \frac{Pr(\mathbf{X} \mid C_k) \times Pr(C_k)}{Pr(\mathbf{X})}. \tag{11.4.4}$$

- C_k is the kth group;
- $\mathbf{X}$ is a matrix of classification features or independent variables;
- $Pr(C_k \mid \mathbf{X})$ is the *posterior* distribution;
- $Pr(C_k)$ is the *prior* distribution;
- $Pr(\mathbf{X} \mid C_k)$ is the *likelihood*; and
- $Pr(\mathbf{X})$ is the *marginal* distribution, sometimes referred to as the *evidence*.

This is sometimes written as

$$posterior = \frac{prior \times likelihood}{evidence}. \tag{11.4.5}$$

The evidence part has nothing to do with the class assignments so it is often ignored. The *prior* distribution, $Pr(C_k)$, is the probability of an object being assigned to a class with no data or evidence to support that assignment. A likely initial prior is just $Pr(C_k) = 1/k$ for each class. This is an *uniformative prior* which I also referred to as "flimsy" or "flat" in Chap. 1. You may have sample data, perhaps training data, that indicates class membership, but these data are not part of $\mathbf{X}$ which is used to do classification; the objects are already classified in the training data. Then $Pr(C_k)$ is simply the class proportions from the training data. This initial prior is adjusted by the data for the class to produce a new probability for classifying the objects; this is the *posterior*. It is possible for new data for the classes to be obtained. In this case, the posterior becomes the prior, the new data for the classes are used for the adjustment, and a new posterior is calculated.

[3] The following draws heavily from https://en.wikipedia.org/wiki/Naive_Bayes_classifier. Last accessed December 1, 2020.

11.4.3 The Naive Adjective: A Simplifying Assumption

Focus on the numerator in Fig. 11.4.4. The product of the prior and the likelihood equals the joint distribution of the class membership and the factors. This equals the product of the probability of each factor conditioned on the other factors and the class membership. That is,[4]

$$Pr(C_k) \times Pr(\mathbf{X} \mid C_k) = Pr(C_k, X_1, X_2, \ldots, X_p) \tag{11.4.6}$$

$$= Pr(X_1 \mid X_2, \ldots, C_k) \times Pr(X_2 \mid X_3, \ldots, C_k) \times \tag{11.4.7}$$

$$\ldots \times Pr(X_n \mid C_k) \times Pr(C_k) \tag{11.4.8}$$

This product is challenging to work with. A simplifying assumption holds that the factors are independent, but each conditioned on the class, C_k. Independence implies that the probabilities of each factor, conditioned on class, are multiplicative. Therefore, you can write:

$$Pr(C_k) \times Pr(\mathbf{X} \mid C_k) = Pr(C_k) \times \prod_{i=1}^{p} Pr(X_i \mid C_k) \tag{11.4.9}$$

This assumption of independence is the reason for the "Naive" adjective for "Naive Bayes." While it is really a simplifying and not a naive assumption, it makes the approach tractable and very useful. The only issue with using the Naive Bayes approach for classification is the nature of the conditional probabilities, $Pr(X_i \mid C_k)$. It is worth repeating that these conditionals are for the features, not the target.

11.4.4 Distribution Assumptions

There are three distributions that can be used for these probabilities:

1. Gaussian;
2. Multinomial; and
3. Bernoulli.

The Gaussian distribution is applicable when features are continuous and the other two when they are discrete. The Multinomial distribution is used when the features have multiple possible level. It is commonly used in text analysis.

---

[4] See https://en.wikipedia.org/wiki/Naive_Bayes_classifier.

The Bernoulli distribution is used for binary features. The Gaussian is the most commonly used.

A problem with these distributions is that they hold for all the features in a training set. Consider the situation where some features are continuous while others are binary. You might be tempted to use the Bernoulli distribution since some features are binary. This is inappropriate because the Bernoulli only applies to a portion of them. What about the other portion? On the other hand, you might be tempted to use the Gaussian distribution since some features are continuous. Using this distribution is equally inappropriate. You cannot apply a Bernoulli distribution to continuous data nor can you apply a Gaussian distribution to binary data. In short, you cannot mix distributions in a single call to the Naive Bayes *fit()* method (which I explain below) for two types of data.

Several remedies have been proposed.[5] Suppose you have a mix of continuous and multinomial data. You can bin the continuous data using the Pandas *cut* and *qcut* functions I outlined in Sect. 5.2.4, thus creating new multinomial features. Then use the Multinomial distribution on all the multinomial features. If you have continuous and binary features, you could bin the continuous ones into two bins: "top tier" and "bottom tier." For example, the top 5% and bottom 95% as well as apply the sklearn *Binarizer* function I discussed in Chap. 5. Then use the Bernoulli distribution. Finally, if you have a mix of multinomial and binary data, then transform the multinomial to binary in a manner comparable to handling a five-point Likert Scale transformation. Overall, the suggestion for this transformation option is to transform from high frequency data to low frequency data; you cannot go in the other direction.

The problem with the transformation option is that you lose information. When you bin a continuous feature, all the variations in the data are hidden when the values are placed in bins; the data are "dumbed down." A second suggestion for handling a mixture of features is to use all the features of one type in one Naive Bayes estimation and all the features of another type in a second estimation. For instance, assuming continuous and binary features, estimate a Gaussian Naive Bayes for the continuous features and a Bernoulli Naive Bayes for the binary features. Then estimate the posterior probabilities for each, add them to your feature set, and then estimate a final Gaussian Naive Bayes on just these two new (probability) features. The Gaussian Naive Bayes is appropriate for this final estimation since the conditional probabilities are continuous.

There is a third option that is a variant of the second. Calculate the two likelihoods, one for the Gaussian features and one for the discrete features, then multiply them along with the prior. That is, if $Pr_G(X_i \mid C_k)$ is the Gaussian likelihood for the continuous data, $Pr_B(X_i \mid C_k)$ is the Bernoulli likelihood function for the binary data, and $Pr(C_k)$ is the prior, then you could simply do

[5] See the discussions on StackOverFlow: https://stackoverflow.com/questions/14254203/mixing-categorial-and-continuous-data-in-naive-bayes-classifier-using-scikit-lea and https://stackoverflow.com/questions/39173169/hybrid-naive-bayes-how-to-train-naive-bayes-classifer-with-numeric-and-category?noredirect=1&lq=1. Both last accessed December 4, 2020.

$Pr_G(X_i \mid C_k) \times Pr_B(X_i \mid C_k) \times Pr(C_k)$. This is a reasonable approach because of the independence assumption. Normalizing this product by the "evidence" gives you the relevant posterior. This approach is implemented in the Python package *mixed-naive-bayes*. Note that you cannot simply estimate a Gaussian Naive Bayes and a Bernoulli Naive Bayes and then multiply the estimated probabilities from each. Those probabilities each result from adjusting the respective likelihoods by the prior and the evidence so multiplying the posteriors will result in a new posterior that is off by factor of the prior over the evidence. The functions in the *mixed-naive-bayes* package handle this correctly. You can install the *mixed-naive-bayes* package using *pip install mixed-naive-bayes*.

11.4.5 Case Study: Naive Bayes Training

I will demonstrate the Naive Bayes (*NB*) for the transactions data set. First, I will demonstrate the Gaussian *NB* for the continuous variables which are the price and discounts. I split the master data set of aggregated transactions into training and testing sets. Then I fit a Gaussian *NB* using the top-two box customer satisfaction and five continuous features. I summarize the results in Fig. 11.15. The accuracy score is 0.678 so about two-thirds of the cases are accurately predicted. The Bernoulii *NB* is shown in Fig. 11.16. The accuracy score is 0.682, slightly better. Finally, I show the Mixed *NB* in Fig. 11.17. The accuracy score for this one is 0.671, slightly below the other two. The lower score for the Mixed *NB* may be due to the ineffectiveness of the categorical variables to explain any of the variation in customer satisfaction.

11.5 Decision Trees for Classification

A *decision tree* is an alternative for identifying key drivers for a target variable. But it can also be used to classify objects based on how important those objects are for determining or explaining the target. My focus is the latter although I will make comments about the former.

Decision trees have several advantages:

1. There is no need to transform the independent variables or features using, for example, logs. Skewed data do not affect trees. The same holds for outliers.
2. They automatically identify local interactions among the features. This is based on the way nodes are split as I describe below.
3. Missing data can be an issue, but generally they are not that serious. The extent of missingness determines the severity of an impact.
4. They are easy for management and clients to understand and interpret.

```
##
## NB with continuous data only: Gaussian NB
##
gcols = [ 'Pprice', 'Ddisc', 'Odisc', 'Cdisc', 'Pdisc', 'sat_t2b' ]
df_gnb = df_agg[ gcols ].copy()
##
## do train/test split
##
df_gnb_train, df_gnb_test = train_test_split( df_gnb, test_size = 0.33, random_state = 42 )
```

```
##
## Create X and y sets
##
gcols = [ 'Pprice', 'Ddisc', 'Odisc', 'Cdisc', 'Pdisc' ]
X = df_gnb_train[ gcols ].copy()
y = df_gnb_train.sat_t2b
##
## ===> Step 1: Instantiate a Gaussian Classifier <===
##
gnb = GaussianNB()
##
## ===> Step 2: Train the model using the training sets <===
##
gnb.fit( X, y )
##
## ===> Step 3: Predict the response for test dataset <===
##
y_pred = gnb.predict( df_gnb_test[ gcols ] )
##
## ===> Step 4: Print predicted values <===
## NOT RUN
## boldprt( 'Predicted Value:' )
```

```
##
## Categorize as Over/Under/Equal as prediction
##
y_test = df_gnb_test.sat_t2b
data = { 'test':y_test, 'predicted':y_pred }
df_gnb = pd.DataFrame( data )
##
## Define a lambda function for recoding
##
df_gnb[ 'Over/Under' ] = df_gnb.apply( lambda x: 'Over' if ( x.predicted > x.test ) else
                                      ( 'Under' if ( x.predicted < x.test ) else 'Equal' ), axis = 1 )
## NOT RUN
##display( df_gnb.head().style.set_caption( 'Over/Under/Equal Categorization' ).\
##     set_table_styles( tbl_styles ).\
##     hide_index() )
##
## Prediction distribution
##
lst_odr = [ 'Over', 'Equal', 'Under' ]
df_gnb[ 'Over/Under' ] = df_gnb[ 'Over/Under' ].astype( CategoricalDtype( lst_odr, ordered = True ) )
##
## Prediction distribution
##
display( df_gnb.stb.freq( [ 'Over/Under' ], sort_cols = True ).style.set_caption( 'GNB Prediction Distribution' ).\
```

GNB Prediction Distribution

Over/Under	count	percent	cumulative count	cumulative percent
Over	81	31.395349	81	31.395349
Equal	175	67.829457	256	99.224806
Under	2	0.775194	258	100.000000

```
##
## Model Accuracy: How often is the classifier correct?
##
boldprt( 'Accuracy:' )
```

```
Accuracy:
      0.678
```

Fig. 11.15 The Gaussian *NB* was used with continuous classifying variables. The accuracy score was 0.678

Of course, they also have disadvantages:

1. Trees can become large trees so arbitrary/subjective "pruning" may be needed.
2. They assume that all variables interact which is the local interaction I mentioned above. So this could be a blessing or curse.
3. Interactions are local, not global so they are specific to the conditions for a node.
4. Execution performance may be poor for large numbers of features.

Trees belong to the family of supervised learning techniques because of the target variable. The methodology is referred to as a tree methodology because the results are displayed as a tree, albeit one that is upside down. Natural tree terminology applies: there is a root, branches, and leaves. A tree is also grown and pruned to make it smaller and more interpretable. The depth of a tree (i.e., its size) is a hyperparameter. The root of the tree contains all the target data (100%) and the goal is to divide that data into smaller units as you move down the tree. Each final leaf of the tree has a small percentage of the target data. The minimum amount in a leaf can be specified so that no leaf has 0% of the total target data.

```
##
## Same problem but now with loyalty program membership which is binary
## But loyalty data are text: Yes/No
## Region and Buyer Rating are not binary although they could be dummified
##
## Use LabelEncoder on loyalty program
##
bcols = [ 'loyaltyProgram', 'sat_t2b' ]
df_bnb = df_agg[ bcols ].copy()
df_bnb[ 'loyaltyProgram' ] = le.fit_transform( df_bnb.loyaltyProgram )
##
## do train/test split
##
df_bnb_train, df_bnb_test = train_test_split( df_bnb, test_size = 0.33, random_state = 42 )
```

```
##
## Create X and y sets
##
bcols = [ 'loyaltyProgram' ]
X = df_bnb_train[ bcols ].copy()
y = df_bnb_train.sat_t2b
##
## ===> Step 1: Instantiate a Bernoulli Classifier <===
##
bnb = BernoulliNB()
##
## ===> Step 2: Train the model using the training sets <===
##
bnb01 = bnb.fit( X, y )
##
## ===> Step 3: Predict the response for test dataset <===
##
y_pred = bnb01.predict( df_bnb_test[ bcols ] )
##
## ===> Step 4: Print predicted values <===
## NOT RUN
## boldprt( 'Predicted Value:' )
```

```
##
## Categorize as Over/Under/Equal on prediction
##
y_test = df_bnb_test.sat_t2b
data = { 'test':y_test, 'predicted':y_pred }
df_bnb = pd.DataFrame( data )
##
## Define a lambda function for recoding
##
df_bnb[ 'Over/Under' ] = df_bnb.apply( lambda x: 'Over' if ( x.predicted > x.test ) else
                                      ( 'Under' if ( x.predicted < x.test ) else 'Equal' ), axis = 1 )
## NOT RUN
##display( df_bnb.head().style.set_caption( 'Over/Under/Equal Categorization' ).\
##     set_table_styles( tbl_styles ).\
##     hide_index() )
##
## Prediction distribution
##
lst_odr = [ 'Over', 'Equal', 'Under' ]
df_bnb[ 'Over/Under' ] = df_bnb[ 'Over/Under' ].astype( CategoricalDtype( lst_odr, ordered = True ) )
##
## Prediction distribution
##
display( df_bnb.stb.freq( [ 'Over/Under' ], sort_cols = True ).style.set_caption( 'BNB Prediction Distribution' ).\
```

BNB Prediction Distribution

Over/Under	count	percent	cumulative count	cumulative percent
Over	82	31.782946	82	31.782946
Equal	176	68.217054	258	100.000000

```
##
## Model Accuracy: How often is the classifier correct?
##
boldprt( 'Accuracy:' )
```

```
Accuracy:
      0.682
```

Fig. 11.16 The Bernoulli *NB* was used with a binary classifying variable. The accuracy score was 0.682

The nature of the target determines the type of tree grown. If the target is continuous, then a *regression tree* is grown; otherwise, a *classification tree* is grown. In either case, the objects are still classified based on how they explain the target. The names "classification tree" and "regression tree" only refer to the criteria that determine how the tree is actually grown. The final leaves of the tree denote or

```
##
## Same problem but now with loyalty program membership, Region, and Buyer Rating,
## which are categorcial, and with continuous data
##
xcols = [ 'Tprice', 'Ddisc', 'Odisc', 'Cdisc', 'Pdisc', 'loyaltyProgram',\
          'Region', 'buyerRating', 'sat_t2b' ]
df_mnb = df_agg[ xcols ].copy()
##
## Use LabelEncoder for loyalty program, Region, and buyerRating
##
df_mnb[ 'loyaltyProgram' ] = le.fit_transform( df_mnb.loyaltyProgram )
df_mnb[ 'Region' ] = le.fit_transform( df_mnb.Region )
df_mnb[ 'buyerRating' ] = le.fit_transform( df_mnb.buyerRating )
##
## do train/test split
##
```

```
##
## Create X and y sets
##
xcols = [ 'Tprice', 'Ddisc', 'Odisc', 'Cdisc', 'Pdisc', 'loyaltyProgram',\
          'Region', 'buyerRating' ]
X = df_mnb_train[ xcols ].copy()
y = df_mnb_train.sat_t2b
##
## ===> Step 1: Instantiate a Gaussian Classifier <===
## Specify categorical feature as columns 5 and 6 (zero-based indexing)
##
mnb = MixedNB( categorical_features = [ 5, 6, 7 ] )
##
## ===> Step 2: Train the model using the training sets <===
##
mnb.fit( X, y )
##
## ===> Step 3: Predict the response for test dataset <===
##
y_pred = mnb.predict( df_mnb_test[ xcols ] )
##
## ===> Step 4: Print predicted values <===
## NOT RUN
## boldprt( 'Predicted Values:' )
```

```
[2 4 3]
```

```
##
## Categorize as Over/Under/Equal on prediction
##
y_test = df_mnb_test.sat_t2b
data = { 'test':y_test, 'predicted':y_pred }
df_mnb = pd.DataFrame( data )
##
## Define a lambda function for recoding
##
df_mnb[ 'Over/Under' ] = df_mnb.apply( lambda x: 'Over' if ( x.predicted > x.test ) else
                                      ( 'Under' if ( x.predicted < x.test ) else 'Equal' ), axis = 1 )
## NOT RUN
##display( df_mnb.head().style.set_caption( 'Over/Under/Equal Categorization' ).\
##     set_table_styles( tbl_styles ).\
##     hide_index() )
##
## Prediction distribution
##
lst_odr = [ 'Over', 'Equal', 'Under' ]
df_mnb[ 'Over/Under' ] = df_mnb[ 'Over/Under' ].astype( CategoricalDtype( lst_odr, ordered = True ) )
##
## Prediction distribution
##
display( df_mnb.stb.freq( [ 'Over/Under' ], sort_cols = True ).style.set_caption( 'MNB Prediction Distribution' ).\
```

MNB Prediction Distribution

Over/Under	count	percent	cumulative count	cumulative percent
Over	81	31.395349	81	31.395349
Equal	173	67.054264	254	98.449612
Under	4	1.550388	258	100.000000

```
##
## Model Accuracy: How often is the classifier correct?
##
boldprt( 'Accuracy:' )
```

```
Accuracy:
      0.671
```

Fig. 11.17 The Mixed *NB* was used with categorical and continuous classifying variables. The accuracy score was 0.671

identify the classifications. The features used to grow the tree can be continuous or discrete.

Trees are grown by partitioning the features into two groups, that is, it bifurcates each feature. The trees are sometimes called *partitioning trees* because of this. The partition or bifurcation of a feature is the best split possible for that feature. The split is the best based on a criterion function for splitting the feature to explain the target.

Some features have only one natural split (e.g., gender) so there is no "best." Others can be split in many ways (an infinite number of ways for continuous features), but not all splits are the best. The importance of each feature at its optimal split is determined and the features are ranked by those importances. The most important feature explaining the target and its optimal split is at the top of the tree just below the root.

11.5.1 Partitioning by Constants

All that follows is for classification trees. The partitions are constants drawn in the space of the target data. If the target is customer satisfaction and the features are gender, with a natural split into male and female, and age, with a split into young and senior, then two constants (i.e., straight lines) are drawn separating the satisfaction measure. Some of the target objects fall into each region cut off by the constants. The number of objects falling into each region is simply counted by the levels of the target. I show an example in the left portion of Fig. 11.18 in which customer satisfaction is the target which has two levels: Dissatisfied and Satisfied (in alphanumeric order). There are 13 people (i.e. objects) and two features (e.g., gender and age). For this example, it is determined that Feature 1 can be optimally divided at value or level C_1 and Feature 2 can be divided at value or level C_2 but only for values of Feature 1 greater than C_1. This means that Feature 1 can only be optimally split once at C_1. In this example, Feature 2 is also only split once given a Feature 1 split. Three regions result from the two splits.

The regions from the split are displayed as an inverted tree which summarizes how the observations or measures on the target were allocated to the regions defined by the splits. I show the summary tree in the right portion of Fig. 11.18. All the people are at the root level so the percent of all people regardless of their satisfaction is 100%. This is all the people in the left portion of the figure. You can see that 61.5% of them at the root are satisfied so the predicted satisfaction level at the root is "Satisfied." The prediction is based on the level with the highest percentage at that split, so a majority rule is used as for the *KNN* method.

For this example, the tree is first bifurcated at C_1 for Feature 1. This feature was determined to be the most important for explaining the target which is why this is the first to be split. The root is the *parent* of the two portions resulting from the split. These portions are the *children* so there is a *parent-child* relationship. A parent-child relationship exists throughout the tree. Each point of a split in the tree is a *node* which can be internal or terminal. Regardless where a node is located, it contains information about the objects at that point. The information varies by software implementation. I show in Fig. 11.18 the sample size, the percent of the parent node allocated to that child node, the percent of the objects at each level of the target at that child node (the percents obviously sum to 100%), and the predicted level for that child node.

A node contains information, albeit latent, about objects in that node. We can extract that latent information by splitting the node thus gaining useful, realized, or revealed information about what is important for the target. The split of a parent node into the two best children nodes from among the number of possible splits is based on the gain in realized information extracted from the possible splits. The split with the maximum realized information gain is the one that is used. The gain in information is the reduction in the *impurity* at a node resulting from a split, or the gain in purity from the split. Impurity is just the degree of heterogeneity. A node that is pure (i.e., has zero impurity) is homogeneous; one that is impure (i.e., has a degree of impurity) is heterogeneous. The goal is to have a node that is homogeneous. There are two measures for impurity: the *Gini Index* and *Entropy*.

11.5.2 Gini Index and Entropy

The Gini Index is based on the probability of misclassifying an object. Suppose there are C classification labels. For example, existing customers could be buyers or non-buyers of a new product, so $C = 2$ and the labels are "buyer" and "non-buyer." In a training data set, it is known who is a buyer and who is a non-buyer. Randomly select an object and then randomly assign that object to one of the class labels. There are then two random draws each with a probability: one draw for the set of objects and one independent draw from the set of classes. Let p_i be the probability that a random object that actually belongs to class i is selected. The probability of incorrectly assigning that object to class k, $k \neq i$, is $\sum_{k \neq i}^{C} p_k = 1 - p_i$ where C is the number of classes. Since the act of randomly selecting an object and the act of randomly assigning it are independent acts, the probability of selecting and assigning is $p_i \times (1 - p_i)$.[6] The Gini Index at a node is

$$G(N_k) = \sum_{i=1}^{C} p_i \times (1 - p_i) \tag{11.5.1}$$

$$= 1 - \sum_{i=1}^{C} p_i^2. \tag{11.5.2}$$

where N_k is node k and C is the number of classes or labels in node k. If $C = 2$, then the Gini Index is $1 - (p_1^2 + p_2^2)$. The Gini Index for the entire split into two nodes is the weighted average of the individual node Gini Indexes with weights equal to the class proportions from the parent.

Using the data in Fig. 11.18, the Gini Index at the root is $1 - (0.385^2 + 0.615^2) = 0.473$. This root can be split using either Feature 1 or Feature 2. Let us check

[6] Note that $\sum_{k=1}^{C} p_k = 1$ so, if $C = 3$ then $p_1 + p_2 + p_3 = 1$. So, $p_1 + p_3 = 1 - p_2$.

Feature 1 first. The split is for $< C_1$ being 0 and $> C_1$ being 1 where 0 represents Dissatisfied and 1 Satisfied for this example. By the diagram, estimates of the respective probabilities based on frequency counts are (1/4 = 0.25, 3/4 = 0.75) and (4/9 = 0.444, 5/9 = 0.555) where the first proportion in each bracketed term refers to the dissatisfied and the second to the satisfied. The Gini Index is $1-(0.25^2+0.75^2) = 0.375$ and $1-(0.444^2+0.556^2) = 0.494$. A weighted average of these using weights from the parent is $0.385 \times 0.375 + 0.615 \times 0.494 = 0.448$. This is the *Weighted Gini Index* for the node based on Feature 1. The information gain, i.e., the gain in purity from splitting Feature 1 at C_1, beyond the parent is $0.473 - 0.448 = 0.025$.

Now check Feature 2 and split with $< C_2$ being 0 and $> C_2$ being 1. By the diagram, then (2/6 = 0.333, 4/6 = 0.667) and (3/7 = 0.429, 4/7 = 0.571). The Gini Index is $(1 - (0.333^2 + 0.667^2) = 0.444$ and $1 - (0.429^2 + 0.571^2) = 0.490$. The weighted average is 0.472. The information gain from the parent is $0.473 - 0.472 = 0.001$. Clearly, splitting on Feature 1 first gives a better improvement in purity: 0.025 vs 0.001.

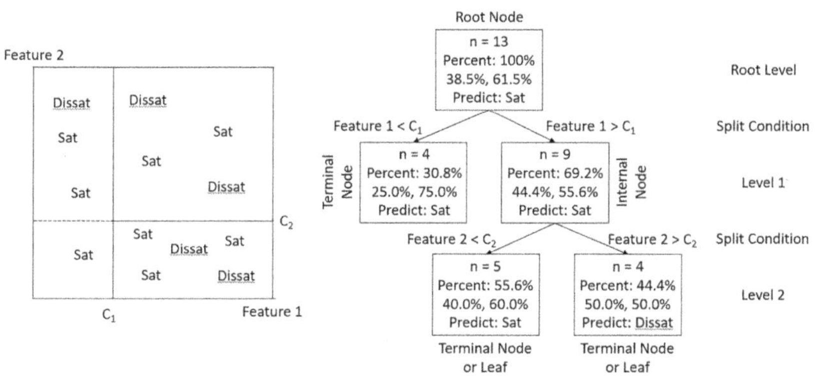

Fig. 11.18 This illustrates two features and there divisions both in feature space and a tree reflecting that space

Notice in Fig. 11.18 that the node for Feature 1 $< C_1$ has only four observations. The sample size is now too small so this node is not split; this is a terminal node or leaf. The node for Feature 1 $< C_1$, however, can be split because the sample size is large enough ($n = 9$), the split now occurring on Feature 2. The Gini Index at C_2 is based on $(0.40, 0.60)$ and $(0.50, 0.50)$ so $1 - (0.40^2 + 0.60^2) = 0.48$ and $1 - (0.50^2 + 0.50^2) = 0.50$ as shown

I show a grown tree based on the data for two features for the example in Fig. 11.19. This splits the parent as the above calculation showed it should. The same holds for the second split of Feature 2. This figure also illustrates the content displayed at each node. I show a typical node's content in Fig. 11.20. At each node, the predicted classification of all its objects is based on the majority rule.

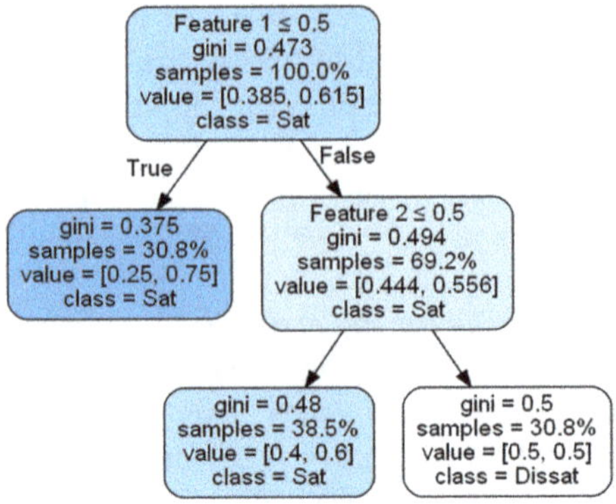

Fig. 11.19 The Gini Index was used to grow the tree illustrated in Fig. 11.18. The values shown match those in the text

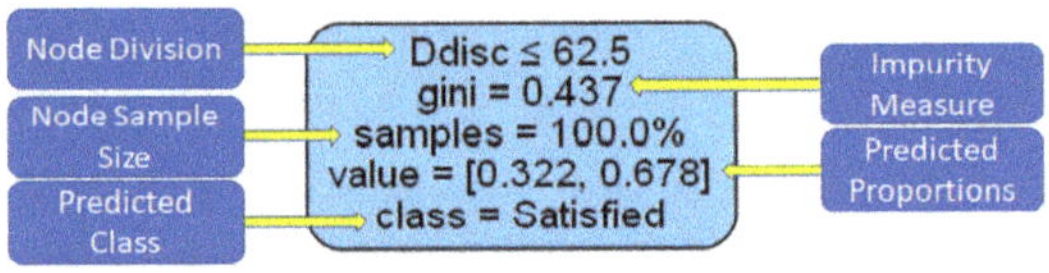

Fig. 11.20 This is the typical content of a tree's nodes. This is for a classification problem

The alternative measure of impurity is Entropy. This is a concept from thermodynamics and refers to the amount of disorder in a system. An entropy value of 0 indicates perfect order and 1 indicates perfect disorder. This concept was carried over to the impurity of a node. An entropy value of 0 means the node is perfectly homogeneous (i.e., no disorder) and 1 means perfectly heterogeneous (i.e., perfect or complete disorder). Entropy is calculated using the probability of observing a class label at a node where the probabilities are based on the proportions of the labels at the node. The measure is

$$E(N_k) = -\sum_{i=1}^{C} p_i \times log_2 p_i \tag{11.5.3}$$

where N_k is node k and C is the number of classes or labels in node k. If $C = 1$, then $p_i = 1$ and $-1 \times log_2 1 = 0$ so the node is perfectly homogeneous (as it should be since there is only one label!). If $C = 2$ and $p_1 = p_2 = 0.5$ so that the labels are uniformly distributed in the node, then $-(0.5 \times log_2 0.5 + 0.5 \times log_2 0.5) = -(log_2 0.5) = 1$ for perfect heterogeneity. But, if $p_1 = 1$ and $p_2 = 0$, then $-(1 \times log_2 1) = 0$ and the node is homogeneous. I show a graph of entropy versus probabilities in Fig. 11.21 and a summary of homogeneity and heterogeneity in Fig. 11.22. I also show the example tree in Fig. 11.19 regrown based on entropy in Fig. 11.23.

```
##
## Graph entropy
##
## Create probabilities
##
p = np.linspace( 0.01, 0.99, 100 )
q = 1 - p
##
## Create DataFrame
##
en =  (-1)*p*np.log2( p ) + (-1)*q*np.log2( q )
data = { 'p':p, 'entropy':en }
df_en = pd.DataFrame( data )
df_en.set_index( 'p', inplace = True )
##
## Plot entropy
##
ax = df_en.plot( legend = False )
ax.set_title( 'Entropy\n$-p \\times log_2( p ) - q \\times log_2( q )$' )
ax.set( xlabel = 'Probability', ylabel = 'Entropy' )
ax.axvline( x = 0.5 );
```

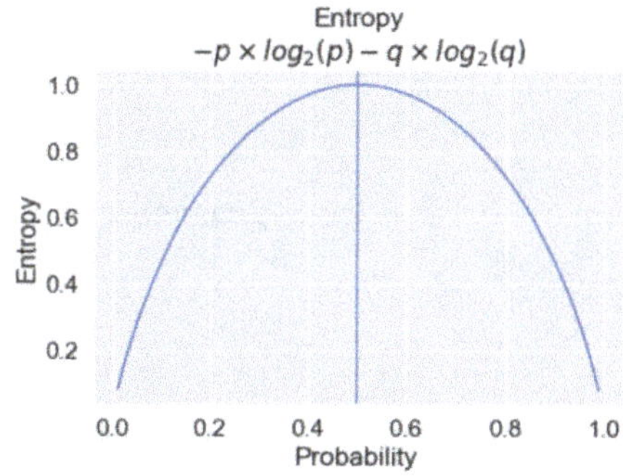

Fig. 11.21 Graph of entropy for a two-class problem

Entropy		
0	1	
Perfect Order	Perfect Disorder	Thermodynamics View
Perfect Homogeneity	Perfect Heterogeneity	Decision Tree View

Fig. 11.22 This shows the relationship between entropy and homogeneity/heterogeneity

My description above is for classification problems, so the trees are classification trees. This is based on the nature of the target, not the features. The target is discrete with levels or labels. If the target is continuous, then a different criterion is used to determine a split. The weighted variance is calculated and the reduction in variance is determined for each node candidate. The feature with the maximum variance reduction is used for a split. In essence, this is a continuous counterpart to information gain.

My examples above were based on discrete features. Continuous features could also be used. This presents a slight challenge because there is a potentially infinite number of split points for a continuous feature. One way to handle this is to identify the unique values in the training data set for the continuous feature and to treat

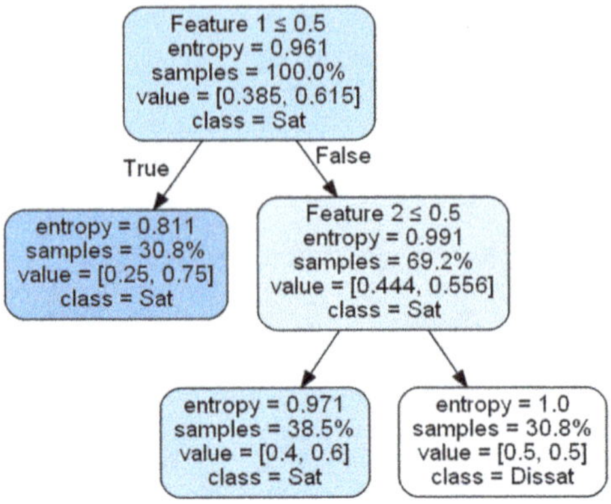

Fig. 11.23 Entropy was used to grow the tree illustrated in Fig. 11.18. Compare this tree to the one in Fig. 11.19

them as "discrete." The disadvantage to this is the large number of potential unique values.

11.5.3 Case Study: Growing a Tree

As an example of growing a tree, consider the transactions data for the living room blinds. Suppose the product manager wants to know the key drivers for customer satisfaction. I show the data preparation in Fig. 11.24. I use the same data I used for the logit example, but make a copy for this example. Since *Region* is categorical with strings for the categories, it cannot be used as it is; it has to be encoded. I used dummy encoding based on the Pandas *get_dummies* function. Once the data were prepared, I then instantiated the *DecisionTreeClassifier* from *sklearn*'s *tree* module which I show in Fig. 11.25. Then I grew a tree as I show in Fig. 11.26.

All the training data are in the root node with 69.7% satisfied, so all the customers are classified as satisfied. The first split is based on the pick-up discount, *Pdics*. All customers with $Pdisc \leq 0.04$ are assigned to the left node (58.2%); otherwise, to the right. (42.8%). On the left, 65.0% are satisfied so all are classified as satisfied. The Gini Index on the left is $1 - (0.35^2 + 0.65^2) = 0.455$ as shown. The Rich Information extracted from the data is that the pick-up discount is not only the most important feature, but also that the discount below and above 4% are the key to customer satisfaction.

```
##
## Create the tree DataFrames
##
tree_train = logit_train.copy()
tree_test = logit_test.copy()
##
## Training data
##
cols = [ 'Pprice', 'Ddisc', 'Odisc', 'Cdisc', 'Pdisc', 'Region' ]
X_train = tree_train[ cols ]
X_test = tree_test[ cols ]
##
## Testing data
##
cols = [ 'sat_t2b' ]
y_train = tree_train[ cols ]
y_test = tree_test[ cols ]
##
## Create dummies
##
dummies = pd.get_dummies( X_train.Region ) ##, prefix = 'dum' )
X_train = pd.concat( [ X_train, dummies ], axis = 1 )
X_train = X_train.drop( columns = [ 'Region' ] )
##
## Apply the label encoder to the test DataFrame
##
dummies = pd.get_dummies( X_test.Region ) ##, prefix = 'dum' )
X_test = pd.concat( [ X_test, dummies ], axis = 1 )
X_test = X_test.drop( columns = [ 'Region' ] )
```

Fig. 11.24 This illustrates the data preparation for growing a decision tree for the furniture Case Study

```
##
## Instantiate the tree
## Specify:
##     - Depth: 3 levels
##     - Minimum sample size for a leaf: 5
##
dtree = tree.DecisionTreeClassifier( random_state = 0, max_depth = 3, min_samples_leaf = 5 )
##
## Fit the tree
##
dtree = dtree.fit( X_train, y_train )
fig, ax = plt.subplots( nrows = 1, ncols = 1, figsize = (18,10) )
feature_names = [ i for i in X_train.columns ]
##
## Display the tree
## Make arrows read (Source: https://stackoverflow.com/questions/62318367/
##                            decision-tree-edges-branches-so-light-that-are-invisible )
##
out = tree.plot_tree( dtree, filled = True, rounded = True, class_names = [ 'Dissatisfied', 'Satisfied' ],\
                feature_names = feature_names, proportion = True, ax = ax );
for o in out:
    arrow = o.arrow_patch
    if arrow is not None:
        arrow.set_edgecolor( 'red' )
```

Fig. 11.25 This illustrates the instantiation of the *DecisionTreeClassifier* function for growing a decision tree for the furniture Case Study

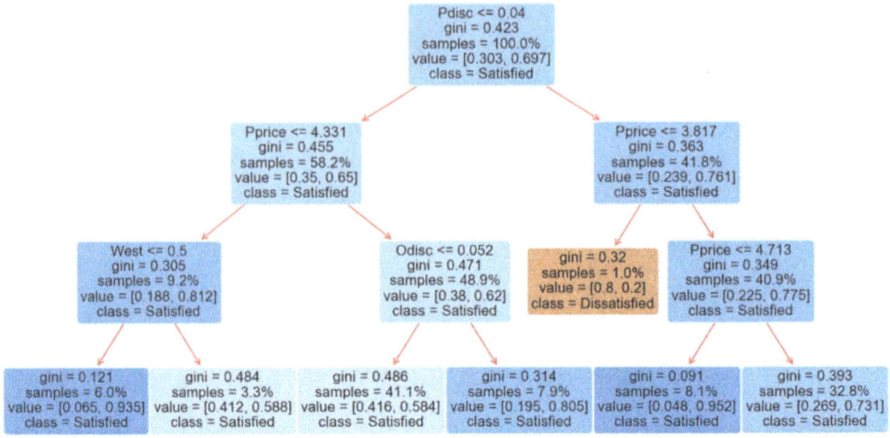

Fig. 11.26 This illustrates the grown decision tree for the furniture Case Study

11.5.4 Case Study: Predicting with a Tree

You can predict the class probabilities for each object using the *predict_proba* method for the tree object or their class assignment using *predict*. The latter is based on the rule that the class assignment is the maximum class probability. I show a tree accuracy report in Figs. 11.27 and 11.28.

```
##
## Check accuracy scores
##
boldprt( 'Accuracy on Training Data:' )
score = dtree.score( X_train, y_train )
print( f'{score:.3f}' )
##
score = dtree.score( X_test, y_test )
boldprt( '\nAccuracy on Testing Data:' )
print( f'{score:.3f}' )
##
## Predict the response for test dataset
##
y_pred = dtree.predict( X_test )
##
## Model Accuracy: How often is the classifier correct?
##
score = round( metrics.accuracy_score( y_test, y_pred ), 3 )
boldprt( 'Accuracy:' )
print( f'\t{score:.3}' )
```

```
Accuracy on Training Data:
0.702

Accuracy on Testing Data:
0.690
Accuracy:
        0.69
```

Fig. 11.27 This illustrates the grown decision tree's accuracy scores for the furniture Case Study

```
##
## Categorize as Over/Under/Equal on prediction
##
df_tree = y_test.copy()
df_tree.rename( columns = {'sat_t2b':'test' }, inplace = True )
df_tree[ 'predicted' ] = y_pred
##
## Define a lambda function for recoding
##
df_tree[ 'Over/Under' ] = df_tree.apply( lambda x: 'Over' if ( x.predicted > x.test ) else
                              ( 'Under' if ( x.predicted < x.test ) else 'Equal' ), axis = 1 )
##
## Prediction distribution
##
lst_odr = [ 'Over', 'Equal', 'Under' ]
df_tree[ 'Over/Under' ] = df_tree[ 'Over/Under' ].astype( CategoricalDtype( lst_odr, ordered = True ) )
##
## Prediction distribution
##
display( df_tree.stb.freq( [ 'Over/Under' ], sort_cols = True ).style.set_caption( 'Decision Tree Prediction Distribution' ).\
    bar( subset = ['count', 'percent' ], align = 'zero', color = 'lightblue' ).set_table_styles( tbl_styles ).hide_index() )
```

Decision Tree Prediction Distribution

Over/Under	count	percent	cumulative_count	cumulative_percent
Over	79	30.620155	79	30.620155
Equal	178	68.992248	257	99.612403
Under	1	0.387597	258	100.000000

Fig. 11.28 This illustrates the grown decision tree's prediction distribution for the furniture Case Study

11.5.5 *Random Forests*

My discussion above was concerned with growing a single tree. You could actually grow many trees to effectively grow a forest. Each tree in the forest is randomly grown so the forest is a random forest. The development of this and the subsequent analysis is beyond the scope of this book. See James et al. (2013) and Hastie et al. (2008) for a technical discussion.

11.6 Support Vector Machines

A final method for classifying objects is the *Support Vector Machine* (*SVM*). This approach is computationally more costly but it also generally produces more accurate classification results. The basic concept is simple. If your data are in two groups so they are binary, then draw a line such that you have the best division of the data points into the two groups. The line is a plane in three dimensions and a hyperplane in more than three dimensions. The concept is the same regardless of the number of dimensions.

First, some terminology. The line dividing the groups is called a *decision line*, *decision surface*, or *decision hyperplane*. It is a decision surface because it allows you to decide which points belong to which class. All points on one side are assigned to one class and all points on the other side are assigned to the other class. This actually reflects the Gestalt *Proximity* and *Similarity Principles*. This is a non-probabilistic approach to class assignment. The other approaches I discussed and illustrated in this chapter are probabilistic with probabilities estimated and a decision rule (i.e., majority rule) based on the probabilities for determining allocation. In this approach, allocation is based on the distance from the decision surface. Distance is the length of a line drawn from a point perpendicular to the surface. A perpendicular line is used because this is the shortest distance to the surface.

The *support vectors* are the data points closest to the decision surface. These close data points are the most difficult to classify because just a small random shock to one of them could move it from one class to another. A random shock, for example, could be just a small measurement error. These points are the ones that "support" the decision surface in the sense of determining where the surface lies. Points far from the surface have little to no impact on where it lies since a small random shock to any of them would have no impact on the surface.

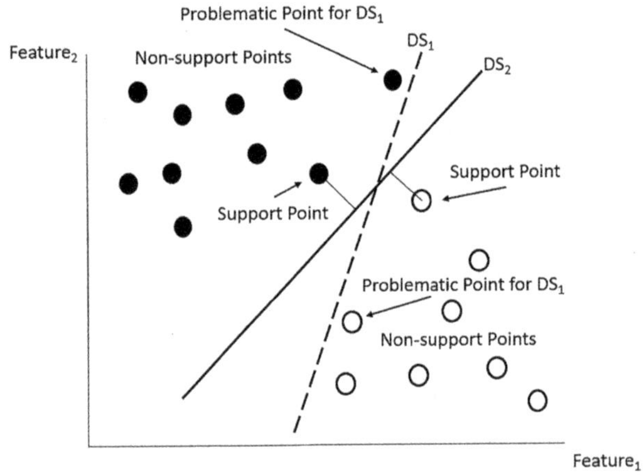

Fig. 11.29 This illustrates data points for a *SVM* problem. Two decision support lines (DS_1 and DS_2) are shown

An example best illustrates the idea. Suppose you have data points for two classes measured on two features as I depict in Fig. 11.29. The two classes are represented by solid and empty circles. A quick inspection of the plot confirms the two classes: the solid circles are all in the top left quadrant and the empty ones are in the lower right quadrant; the two classes are clearly separated. Nonetheless, you want to draw a straight line, a decision surface, that best separates the two groups. I illustrate two possibilities. The one labeled DS_1 separates the two groups but there are two points very close to this line that "support" it, but they are clearly problematic. A small amount of random noise added to either point could change the class assignment.

A second line, DS_2, also separates the two classes but has a greater distance from the points closest to this line. The points closest to it are the support points forming a support vector which is a subset of the data. It is the support vector that is the basis for the decision surface. The distance between the support points and the line forms a channel, sometimes referred to as a "street," around the line. This is formally called a *margin*. The line is in the middle of the margin and is thus measured as the median of the width of the margin. The goal is to find the maximum margin, the widest street, between the support points. The decision surface is the median. The mathematics for this is more challenging. See Deisenroth et al. (2020, Chapter 12) for a technical discussion.

11.6.1 Case Study: SVC *Application*

The *sklearn* package has two modules that support classification and regression problems just as it has two for Naive Bayes and decision trees. The classification module is SVC and the regression module is SVR. I am only concerned with classification in this book. I illustrate the SVC module in Fig. 11.30 for the customer satisfaction classification problem. I will use the same train/test data sets as for the customer satisfaction logit model. The problem is the same: classify people based on their top-two box satisfaction rating. For this SVM application, there is one problem. The logit model used the price and discounts as features, but it also used the marketing region as a feature. The *Region* variable is categorical with categories as words: "Midwest", "Northeast", "South", and "West." These cannot be used directly since words cannot be used in calculations. They were dummified using a function in the formula statement. The *SVM* method has the same problem. In this case, however, there is no formula *per se* so there is no function to create dummy variables. Pandas has a *get_dummies* method to handle this situation. Another approach to handle this is to use the *LabelEncoder* I discussed in Chap. 5. This function converts the categories of a categorical variable to integers. A nice feature of using this encoder is that it has a reverse method so that you can always retrieve the labels associated with the integers. I decided to use the Pandas *get_dummies* function for this problem.

11.6.2 Case Study: Prediction

I provide the fit and accuracy measure in Fig. 11.31. You can do a scenario analysis for the *SVM* classification. I show how you can do this in Fig. 11.32. The set-up is actually just like the other ones I showed in this chapter.

```
##
## Create the SVM DataFrames
##
svm_train = logit_train.copy()
svm_test = logit_test.copy()
##
## Training data
##
cols = [ 'Pprice', 'Ddisc', 'Odisc', 'Cdisc', 'Pdisc', 'Region' ]
X_train = svm_train[ cols ]
X_test = svm_test[ cols ]
##
## Testing data
##
cols = [ 'sat_t2b' ]
y_train = svm_train[ cols ]
y_test = svm_test[ cols ]
##
## Create dummies
##
dummies = pd.get_dummies( X_train.Region, prefix = 'dum' )
X_train = pd.concat( [ X_train, dummies ], axis = 1 )
X_train = X_train.drop( columns = ['Region'] )
##
## Apply the Label encoder to the test DataFrame
##
dummies = pd.get_dummies( X_test.Region, prefix = 'dum' )
X_test = pd.concat( [ X_test, dummies ], axis = 1 )
X_test = X_test.drop( columns = [ 'Region' ] )
```

Fig. 11.30 This illustrates DataFrame setup for a *SVM* problem

```
##
## ===> Step 1: Instantiate a SVM Classifier <===
## Notice that SVC is used for the classifier
##
svmC = svm.SVC( )
##
## ===> Step 2: Fit the model using the training sets <===
##
svm01 = svmC.fit( X_train, y_train )
##
## ===> Step 3: Predict the response for test dataset <===
##
y_pred = svm01.predict( X_test )
##
## ===> Step 4: Model Accuracy: How often is the classifier correct? <===
##
score = round( metrics.accuracy_score( y_test, y_pred ), 3 )
boldprt( 'Accuracy:' )
print( f'\t{score:.3}' )
```

```
Accuracy:
        0.682
```

Fig. 11.31 This illustrates the fit and accuracy measures for a *SVM* problem

```
##
## Set scenario
##
data = { 'Pprice':[ 1, 2 ], 'Ddisc':[ 0.03, 0.03 ], 'Odisc':[ 0.03, 0.03 ], 'Cdisc':[ 0.03, 0.03 ],
         'Pdisc':[ 0.03, 0.04 ], 'Midwest':[ 0, 1 ], 'Northeast':[ 0, 0 ], 'South':[ 0, 0 ],
         'West':[ 1, 0 ]}
scenario = pd.DataFrame( data )
##
## Predict using scenario
##
cols = [ 'Pprice', 'Ddisc', 'Odisc', 'Cdisc', 'Pdisc', 'Midwest', 'Northeast', 'South', 'West' ]
scen_pred = svm01.predict( scenario )
##
## Label predictions
##
scenario[ 'Prediction' ] = [ 'Satisifed' if x == 1 else 'Dissatisfied' for x in list( scen_pred ) ]
scenario.style.set_caption( 'SVC Scenario' ).set_table_styles( tbl_styles ).hide_index()
```

SVC Scenario

Pprice	Ddisc	Odisc	Cdisc	Pdisc	Midwest	Northeast	South	West	Prediction
1	0.030000	0.030000	0.030000	0.030000	0	0	0	1	Satisifed
2	0.030000	0.030000	0.030000	0.040000	1	0	0	0	Satisifed

Fig. 11.32 This illustrates how to do a scenario analysis using a *SVM*

```
##
## Create DataFrame of accuracy measures
##
data = { 'Method':[ 'Logit', 'GNB', 'BNB', 'MNB', 'Tree', 'SVC' ],
         'Accuracy':[ 0.674, 0.678, 0.682, 0.671, 0.690, 0.682 ] }
df_acc = pd.DataFrame( data )
df_acc.set_index( 'Method', inplace = True )
df_acc.style.set_caption( 'Accuracy Comparisons' ).\
    set_table_styles( tbl_styles ).bar( vmin = 0.67, color = 'lightblue' ).\
    format( {'Accuracy':'{:.1%}'} )
```

Accuracy Comparisons

Method	Accuracy
Logit	67.4%
GNB	67.8%
BNB	68.2%
MNB	67.1%
Tree	69.0%
SVC	68.2%

Fig. 11.33 This illustrates the fit and accuracy measure for a *SVM* problem

11.7 Classifier Accuracy Comparison

The accuracy of the six methods can be compared. I show this in Fig. 11.33. You can see that for this problem, the decision tree has the highest accuracy rate, but the measures are all close. My recommendation in this case is to use the decision tree because of the graphical output.

Chapter 12
Grouping with Unsupervised Learning Methods

I will turn my attention to *unsupervised learning* methods in this chapter. Recall that these methods do not have a target variable that guides them to learn from a set of features. There are still features, but without a target another approach is needed to extract the information buried inside your data. Supervised learning methods have a target and a set of parameters that indicate how the features relate to the target. The learning, therefore, is the identification of the parameters so that the relationship can also be identified. There are no parameters without a target because there is no relationship to something that does not exist. The only problem is a relationship among the features as standalone entities. Unlike supervised learning methods which use estimation procedures, unsupervised learning methods use *algorithms* to identify the relationship among the features. Algorithms, widely used in computer science, machine learning, and in other quantitative areas where estimation is impossible, are heuristics that produce a result. See Cormen et al. (2009) for a classic treatment of algorithms.

The unsupervised learning algorithms I will describe in this chapter are concerned with grouping objects. This is not unlike the goal of the classification methods I reviewed in the previous chapter. Those methods were concerned with classifying from a prediction point of view. The classes are known and the question is which class does a new, previously unclassified object belong. Now objects are not predicted to belong to one group or another since groups are unknown. Instead, the groups are created from the data. The question is: "*What are the groups?*" Algorithms group objects so that they belong to one group or another; predictions for new objects are not involved. Predictions are, in fact, impossible because classes are unknown a priori. Assignment to groups and prediction of class membership are different activities.

We had a number of measures to assess prediction accuracy because we had a target for making those assessments. Now we are grouping objects. What makes a good group? This is a difficult, if not impossible, question to answer because there is no basis for an assessment. Hand et al. (2001, p. 295, emphasis in original) stated

W. R. Paczkowski, *Business Analytics*,
https://doi.org/10.1007/978-3-030-87023-2_12

that there is "direct notion of *generalization* to a test data set", which means "the validity of a clustering is often in the eye of the beholder."

I will cover the following two types of grouping methods of unsupervised learning:

- Clustering; and
- Mixture Models.

Clustering is a complex topic, and one not without controversy. Someone once said there is an infinite number of clustering solutions. That person probably underestimated! This is due, in part, to the number of methods and their variations available. I will look at two broad clustering methods:

- Hierarchical Clustering
- K-Means Clustering

Mixture models are in the class of clustering methods, but I will treat them separately in this chapter because they are probabilistic, more in line with logistic regression, Naive Bayes, and classification decision trees. Decision trees are probabilistic because the assignments to a class in a node (e.g., satisfied, dissatisfied) are based on the proportions of the objects in the node. The same holds for K-Nearest Neighbor methods. In fact, decision trees and K-Nearest Neighbor methods both use a majority wins strategy for assigning class labels which means they both rely on proportions: the label with the highest proportion wins.

Proportions are unbiased maximum likelihood estimators of probabilities. Logistic regression and Naive Bayes begin with a probability statement and end with a probability for predictions. Decision trees and K-Nearest Neighbor do not begin with a probability statement but end with one.

12.1 Training and Testing Data Sets

The supervised approaches I described in the previous chapters all had a target variable; that is the chief characteristic of supervised learning. This target is important because it is in the master data set which gets split in two parts: training and testing. The target is carried along with the split so it appears in both subsets. You can use a trained model to make predictions of the target and then test how well the model predicts because the target is in the testing data set.

You cannot do this testing with unsupervised clustering because there is no target. You can always split a master data set into training and testing parts. The split function does not know about a target. This is not completely correct because the *train_test_split* function has parameters for X and y (note the cases) where X is the set of features and y is the target. The y parameter is option; the X is required. I chose to use only X as a DataFrame in what I presented because doing otherwise requires eventually merging the features and target data sets. Passing just X avoids this extra step.

The problem comes in when you make a prediction. Predict what? All you have is a series of features. You do not know how the objects (e.g., customers) are grouped before you do a clustering of them so you cannot test how well the algorithm clusters the objects. The notion of using training and testing data sets to do predictions. In fact, the *sklearn* hierarchical clustering function, which I discuss below, does not have a predict method. The *KMeans* function does but only for assigning to the nearest cluster. However, the methods do have hyperparameters so tuning can be done to set them. You then need a tuning data set which is how the train-test-split is used.

12.2 Hierarchical Clustering

Hierarchical clustering works by creating, as its name suggests, a hierarchy of groups of objects. The objects are measured on several features that can be continuous, ordinal, or nominal. The latter two have to be appropriately encoded, usually dummy encoded. Character features can be used but, of course, they also have to be appropriately coded. The hierarchy is typically displayed as a tree, called a *dendrogram*, with all the individual objects as separate individual leaves at one end of the tree and a full collection of the objects at the opposite, or *root*, end. In this regard, the hierarchical clustering dendrogram is like a decision tree. The difference is how the dendrogram is created.

The decision tree approach uses an impurity measure, either Gini Index or Entropy, to group values of the features so that the objects in a group are homogeneous. For the customer satisfaction example I used, customers are grouped based on the values of their features. The final groupings of customers are the terminal leaves on the tree. Each one is a homogeneous group (i.e., cluster) of customers and each of these is defined or identified by reading up the tree from a terminal leaf. The target class is predicted at each terminal leaf based on the maximum probability of belonging to a class. The root is completely heterogeneous.

For a hierarchical clustering dendrogram, the root is also completely heterogeneous since it contains all the objects. Each terminal leaf is homogeneous since it contains only one object. The groups are formed based on a distance measure which has the sense of an impurity measure. Impurity is how heterogeneous are the objects, distance is how similar (i.e., homogeneous).

12.2.1 Forms of Hierarchical Clustering

There are two forms of hierarchical clustering:

1. Agglomerative in which all objects start as their own cluster (i.e., each initial cluster has only one object) and the hierarchy is built upward from that point.

Higher-level clusters are built from lower-level clusters. The dendrogram is thus built from the bottom up; this is a *bottoms-up* approach.

2. Divisive in which all objects are in one initial cluster (i.e., the initial cluster has all the objects) which is the root. The objects are then successively pulled from this cluster to form new clusters below the high-level cluster. The dendrogram is thus built from the top down; this is a *top-down* approach.

The Divisive approach is more compute-intensive because you have to examine all cases to determine which ones to pull out, one object at a time. The Agglomerative approach is more efficient and is the one typically used. This is the one I will describe below. See Paczkowski (2016) for a brief comparison of the two approaches and (Everitt et al., 2001) for more details.

12.2.2 Agglomerative Algorithm Description

Hierarchical clustering uses an algorithmic approach to grouping objects. The objects are individual objects or clusters of objects. In fact, at some point the objects are always other clusters. The clusters are formed so that their members are homogeneous based on the features, although the degree of homogeneity declines as you move up the tree. The root is a completely heterogeneous grouping and the terminal leaves at the bottom are perfectly homogeneous.

The hierarchical clustering algorithm groups objects by measuring how *similar* or *dissimilar* they are based on a distance measure that is a function of the features. Objects that are "close" are considered similar and so are joined into a group; the objects are *linked* together. The process is iterative. Once a group is created, that group becomes a new object (so a group and an object become the same thing) and the process is repeated until all objects form one group: the root.

This iterative process has four steps:

Step 1: Create initial clusters with each object as its own cluster.

- If there are 5 objects, there are 5 initial clusters.
- When starting, all objects are leaves or terminal nodes.
- These clusters are *singleton clusters*.
- Each cluster is perfectly homogeneous.

Step 2: Calculate a distance measure for all pairs of clusters.

- The distance shows the degree of similarity: small distance apart $\Rightarrow$ similar.

Step 3: Merge or link two clusters that are most similar to form a new cluster. The new cluster is homogeneous based on the similarity of the objects in the new cluster.

- The degree of homogeneity is reduced.

Step 4: Repeat starting with Step 2, otherwise stop if all objects are grouped.

- When stop, all objects are at the root.
- The degree of homogeneity is zero.

This is an algorithm, not an estimation process because there are no parameters involved; just a series of steps or rules are used. To implement this clustering algorithm, you need two things: a distance *metric* between pairs of clusters and a rule for how clusters are joined or *linked* based on the distance metric.

12.2.3 Metrics and Linkages

The metric is how the distances are calculated based on a set of features. There is a wide array of metrics in *scipy*. The most commonly used are: Euclidean distance (*L2*), Manhattan distance, and cosine distance. The cosine similarity is the cosine of the angle between two feature vectors. The Euclidean is the *scipy* default.

At the very bottom of the dendrogram, all the clusters are singletons; all the clusters have exactly one object since each object is its own cluster. Finding the distance between all pairs is using easy a metric, such as Euclidean distance. As you move up the dendrogram, however, the clusters have more than one object in them. This is the idea of a cluster: a collection of homogeneous objects. Once clusters consist of two or more objects, the problem becomes finding the distance between pairs of clusters. What point in a cluster (the points being the objects that comprise the cluster) is used to calculate a new cluster distance? In other words, "*From where in a cluster do you measure distance*"*?* The center point? The most distant or nearest point? The average or median point? This is where the *linkage* comes in. The linkage method is based on the points inside a cluster. The Python package *scipy* has seven linkage methods that determine how the distance between two clusters are determined when those clusters consist of several objects. The linkage methods are:

1. Ward's minimum variance linkage;
2. Maximum or complete linkage;
3. Average linkage;
4. Single linkage;
5. Weighted linkage;
6. Centroid linkage; and
7. Median linkage.

The single linkage, also referred to as the "Nearest Point Algorithm," is the default in *scipy*. It uses the minimum of all the distances of pairs of points from one cluster to another. The average distance method uses the average distance of all the pairs. Ward's method is the most commonly used, even though it is not the default in *scipy*.

Those clusters that are closest, i.e., most similar or least dissimilar, based on the linkage and distance metric you selected, are joined to form a new cluster. Once that new cluster is formed, the component clusters are then deleted from further consideration; they are now in the new cluster. This process continues until all the objects are joined at the root.

12.2.4 Preprocessing Data

Some objects will have a large impact on the distance calculations because of their scale. This is the exact same concept as outliers affecting the mean. In that case, standardization reduces the impact of those points. Standardization is necessary for hierarchical clustering because scales can have an adverse effect on the distance calculations and thus distort results. Mean centering and scaling by the standard deviation are typical. I discussed these in Chap. 5.

Categorical variables also present a problem when the categories are strings. Regions, defined as Midwest, Northeast, South, and West, is an example. They are usually labeled encoded with nominal values based on the sorted levels.

Finally, missing values for any feature must be handled. A distance for joining clusters cannot be calculated with missing values. They have to be filled in or the whole record containing at least one missing value must be deleted. If you impute a missing value, you have to be careful that the imputed value is not based on other values that are too representative of the entire sample. If they are, then you run the risk of not getting a good clustering solution because the imputed value itself could distort results. Interpolation using a small window around the missing observation is best.

12.2.5 Case Study Application

I will continue with the furniture Case Study, focusing now on clustering customers based on seven features: Region, unit sales, pocket price, and the four discounts. The Data Cube was collapsed on the time dimension so some features had to be aggregated. In particular, unit sales were summed and the price and discounts were each averaged. Region, of course, was left untouched since it is unique for each local boutique retailer. I show some of the aggregated data in Fig. 12.1.

Once the Data Cube is subsetted and the subset data are appropriately aggregated, they have to be preprocessed. I did this in two steps. First, I standardized the total sales, the average pocket price, and each of the four discounts. I used the *StandardScaler*. I checked the descriptive statistics to make sure that each mean is zero and each standard deviation is 1.0. I show the code for this in Fig. 12.2. I also label encoded the Region variable. I used the *labelEncoder* function as I described in Chap. 5. I show this in Fig. 12.3

```
##
## Subset the df_agg data
##
cols = [ 'Region', 'totalUsales', 'meanPprice', 'meanDdisc', 'meanCdisc', 'meanPdisc', 'meanOdisc' ]
df_hclusters = df_agg[ cols ]
##
df_hclusters.head().style.set_caption( 'Subset Data for Clustering' ).\
    set_table_styles( tbl_styles ).\
    hide_index()
```

Subset Data for Clustering

Region	totalUsales	meanPprice	meanDdisc	meanCdisc	meanPdisc	meanOdisc
Northeast	3461	5.400400	0.131820	0.068248	0.038241	0.051113
West	1001	5.797762	0.128930	0.071814	0.036907	0.050093
West	787	5.744514	0.136357	0.071179	0.040036	0.046714
West	1873	5.702487	0.136744	0.071500	0.040564	0.050641
West	2350	5.722586	0.135367	0.070184	0.041439	0.049990

Fig. 12.1 This is a sample of the aggregated data for the furniture Case Study hierarchical clustering of customers

```
##
## Standardize the six numeric variables.
##
cols = [ 'totalUsales', 'meanPprice', 'meanDdisc', 'meanCdisc', 'meanPdisc', 'meanOdisc' ]
##
## Extract the variables as a temporary DataFrame
##
tmp = df_hclusters[ cols ]
##
## Standardize and horizontally concatenate with the Region variable;
## concatenate on axis = 1
##
tmp_standard = pd.DataFrame( StandardScaler().fit_transform( tmp ), columns = cols )
df_hclusters = pd.concat( [ df_hclusters[ 'Region'], tmp_standard ], axis = 1 )
##
display( df_hclusters.head().style.set_caption( 'Standardized Data for Clustering' ).\
    set_table_styles( tbl_styles ).\
    hide_index() )
display( df_hclusters.describe().T.style.set_caption( 'Descriptive Statistics' ).\
    set_table_styles( tbl_styles ) )
```

Standardized Data for Clustering

Region	totalUsales	meanPprice	meanDdisc	meanCdisc	meanPdisc	meanOdisc
Northeast	0.244311	0.216148	0.542271	-0.609569	-1.088793	0.524922
West	-0.659626	0.739036	0.428604	0.627404	-1.946044	0.026226
West	-0.738261	0.668967	0.720783	0.406992	0.065101	-1.626092
West	-0.339206	0.613664	0.735987	0.518495	0.404748	0.294217
West	-0.163930	0.640112	0.681844	0.061867	0.966986	-0.024256

Descriptive Statistics

	count	mean	std	min	25%	50%	75%	max
totalUsales	779.000000	0.000000	1.000642	-0.928602	-0.624534	-0.321568	0.351424	13.534752
meanPprice	779.000000	-0.000000	1.000642	-1.985026	-0.683220	0.051265	0.714262	1.969991
meanDdisc	779.000000	0.000000	1.000642	-1.672922	-1.450223	0.559384	0.708754	1.287903
meanCdisc	779.000000	-0.000000	1.000642	-3.945506	-0.561073	0.022167	0.523411	4.582131
meanPdisc	779.000000	0.000000	1.000642	-4.396222	-0.573147	0.056118	0.580254	3.256133
meanOdisc	779.000000	-0.000000	1.000642	-4.525364	-0.552308	0.031875	0.570746	3.784331

Fig. 12.2 This shows the standardization of the aggregated data for the furniture Case Study

```
##
## Convert character Labels to numerics
##
x = le.fit_transform( df_hclusters.Region )
df_hclusters[ 'Region' ] = x
##
df_hclusters.head().style.set_caption( 'Character Labels as Numerics' ).\
    set_table_styles( tbl_styles ).\
    hide_index()
```

Character Labels as Numerics

Region	totalUsales	meanPprice	meanDdisc	meanCdisc	meanPdisc	meanOdisc
1	0.244311	0.216148	0.542271	-0.609569	-1.088793	0.524922
3	-0.659626	0.739036	0.428604	0.627404	-1.946044	0.026226
3	-0.738261	0.668967	0.720783	0.406992	0.065101	-1.626092
3	-0.339206	0.613664	0.735987	0.518495	0.404748	0.294217
3	-0.163930	0.640112	0.681844	0.061867	0.966986	-0.024256

Fig. 12.3 This shows the label encoding of the Region variable for the furniture Case Study

Once the data were properly aggregated and preprocessed, I was ready to create the clusters. This is a iterative process: try several solutions to see which gives the best results for your problem. In this example, I only created one solution. I used the *scipy* package hierarchical clustering functions which I imported in my Best Practices section of a Jupyter notebook using the command *import scipy.cluster.hierarchy as shc*. The *sklearn* package also has a hierarchical clustering set of functions, but *scipy* has more functionality. I provide the code to create the clusters and draw the dendrogram in Fig. 12.4. I divided this into four steps:

1. Instantiate the function. I used Ward's linkage method and the default Euclidean distance metric.
2. Create the figure space. This is where the dendrogram will be plotted. This is not necessary if you accept the default graph size.
3. Create the dendrogram. I used the *dendrogram* function.
4. Document the dendrogram graph. I specified a coordinate at a distance of 23 for drawing a horizontal line as a cut-off line which I will explain below. I also included boxes to highlight the clusters based on the cut-off line.

I show the resulting dendrogram in Fig. 12.5. First, notice that all the customers are at the bottom of the dendrogram with each one as an individual leaf; each customer is his/her own cluster as I noted above. You can also see that the terminal leaves are comparable to the terminal leaves of a decision tree except for the fact that the dendrogram has sample size 1 for a leaf whereas the decision tree had a sample size greater than 1. Finally, you can see that the root is at the top of the dendrogram with all the customers, so the sample size is 100%; this is the same for a decision tree. The difference between the dendrogram and the decision tree, as I noted above,

```
##
## ===> Step 1: Instantiate the linkage method <===
##
## Use Ward's minimum variance linkage method and the default
## Euclidean distance method
##
ward = shc.linkage( df_hclusters, method = 'ward' )
##
## ===> Step 2: Create the plot figure space <===
##
plt.figure( figsize = ( 10, 7  ) )
##
## ===> Step 3: Create the dendrogram <===
##
shc.dendrogram( ward )
##
## ===> Step 4: Document the dendrogram <===
##
plt.title( 'CID Clustering\nHierarchical Clustering Dendrogram\nWard\'s Method', fontsize = font_title )
plt.xlabel( 'Customer (CID)' )
plt.ylabel( 'Distance' )
plt.text( 2500, 23.5, 'Cut-off Line' )
##
max_dist = 23  ## for horizontal cut-off line
plt.axhline( y = max_dist, c = 'black', ls = '-', lw = 1.5 );
##
rec_lst = [ [ (20, 14 ), 500, 4 ], [ (800, 17 ), 1200, 3 ], [ (3600, 17 ), 1350, 4 ],
          [ (5900, 18 ), 1600, 4 ]  ]
txt_lst = [ 100, 1100, 4000, 6300 ]
for i in range( len( rec_lst ) ):
    x = rec_lst[ i ]
    rectangle = plt.Rectangle( x[ 0 ], x[ 1 ], x[ 2 ], fill = None, ec = "black", lw = 3 )
    plt.gca().add_patch(rectangle)
    plt.text( txt_lst[ i ], 23.5, 'Cluster ' + str( i + 1 ) )
```

Fig. 12.4 This shows the code for the hierarchical clustering for the furniture Case Study

is that the former does not tell you why the clusters were formed except for the fact that the objects are similar (i.e., they are close).

The customers are grouped using the step-by-step iterative algorithmic procedure I described above. If you follow up the dendrogram, you can see the clusters as they are formed. But which clusters do you use? Where do you "draw the line" in identifying clusters? I literally drew a line at a distance of 23 as you can see in Fig. 12.4 (see the variable *max_dist*). Clusters that are formed just below this line are the clusters to study. Those above the line are not clusters to study. Certainly the root signifies a cluster, is above the line, and certainly is not worth studying.

My cut-off line at 23 is arbitrary. It is a hyperparameter. At this level, there are four clusters to study. I highlighted these with boxes. Can you go further now and identify the customers in each of these clusters? You can flatten the clustering data using the *scipy* function *fcluster* which has the variable from the linkage (e.g., "ward" in this problem), the maximum distance (i.e., *max_dist*), and a criterion for applying the maximum distance as parameters. A flattened cluster file is just a 2D flat file of the hierarchical data. I show an example in Fig. 12.6. Notice that this gives the cluster assignment of each customer. The cluster assignments depend on the *max_dist* value; a lower value will yield more clusters. You can now examine these clusters, say, using frequency distributions, boxplot, and a table of cluster means. I show some possibilities in Figs. 12.7, 12.8, and 12.9.

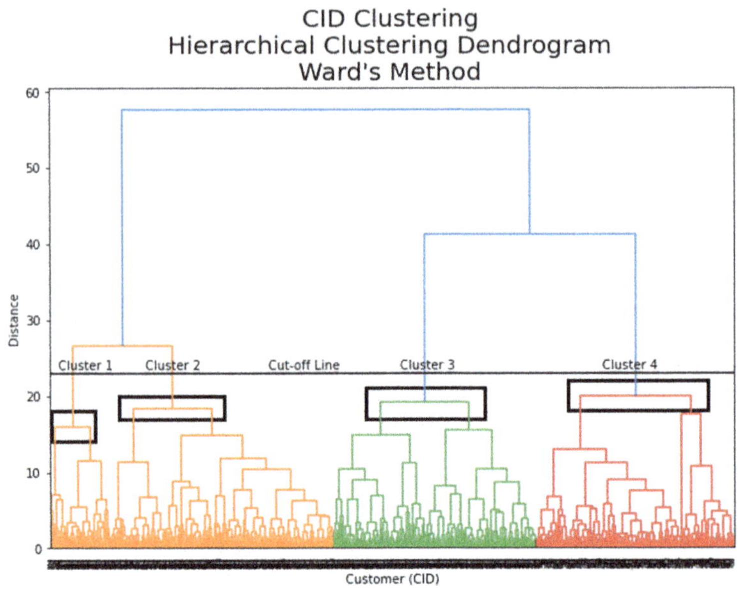

Fig. 12.5 This shows the dendrogram for the hierarchical clustering for the furniture Case Study. The horizontal line at distance 23 is a cut-off line: clusters formed below this line are the clusters we will study

```
##
## Identify the CIDs in each cluster
## Consider any cluster grouping formed below 23
##
cluster_labels = shc.fcluster( ward, max_dist, criterion = 'distance' )
df_hclusters[ 'Cluster_Number' ] = cluster_labels
##
df_hclusters.head().style.set_caption( 'DataFrame with Cluster Assignment Number' ).\
    set_table_styles( tbl_styles ).\
    hide_index()
```

DataFrame with Cluster Assignment Number

Region	totalUsales	meanPprice	meanDdisc	meanCdisc	meanPdisc	meanOdisc	Cluster_Number
1	0.244311	0.216148	0.542271	-0.609569	-1.088793	0.524922	2
3	-0.659626	0.739036	0.428604	0.627404	-1.946044	0.026226	4
3	-0.738261	0.668967	0.720783	0.406992	0.065101	-1.626092	4
3	-0.339206	0.613664	0.735987	0.518495	0.404748	0.294217	4
3	-0.163930	0.640112	0.681844	0.061867	0.966986	-0.024256	4

Fig. 12.6 This is the flattened hierarchical clustering solution. Notice the cluster numbers

```
##
## Examine the cluster size distribution
##
df_hclusters[ 'Cluster' ] = [ 'Cluster ' + str( x ) for x in df_hclusters.Cluster_Number ]
df_hclusters.stb.freq( [ 'Cluster' ] ).style.set_caption( 'Cluster Distribution' ).\
    set_table_styles( tbl_styles ).\
    bar( subset = [ 'count' ], align='mid', color = 'lightblue').hide_index()
```

Cluster Distribution

Cluster	count	percent	cumulative_count	cumulative_percent
Cluster 2	254	32.605905	254	32.605905
Cluster 3	230	29.525032	484	62.130937
Cluster 4	225	28.883184	709	91.014121
Cluster 1	70	8.985879	779	100.000000

Fig. 12.7 This is a frequency distribution for the size of the clusters for the hierarchical clustering solution

```
##
## Create a boxplot for each cluster for Order Discount
##
ax = sns.boxplot( x = 'Cluster_Number', y = 'meanOdisc', data = df_hclusters )
ax.set_title( 'Order Discount\nby Clusters\nHierarchical Clustering', fontsize = font_title )
ax.set( xlabel = 'Clusters', ylabel = 'Order Discount' )
base = 'Base: All data; n = ' + str( df_hclusters.shape[ 0 ] )
footer();
```

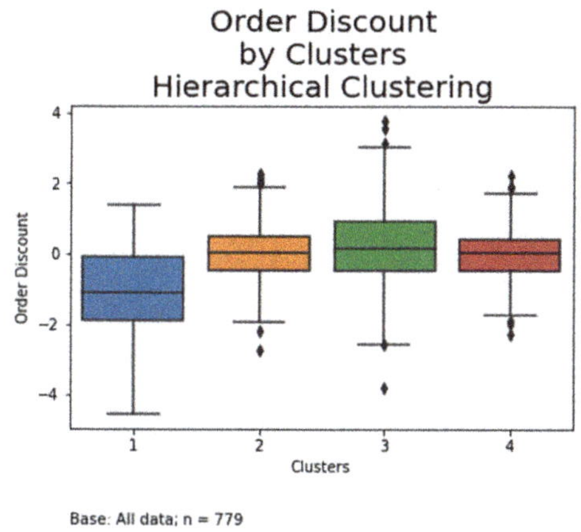

Fig. 12.8 This are the boxplots for the size of the clusters for the hierarchical clustering solution

12.2.6 Examining More than One Solution

It should be clear that there are many hierarchical clustering solutions you can create. You can create these by varying these hyperparameters:

1. distance metric;
2. linkage method;

```
##
## Cluster means summary
##
x = [ 'totalUsales', 'meanPprice', 'meanDdisc', 'meanCdisc', 'meanPdisc', 'meanOdisc' ]
df_pv = pd.pivot_table( df_hclusters, index = [ 'Cluster' ], values = x )
df_pv.style.set_caption( 'Cluster Means Summary' ).set_table_styles( tbl_styles ).format( '{:.3f}' )
```

Cluster Means Summary

Cluster	meanCdisc	meanDdisc	meanOdisc	meanPdisc	meanPprice	totalUsales
Cluster 1	1.494	0.293	-1.002	0.151	0.195	-0.576
Cluster 2	-0.283	0.665	0.056	-0.077	-0.238	0.113
Cluster 3	-0.198	-1.455	0.213	0.106	0.435	-0.333
Cluster 4	0.056	0.646	0.031	-0.068	-0.237	0.392

Fig. 12.9 This is a summary of the cluster means for the hierarchical clustering solution

3. cut-off line; and
4. features.

How do you judge which solution is the best? One thing you could look at it cluster size. You should want clusters with more than a few objects (e.g., customers). This form of clustering is often used for market segmentation and marketing managers want an even distribution across the segments. You could also look at the number of clusters. You do not want a number that is too unmanageable, nor should you want two or three. For market segmentation, I often recommend four or five. But ultimately, as stressed by Hand et al. (2001, p. 295), the solution is in the eye of the beholder.

12.3 K-Means Clustering

K-Means clustering is not hierarchical so a dendrogram is not produced. Also, it requires that you specify the number of clusters, k; setting $k = 3$ means you want 3 clusters. You can try different values of k to get different solutions. The k is a hyperparameter.

12.3.1 Algorithm Description

The algorithm is iterative as it is for hierarchical clustering. In this case, objects are successively joined based on the means of the features. This immediately suggests that the features must be at least at the interval level so that means can be calculated. The algorithm is:

1. Create k initial clusters, sometimes called *seed clusters* or *seed points*.
2. Group objects with each of the k seeds based on their shortest distance from the seeds.

3. Create new seeds as the means (or *centroids*) of the groups.
4. Merge an object into each group based on the shortest distance between the object and the centroid.
5. Repeat starting with Step 3, otherwise stop when all objects are assigned to the k groups.

   So, the metric is the shortest distance and the linkage is the centroid.

12.3.2 Case Study Application

I show the initial data setup for the furniture Case Study in Fig. 12.10. Notice that this is the same as for the hierarchical data in Fig. 12.1. The data for that clustering algorithm had to be standardized before using the algorithm. The data for the K-Means algorithm must be *whitened*, which means each feature must be scaled by its standard deviation. The features are not mean centered because the algorithm works by calculating the mean of each feature; centering sets all means to zero. The algorithm, of course, would not produce anything meaningful, if it produces anything at all, since all means would be the same: zero. You use the *whiten* function in *scipy*. This function uses the feature data as an argument. After the data are whitened, use the *sklearn* function *KMeans* to link the centroids and get the cluster assignments. I use *sklearn* for this application because it has more functionality. I show the code for this in Fig. 12.11. The algorithm begins with a random selection of observations as the initial centroids. You can specify a random seed for reproducibility. I used 42. Once you have the cluster assignments, you can append them to your DataFrame of original data and analyze the data by clusters. For example, you could create a frequency table as in Fig. 12.12 and the cluster means as in Fig. 12.13.

```
##
## Set up data
##
cols = [ 'Region', 'totalUsales', 'meanPprice', 'meanDdisc', 'meanCdisc', 'meanPdisc', 'meanOdisc' ]
df_kclusters = df_agg[ cols ].copy()
##
df_kclusters.head().style.set_caption( 'K-Means Clustering Data' ).\
    set_table_styles( tbl_styles ).\
    hide_index()
```

K-Means Clustering Data

Region	totalUsales	meanPprice	meanDdisc	meanCdisc	meanPdisc	meanOdisc
Northeast	3461	5.400400	0.131820	0.068248	0.038241	0.051113
West	1001	5.797762	0.128930	0.071814	0.036907	0.050093
West	787	5.744514	0.136357	0.071179	0.040036	0.046714
West	1873	5.702487	0.136744	0.071500	0.040564	0.050641
West	2350	5.722586	0.135367	0.070184	0.041439	0.049990

Fig. 12.10 This is a sample of the aggregated data for the furniture Case Study for K-Means clustering of customers

```
##
## Subset the data for all numerics
##
cols = [ 'totalUsales', 'meanPprice', 'meanDdisc', 'meanCdisc', 'meanPdisc', 'meanOdisc' ]
tmp = df_kclusters[ cols ]
##
## ===> Step 1: Scale by the standard deviation <===
##
whitened = whiten( tmp )
##
## ===> Step 2: Instantiate the KMeans function <===
##
n_clusters = 4
kmeans = KMeans( n_clusters, random_state = 42  )
##
## ===> Step 3: Fit the centroids <===
##
centroids = kmeans.fit_transform( whitened )
##
## ===> Step 4: Fit the cluster assignments <===
##
clusters = kmeans.fit_predict( whitened )
##
df_kclusters[ 'Cluster_Number' ] = clusters
df_kclusters.head()
```

	Region	totalUsales	meanPprice	meanDdisc	meanCdisc	meanPdisc	meanOdisc	Cluster_Number
0	Northeast	3461	5.400400	0.131820	0.068248	0.038241	0.051113	1
1	West	1001	5.797762	0.128930	0.071814	0.036907	0.050093	1
2	West	787	5.744514	0.136357	0.071179	0.040036	0.046714	0
3	West	1873	5.702487	0.136744	0.071500	0.040564	0.050641	1
4	West	2350	5.722586	0.135367	0.070184	0.041439	0.049990	1

Fig. 12.11 This are the setup for a K-Means clustering. Notice that the random seed is set at 42 for reproducibility

```
##
## Examine the cluster size distribution
##
df_kclusters[ 'Cluster' ] = [ 'Cluster ' + str( x ) for x in df_kclusters.Cluster_Number ]
df_kclusters.stb.freq( [ 'Cluster' ] ).style.set_caption( 'Cluster Distribution' ).\
    set_table_styles( tbl_styles ).\
    bar( subset = [ 'count' ], align='mid', color = 'lightblue').hide_index()
```

Cluster Distribution

Cluster	count	percent	cumulative_count	cumulative_percent
Cluster 1	283	36.328626	283	36.328626
Cluster 2	213	27.342747	496	63.671374
Cluster 3	183	23.491656	679	87.163030
Cluster 0	100	12.836970	779	100.000000

Fig. 12.12 This is an example frequency table of the K-Means cluster assignments from Fig. 12.11

```
##
## Cluster means summary
##
x = [ 'totalUsales', 'meanPprice', 'meanDdisc', 'meanCdisc', 'meanPdisc', 'meanOdisc' ]
df_pv = pd.pivot_table( df_kclusters, index = [ 'Cluster' ], values = x )
df_pv.style.set_caption( 'Cluster Means Summary' ).set_table_styles( tbl_styles ).format( '{:.3f}' )
```

Cluster Means Summary

Cluster	meanCdisc	meanDdisc	meanOdisc	meanPdisc	meanPprice	totalUsales
Cluster 0	0.073	0.123	0.047	0.040	5.370	1291.730
Cluster 1	0.069	0.135	0.051	0.040	5.471	2057.806
Cluster 2	0.069	0.080	0.051	0.040	5.567	1885.568
Cluster 3	0.070	0.134	0.050	0.040	4.415	5819.798

Fig. 12.13 This is a summary of the cluster means for the K-Means cluster assignments from Fig. 12.11

12.4 Mixture Model Clustering

Mixture models arise when your data are draws from several distributions. You learned in an elementary statistics course that your data are draws from a single distribution, usually a normal distribution. The data, however, may come from several distributions. This is evident if a histogram is skewed and/or multimodal. Multimodality is *always* due to a mixture of two or more unimodal distributions reflecting different populations. The individual unimodal distributions are weighted. See Paczkowski (2016) for a discussion and example.

K-Means clustering has two disadvantages:

- It lacks flexibility in handling different shapes of clusters. It places a circle or hypersphere around each cluster, but the clusters may not be spherical.
- It lacks of probabilistic cluster assignment. But the assignment to a cluster is not always certain. There is uncertainty in most assignments which is a probabilistic concept.

Mixture models deal with these issues. The most common distribution for continuous data is the normal, or *Gaussian*, distribution. I show the set-up and the results for this clustering approach in Fig. 12.14. I also show two types of summaries in Figs. 12.15 and 12.16 that are comparable to what I showed for the other two clustering approaches.

```
##
## Set up data
##
x = [ 'Region', 'totalUsales', 'meanPprice', 'meanDdisc', 'meanCdisc', 'meanPdisc', 'meanOdisc' ]
df_mclusters = df_agg[ x ].copy()
##
x = [ 'totalUsales', 'meanPprice', 'meanDdisc', 'meanCdisc', 'meanPdisc', 'meanOdisc' ]
tmp = df_mclusters[ x ]
##
## ===> Step 1: Instantiate the model <===
##
n_components = 4
gmm = GaussianMixture( n_components )
##
## ===> Step 2: Fit the model <===
##
mixture = gmm.fit( tmp )
##
## ===> Step 3: Add cluster labels to main cluster DataFrame <===
##
cluster_number = mixture.predict( tmp )
df_mclusters[ 'Cluster_Number'] = cluster_number
##
df_mclusters.head().style.set_caption( 'DataFrame with Mixture Cluster Assignments' ).\
    set_table_styles( tbl_styles ).\
    hide_index()
```

DataFrame with Mixture Cluster Assignments

Region	totalUsales	meanPprice	meanDdisc	meanCdisc	meanPdisc	meanOdisc	Cluster_Number
Northeast	3461	5.400400	0.131820	0.068248	0.038241	0.051113	2
West	1001	5.797762	0.128930	0.071814	0.036907	0.050093	1
West	787	5.744514	0.136357	0.071179	0.040036	0.046714	1
West	1873	5.702487	0.136744	0.071500	0.040564	0.050641	2
West	2350	5.722586	0.135367	0.070184	0.041439	0.049990	2

Fig. 12.14 This are the setup for a Gaussian mixture clustering

```
##
## Examine the cluster size distribution
##
df_mclusters[ 'Cluster' ] = [ 'Cluster ' + str( x ) for x in df_mclusters.Cluster_Number ]
df_mclusters.stb.freq( [ 'Cluster' ] ).style.set_caption( 'Cluster Distribution' ).\
    set_table_styles( tbl_styles ).\
    bar( subset = [ 'count' ], align='mid', color = 'lightblue').hide_index()
```

Cluster Distribution

Cluster	count	percent	cumulative_count	cumulative_percent
Cluster 3	375	48.138639	375	48.138639
Cluster 0	236	30.295250	611	78.433890
Cluster 1	167	21.437741	778	99.871630
Cluster 2	1	0.128370	779	100.000000

Fig. 12.15 This is an example frequency table of the Gaussian Mixture cluster assignments from Fig. 12.14

```
##
## Cluster means summary
##
x = [ 'totalUsales', 'meanPprice', 'meanDdisc', 'meanCdisc', 'meanPdisc', 'meanOdisc' ]
df_pv = pd.pivot_table( df_mclusters, index = [ 'Cluster' ], values = x )
df_pv.style.set_caption( 'Cluster Means Summary' ).set_table_styles( tbl_styles ).format( '{:.3f}' )
```

Cluster Means Summary

Cluster	meanCdisc	meanDdisc	meanOdisc	meanPdisc	meanPprice	totalUsales
Cluster 0	0.070	0.080	0.050	0.040	5.556	1864.165
Cluster 1	0.070	0.134	0.050	0.040	5.472	977.814
Cluster 2	0.070	0.135	0.050	0.040	4.931	4094.168
Cluster 3	0.070	0.131	0.050	0.040	4.838	39630.000

Fig. 12.16 This is a summary of the cluster means for the Gaussian Mixture cluster assignments from Fig. 12.14

..ography

Adler, M.J. and C.V. Doren. 1972. *How to Read a Book: The Classic Guide to Intelligent Reading*. Revised ed. Touchstone.

Adriaans, P. 2019. Information. In *The Stanford Encyclopedia of Philosophy*, ed. E.N. Zalta (Spring 2019 ed.). Stanford: Metaphysics Research Lab, Stanford University.

Agresti, A. 2002. *Categorical Data Analysis*. 2nd ed. New York: Wiley.

Akoglu, H. 2018. User's guide to correlation coefficients. *Turkish Journal of Emergency Medicine* 18(3): 91–93.

Andrienko, G., N. Andrienko, and A. Savinov. 2001. Choropleth maps: Classification revisited. In *Proceedings of the 20th International Cartographic Conference (ICA 2001)*, 1209–1219. https://www.researchgate.net/publication/228959242_Choropleth_maps_Classification_revisited.

Arezki, R., V.A. Ramey, and L. Sheng. 2015. News shocks in open economies: Evidence from giant oil discoveries. IMF Working Paper.

Baltagi, B.H. 1995. *Econometric Analysis of Panel Data*. New York: Wiley.

Barsky, R., and E. Sims. 2011. News shocks and business cycles. Unpublished working paper.

Beaudry, P., and F. Portier. 2006. Stock prices, news, and economic fluctuations. *American Economic Review* 96(4), 1293–1307.

Belsley, D.A., E. Kuh, and R.E. Welsch. 1980. *Regression Diagnostics: IdentifyingInfluential Data and Sources of Collinearity*. New York: Wiley.

Berman, J. and J. Pfleeger. 1997. Which industries are sensitive to business cycles? *Monthly Labor Review*, 120: 19–25.

Bilder, C.R. and T.M. Loughin. 2015. *Analysis of Categorical Data with R*. Boca Raton: CRC Press.

Blumberg, A.E. 1976. *Logic: A First Course*. New York: Alfred A. Knopf.

Box, G. and D. Cox. 1964. An analysis of transformations. *Journal of the Royal Statistical Society*, 26: 211–252. Series B.

Box, G., W. Hunter, and J. Hunter. 1978. *Statistics for Experimenters: An Introduction to Design, Data Analysis, and Model Building*. New York: Wiley.

Box, G., G. Jenkins, and G. Reinsel. 1994. *Time Series Analysis, Forecasting and Control*. 2nd ed. Englewood: Prentice Hall.

Brill, J.E. 2008. Likert scale. In *Encyclopedia of Survey Research Methods*, ed. P.J. Lavrakas, 428–429. New York: SAGE Publications Inc.

Buckingham, W. 2010. *Encyclopedia of Geography*. Chapter Choropleth Maps, 407–408. SAGE Publications Inc.

Capurro, R. and B. Hjorland. 2003. *The Concept of Information*. Vol. 37. Chapter 8, 343–411.

W. R. Paczkowski, *Business Analytics*,
https://doi.org/10.1007/978-3-030-87023-2

Carr, D.B., R.J. Littlefield, W.L. Nicholson, and J.S. Littlefield. 1987. Scatterplot matrix techniques for large n. *Journal of the American Statistical Association* 82(398), 424–436.

Carroll, R.J. and D. Ruppert. 1988. *Transformation and Weighting in Regression*. London: Chapman and Hall.

Celko, J. 2000. *SQL for Smarties: Advanced SQL Programming*. 2nd ed. London: Academic Press.

Checkland, P. and S. Howell. 1998. *Information, Systems and Information Systems: Making Sense of the Field*. New York: Wiley.

Cleveland, W.S. 1979. Robust locally weighted regression and smoothing scatterplots. *Journal of the American Statistical Association* 74(368), 829–836.

Cleveland, W.S. 1981. Lowess: A program for smoothing scatterplots by robust locally weighted regression. *The American Statistician* 35(1), 54.

Coad, A. 2009. On the distribution of product price and quality. *Journal of Evolutionary Economics* 19, 589–604.

Cochrane, W.G. 1963. *Sampling Techniques*. 2nd ed. New York: Wiley.

Collette, A. 2014. *Python and HDF5: Unlocking Scientific Data*. Newton: O'Reilly Media Inc.

Cormen, T.H., C.E. Leiserson, R.L. Rivest, and C. Stein. 2009. *Introduction to Algorithms* 3rd ed. New York: MI Press.

Coveney, P. and R. Highfield. 1990. *The Arrow of Time: A Voyage Throgh Science to Solve Time's Greatest Mystery*. New York: Ballantine Books.

Cramer, H. 1946. *Mathematical Methods of Statistics*. Princeton: Princeton University.

D'Agostino, R.B., A. Belanger, and J. Ralph B. D'Agostino. 1990. A suggestion for using powerful and informative tests of normality. *American Statistician* 44(4), 316–321.

Davies, P. 1995. *About Time: Einstein's Unfinished Revolution*. New York: Simon and Schuster.

Deisenroth, M.P., A.A. Faisl, and C.S. Ong. 2020. *Mathematics for Machine Learning*. Cambridge: Cambridge University.

Dershowitz, N. and E.M. Reingold. 2008. *Calendrical Calculations*. 3rd ed. Cambridge: Cambridge University.

Dhrymes, P.J. 1971. *Distributed Lags: Problems of Estimation and Formulation*. Mathematical Economics Texts. San Francisco: Holden-Day, Inc.

Diamond, W. 1989. *Practical Experiment Design for Engineers and Scientists*. New York: Van Nostrand Reinhold.

Doane, D.P. and L.E. Seward. 2011. Measuring skewness: A forgotten statistic? *Journal of Statistics Education* 19(2), 1–18.

Dobbin, K.K. and R.M. Simon. 2011. Optimally splitting cases for training and testing high dimensional classifiers. *BMC Medical Genomics* 4(1), 1–8.

Dobson, A.J. 2002. *An Introduction to Generalized Linear Models*. 2nd ed. Texts in Statistical Science. London: Chapman & Hall/CRC.

Dougherty, C. 2016. *Introduction to Econometrics*. 5th ed. Oxford: Oxford University.

Draper, N. and H. Smith. 1966. *Applied Regression Analysis*. New York: Wiley.

Dudewicz, E.J. and S.N. Mishra. 1988. *Modern Mathematical Statistics*. New York: Wiley.

Duff, I.S., A. Erisman, and J.K. Reid. 2017. *Direct Methods for Sparse Matrices*. 2nd ed. Numerical Mathematics and Scientific Computation. Oxford: Oxford University.

Emerson, J.D. and M.A. Stoto. 1983. *Understanding Robust and Exploratory Data Analysis*, Chapter Transforming Data, 97–128. New York: Wiley.

Enders, C.K. 2010. *Applied Missing Data Analysis*. New York: The Guilford Press.

Everitt, B.S., S. Landau, and M. Leese. 2001. *Cluster Analysis*. 4th ed. London: Arnold Publishers.

Fan, J., R. Samworth, and Y. Wu. 2009. Ultrahigh dimensional feature selection: Beyond the linear model. *Journal of Machine Learning Research* 10, 2013–2038.

Faraway, J.J. 2016. Does data splitting improve prediction? *Statistics and Computing* 26, 49–60.

Floridi, L. 2010. *Information: A Very Short Introduction*. Oxford: Oxford University.

Fox, J. 2019. *Regression Diagnostics: An Introduction*. 2nd ed. *Quantitative Applications in the Social Sciences Book*. Vol. 79. New York: SAGE Publications Inc.

Freedman, D., R. Pisani, and R. Purves. 1978. *Statistics*. New York: W.W. Norton & Company.

—, J.L. and F.J. Williams. 1969. *Modern Business Statistics*. Englewood: Prentice-Hall, Inc. Revised edition by Benjamin Perles and Charles Sullivan.

Frigge, M., D.C. Hoaglin, and B. Iglewicz. 1989. Some implementations of the boxplot. *The American Statistician 43*(1), 50–54.

Gelman, A. and J. Hill. 2007. *Data Analysis Using Regression and Multilevel/Hierarchical Models*. Cambridge: Cambridge University.

Gelman, A., J. Hill, and A. Vehtari. 2021. *Regression and Other Stories*. Cambridge: Cambridge University.

Georgantzas, N.C. and W. Acar. 1995. *Scenario-Driven Planning: Learning to Manage Strategic Uncertainty*. Westport: Quorum Books.

Goldberger, A.S. 1964. *Econometric Theory*. New York: Wiley.

Granger, C.W.J. 1979. Seasonality: Causation, interpretation, and implications. Technical report, NBER. http://www.nber.org/chapters/c3896. This PDF is a selection from an out-of-print volume from the National Bureau of Economic Research.

Greene, W.H. 2003. *Econometric Analysis*. 5th ed. Englewood: Prentice Hall.

Gujarati, D. 2003. *Basic Econometrics*. 4th ed. New York: McGraw-Hill/Irwin.

Hand, D., H. Mannila, and P. Smyth. 2001. *Principles of Data Mining*. Cambridge: The MIT Press.

Hastie, T., R. Tibshirani, and J. Friedman. 2008. *The Elements of Statistical Learning: Data Mining, Inference, and Prediction*. 2nd ed. Berlin: Springer.

Hausman, J. and C. Palmery. 2012. Heteroskedasticity-robust inference in finite samples. *Economic Letters* 116(2), 232–235.

Healey, C.G. and A.P. Sawant. 2012. On the limits of resolution and visual angle in visualization. *ACM Transactions on Applied Perception* 9(4), 1–21.

Hernandez, M.J. and J.L. Viescas. 2000. *SQL Queries for Mere Mortals: A Hands-on Guide to Data Manipulation in SQL*. Reading: Addison-Wesley.

Hildebrand, D.K., R.L. Ott, and J.B. Gray. 2005. *Basic Statistical Ideas for Managers*. 2nd ed. Mason: Thomson South-Western.

Hill, R.C., W.E. Griffiths, and G.C. Lim. 2008. *Principles of Econometrics*. 4th ed. New York: Wiley.

Hirshleifer, J. and J.G. Riley. 1996. *The Analytics of Uncertainty and Information*. Cambridge: Cambridge University.

Hocking, R.R. 1996. *Methods and Applications of Linear Models: Regression and the Analysis of Variance*. New York: Wiley.

Hoffmann, E. 1980. Defining information: An analysis of the information content of documents. *Information & Processing Management* 16, 291–304.

Hsiao, C. 1986. *Analysis of Panel Data*. Cambridge: Cambridge University.

Huber, P.J. 1994. Huge data sets. In *Compstat*, ed. R. Dutter and W. Grossmann. Heidelberg: Physica.

Hubert, M. and S.V. der Veeken. 2008. Outlier detection for skewed data. *Journal of Chemometrics* 22(3), 235–246.

Hunt, J. 2019. *Advanced Guide to Python 3 Programming*. Berlin: Springer.

James, G., D. Witten, T. Hastie, and R. Tibshirani. 2013. *An Introduction to Statistical Learning: With Applications in R*. New York: Springer Science+Business Media.

Joanes, D.N. and C.A. Gill. 1998. Comparing measures of sample skewness and kurtosis. *Journal of the Royal Statistical Society. Series D* 47(1), 183–189.

Jobson, J. 1992. *Applied Multivariate Data Analysis*. Categorical and Multivariate Methods. Vol. II. Berlin: Springer.

Johnston, J. 1972. *Econometric Methods*. 2nd ed. New York: McGraw-Hill Book Company.

Kennedy, P. 2003. *A Guide to Econometrics*. 5th ed. Cambridge: MIT Press.

Kmenta, J. 1971. *The Elements of Econometrics*. New York: The MacMillan Company.

Knight, F.H. 1921. *Risk, Uncertainty, and Profit*. Boston: Houghton Mifflin.

Kosslyn, S.M. 2006. *Graph Design for the Eye and Mind*. Oxford: Oxford University.

Kreft, I.G. and J. de Leeuw. 1998. *Introducing Multilevel Modeling*. 1st ed. New York: SAGE Publications Ltd.

Kwiatkowski, D., P.C. Phillips, P. Schmidt, and Y. Shin. 1992. Testing the null hypothesis of stationarityagainst the alternative of a unit root. *Journal of Econometrics* 54, 159–178.

Lay, D.C. 2012. *Linear Algebra and Its Applications*. 4th ed. London: Pearson Education.

Lemahieu, W., B. Baesens, and S. vanden Broucke. 2018. *Principles of Data Management: The Practical Guide to Storing, Managing and Analyzing Big and Small Data*. 1st ed. Cambridge: Cambridge University.

Levy, P.S. and S. Lemeshow. 2008. *Sampling of Populations: Methods and Applications*. 4th ed. New York: Wiley.

Lewin-Koh, N. 2020, March. Hexagon binning: An overview. techreport. http://cran.r-project.org/web/packages/.

Luke, D.A. 2004. *Multilevel Modeling*. Quantitative Applications in the Social Sciences. New York: SAGE Publications. Series/Number 07-143.

MacKinnon, J.G. and H. White. 1985. Some heteroskedasticity-consistent covariance matrix estimators with improved finite sample properties. *Journal of Econometrics* 29(3), 305–325.

Mangiafico, S.S. 2016. Summary and analysis of extension program evaluation in r. Version 1.9.0. http://rcompanion.org/handbook/. Last accessed October 15, 2017.

McCullagh, P. and J.A. Nelder. 1989. *Generalized Linear Models*. 2nd ed. London: Chapman & Hall.

McKinney, W. 2018. *Python for Data Analysis: Data Wrangling with Pandas, Numpy, and ipython*. 2nd ed. Newton: O'Reilly.

Mingers, J. and C. Standing. 2018. What is information? toward a theory of information as objective and veridical. *Journal of Information Technology* 33, 85–104.

Moore, D.S. and W.I. Notz. 2017. *Statistics: Concepts and Controversies*. 9th ed. San Francisco: W.H. Freeman & Company.

Morgenstern, O. 1965. *On the Accuracy of Economic Observations*. 2nd Revised ed. Princeton: Princeton University.

Morgenthaler, S. 1997. *The Practice of Data Analysis: Essays in Honor of John W. Tukey*, Chapter Gaussianizing Transformations and Estimation, 247–259. Princeton: Princeton University.

Mosteller, F. and J.W. Tukey. 1977. *Data Analysis and Regression: A Second Course in Statistics*. Reading: Addison-Wesley Publishing Company.

Mulligan, K. and F. Correia. 2020. Facts. In *The Stanford Encyclopedia of Philosophy*, ed. E.N. Zalta. California: Metaphysics Research Lab, Stanford University. https://plato.stanford.edu/archives/win2020/entries/facts/.

Nelson, C.R. 1973. *Applied Time Series Analysis for Managerial Forecasting*. San Francisco: Holden-Day, Inc.

Neter, J., W. Wasserman, and M.H. Kutner. 1989. *Applied Linear Regression Models*. 2nd ed. Homewood: Richard D. Irwin, Inc.

Paczkowski, W.R. 2016. *Market Data Analysis Using JMP*. Bengaluru: SAS Press.

Paczkowski, W.R. (2018. *Pricing Analytics: Models and Advanced Quantitative Techniques for Product Pricing*. London: Routledge.

Paczkowski, W.R. 2020. *Deep Data Analytics for New Product Development*. London: Routledge.

Paczkowski, W.R. 2021a. *Business Analytics: Data Science for Business Problems*. Berlin: Springer.

Paczkowski, W.R. 2021b. *Modern Survey Analysis: Using Python for Deeper Insights*. Berlin: Springer.

Parzen, E. 1962. *Stochastic Processes*. San Francisco: Holden-Day, Inc.

Peebles, D. and N. Ali. 2015. Expert interpretation of bar and line graphs: The role of graphicacy in reducing the effect of graph format. *Frontiers in Psychology* 6, Article 1673.

Peterson, M.P. 2008. *Encyclopedia of Geographic Information Science*. Chapter Choropleth Map, 38–39. New York: SAGE Publications, Inc.

Picard, R.R. and K.N. Berk. 1990. Data splitting. *The American Statistician* 44(2), 140–147.

Pinker, S. 1990. *Artificial Intelligence and the Future of Testing*. Chapter A theory of graph comprehension, 73–126. Hove: Psychology Press.

Pinker, S. 2014. *The Sense of Style: The Thinking Person's Guide to Writing in the 21st Century*. Baltimore: Penguin Books.

Popper, K.R. 1972. *Objective Knowledge: An Evolutionary Approach*. Oxford: Oxford University.

Ray, J.-C. and D. Ray. 2008. Multilevel modeling for marketing: a primer. *Recherche et Applications en Marketing* 23(1), 55–77.

Reitermanov, Z. 2010. Data splitting. In *WDS'10 Proceedings of Contributed Papers*, 31–36. Part I.

Ripley, B.D. 1996. *Pattern Recognition and Neural Networks*. Cambridge: Cambridge University.

Romano, J.P. and C. DiCiccio. 2019. Multiple data splitting for testing. techreport Technical Report No. 2019-03. Stanford: Stanford University.

Russell, S. and P. Norvig. 2020. *Artificial Intelligence: A Modern Approach* 4th ed. Pearson Series in Artifical Intelligence. London: Pearson.

Samuelson, P.A. 1973. *Economics* 9th ed. New York: McGraw-Hill.

Savage, L.J. 1972. *The Foundations of Statistics* 2nd Revised ed. New York: Dover Publications, Inc.

Schmittlein, D.C., D.G. Morrison, and R. Colombo. 1987. Counting your customers: Who are they and what will they do next? *Management Science* 33(1), 1–24.

Sedgewick, R., K. Wayne, and R. Dondero. 2016. *Inroduction to Python Programming: An Interdisciplinary Approach*. London: Pearson.

Shao, J. 2003. *Mathematical Statistics*. 2nd ed. Berlin: Springer.

Silverman, B. 1986. *Density Estimation for Statistics and Data Analysis*. London: Chapman and Hall.

Snee, R.D. 1977. Validation of regression models: Methods and examples. *Technometrics* 19(4), 415–428.

Snijders, T.A. and R.J. Bosker. 2012. *Multilevel Analysis: An Introduction to Basic and Advanced Multilevel Modeling*. 2nd ed. New York: SAGE.

Spurr, W.A. and C.P. Bonini. 1968. *Statistical Analysis for Business Decisions*. Homewood: Richard D. Irwin, Inc. Second Printing.

Stevens, S.S. 1946. On the theory of scales of measurement. *Science* 103(2684), 677–680.

Stewart, I. 2019. *Do Dice Play God? The Mathematics of Uncertainty*. London: Profile Books LTD.

Strang, G. 2006. *Linear Algebra and Its Applications*. 4th ed. Boston: Thomson Brooks/Cole.

Stross, R. 2010. Failing like a buggy whip maker? better check your simile. New York: New York Times. Jan. 10, 2010, Section BU, Page 4.

Stupak, J.M. 2019. Introduction to US economy: The business cycle and growth. *In Focus*. Congressional Research Service.

Thompson, S.K. 1992. *Sampling*. New York: Wiley.

Tufte, E.R. 1983. *The Visual Display of Quantitative Information*. Cheshire: Graphics Press.

Tukey, J.W. 1957. On the comparative anatomy of transformations. *Annuals of Mathematical Statistics* 28(3), 602–632.

Tukey, J.W. 1977. *Exploratory Data Analysis*. London: Pearson.

VanderPlas, J. 2017. *Python Data Science Handbook: Essential Tools for Working with Data*. Newton: O'Reilly Media.

Vanderplas, S., D. Cook, and H. Hofmann. 2020. Testing statistical charts: What makes a good graph? *The Annual Review of Statistics and Its Application* 7, 61–88.

Velleman, P.F. and L. Wilkinson. 1993. Nominal, ordinal, interval, and ratio typologies are misleading. *The American Statistician* 47(1), 65–72.

Wan, X., W. Wang, J. Liu, and T. Tong. 2014. Estimating the sample mean and standard deviation from the sample size, median, rangeand/or interquartile range. *BMCMedical Research Methodology 14*(1), 1–13.

Weglarczyk, S. 2018. Kernel density estimation and its application. In *ITM Web of Conferences*. Vol. 23, 37.

Wegman, E.J. 1990. Hyperdimensional data analysis using parallel coordinates. *Journal of the American Statistical Association* 85(411), 664–675.

Wegman, E.J. 2003. Visual data mining. *Statistics in Medicine* 22, 1383–1397.

Wei, W.W. 2006. *Time Series Analysis: Univariate and Multivariate Methods*. 2nd ed. London: Pearson.

Weisberg, S. 1980. *Applied Linear Regression*. New York: Wiley.

Weiss, N.A. 2005. *Introductory Statistics*. 7th ed. London: Pearson Education, Inc.

White, H. 1980. A heteroskedasticity-consistent covariance matrix estimator and a direct test for heteroskedasticity. *Econometrica* 48, 817–838.

Wilder-Jame, E. 2016. Breaking down data silos. In *Harvard Business Review*. https://hbr.org/2016/12/breaking-down-data-silos#comment-section

Witten, I.H., E. Frank, and M.A. Hall. 2011. *Data Mining: Practical Machine Learning Tools and Techniques*. 3rd ed. Amsterdam: Elsevier Inc.

Wooldridge, J.M. 2002. *Econometric Analysis of Cross Section and Panel Data*. Cambridge: MIT Press.

Wooldridge, J.M. 2003. Cluster-sample methods in applied econometrics. *American Economic Review* 93(2), 133–138.

Yeo, I.-K. and R.A. Johnson. 2000. A new family of power transformations to improve normality or symmetry. *Biometrika* 87(4), 954–959.

Zarembka, P. 1974. *Frontiers in Econometrics*. Chapter Transformation of Variables in Econometrics, 81–104. New York: Academic Press.

Zarnowitz, V. 1992. The regularity of business cycles. In *Business Cycles: Theory, History, Indicators, and Forecasting*, ed. V. Zarnowitz. Chapter 8, 232–264. Chicago: University of Chicago.

Zeckhauser, R. 2006. Investing in the unknown and unknowable. *Capitalism and Society* 1(2), 1–39.

Zhao, N., Q. Xu, M.-L. Tang, B. Jiang, Z. Chen, and H. Wang. 2020. High-dimensional variable screening under multicollinearity. *Stats* 9(1), e272.

Index

W. R. Paczkowski, *Business Analytics*,
https://doi.org/10.1007/978-3-030-87023-2

GPSR Compliance
The European Union's (EU) General Product Safety Regulation (GPSR) is a set of rules that requires consumer products to be safe and our obligations to ensure this.

If you have any concerns about our products, you can contact us on

ProductSafety@springernature.com

In case Publisher is established outside the EU, the EU authorized representative is:

Springer Nature Customer Service Center GmbH
Europaplatz 3
69115 Heidelberg, Germany

www.ingramcontent.com/pod-product-compliance
Lightning Source LLC
Chambersburg PA
CBHW072113160625
28292CB00003B/66
* 9 7 8 3 0 3 0 8 7 0 2 2 5 *